Functional Carbohydrates

Development, Characterization, and Biomanufacture

Functional Carbohydrates

Development, Characterization, and Biomanufacture

Edited by
Jian Chen
Yang Zhu
Song Liu

CRC Press
Taylor & Francis Group
Boca Raton London New York

CRC Press is an imprint of the
Taylor & Francis Group, an **informa** business

CRC Press
Taylor & Francis Group
6000 Broken Sound Parkway NW, Suite 300
Boca Raton, FL 33487-2742

International Standard Book Number-13: 978-1-4987-1877-6 (Hardback)

Library of Congress Cataloging-in-Publication Data

Names: Chen, Jian, 1962- | Zhu, Yang, 1955- | Liu, Song (Biotechnologist)
Title: Functional carbohydrates : development, characterization, and biomanufacture / [edited by] Jian Chen, Yang Zhu, Song Liu.
Description: Boca Raton : CRC Press, 2017. | Includes bibliographical references and index.
Identifiers: LCCN 2017016934| ISBN 9781498718776 (hardback : alk. paper) | ISBN 9781315371061 (e-book) | ISBN 9781315353890 (e-book) | ISBN 9781498718783 (e-book) | ISBN 9781315334837 (ebook)
Subjects: LCSH: Functional foods. | Carbohydrates in human nutrition.
Classification: LCC QP144.F85 F8216 2017 | DDC 613.2--dc23
LC record available at https://lccn.loc.gov/2017016934

Visit the Taylor & Francis Web site at
http://www.taylorandfrancis.com

and the CRC Press Web site at
http://www.crcpress.com

FSC www.fsc.org MIX Paper from responsible sources FSC® C013056

Printed and bound in Great Britain by
TJ International Ltd, Padstow, Cornwall

Contents

Preface

Before understanding the structures and chemical properties of carbohydrates, people have widely used carbohydrates in daily life. For example, those plants rich in carbohydrates (rice and wheat) served as staple foods that could be further made into fermented beverages. Traditional carbohydrates such as glucose and starch were generally used as the energy-supplying materials of human bodies. "Functional carbohydrates" is the term used to describe those carbohydrates that play an important role in strengthening immunity, decreasing the level of blood-lipid, and regulating the intestinal flora of humans. To date, functional carbohydrates mainly covered dietary fiber, functional polysaccharides, functional oligosaccharides, sugar alcohols, and other functional monosaccharides. To facilitate tracking the important progress in functional carbohydrates, the book *Functional Carbohydrates: Development, Characterization, and Biomanufacture* was thus proposed by Prof. Chen Jian (Jiangnan University, China), Prof. Zhu Yang (Wageningen University, Netherlands), and Associate Prof. Liu Song (Jiangnan University).

This book addresses the history and recent developments of a selected number of important functional carbohydrates (Chapter 1). In Chapters 2 through 9, we introduce the source, properties, and applications of functional acid (hyaluronic acid [Chapter 2]), saccharides (curdlan oligosaccharides [Chapter 3], α-cyclodextrin [Chapter 4], galactooligosaccharides [Chapter 5], hemicellulose hydrolysates [Chapter 6], and *Ganoderma lucidum* polysaccharide [Chapter 7]), and some saccharide derivatives (glucosamine, N-acetylglucosamine [Chapter 8], and sugar alcohols [Chapter 9]). Especially, we describe in detail the biomanufacture of these carbohydrates based on fermentation or enzyme catalysis, including the strain screening and improvement, optimization of fermentation process, and product downstream processing.

The authors of this book are from Wageningen University (Netherlands), Jiangnan University (China), Shandong University (China), Nanjing Forestry University (China), Shanghai Jiao Tong University (China), and Chinese Academy of Sciences, all well known for their achievements in the field of enzyme and fermentation technology in general and functional carbohydrates in particular. We are particularly grateful to all authors for their indispensable contributions.

Editors

Jian Chen, PhD, is a professor of food biotechnology at Jiangnan University. Professor Chen received his PhD from Jiangnan University (former Wuxi University of Light Industry) in 1990. His research interests include stress tolerance and the response of food microorganisms, the production of food additives by biotechnology, and food safety issues in fermented foods. Using lactic bacteria and yeast as models, he investigated the roles of key genes and antioxidants on tolerance and response of food microorganisms to cold, acidic, oxidative, and osmotic stresses. He has successfully achieved the industrial-scale production of more than 30 different food additives by metabolic engineering and process optimization. He also screened and improved a series of food enzymes, which have been widely applied in juice processing and bakery areas. He is working on mechanisms inside the accumulation of nitrogen-containing small molecule harmful components during the production of fermented foods, such as ethylcarbamate and bioamines. Dr. Chen has published over 300 research articles and 16 invited reviews. He currently serves on the editorial board of eight peer-reviewed journals, including the *Journal of Agricultural and Food Chemistry* and *Process Biochemistry.* Professor Chen was named as a fellow of the International Academy of Food Science & Technology (IA FoST) in 2012.

Yang Zhu, PhD, studied biochemical engineering at East China University of Science and Technology, where he worked also as a faculty member in the Department of Food Science and Technology, and the Department of Biochemical Engineering after his graduation. He was also a visiting scholar at the Institute of Microbiology of the University of Münster, Germany. Since 1993, he has been at TNO Nutrition and Food Research, the Netherlands and Wageningen University and Research, the Netherlands. He received his PhD in bioprocess engineering in 1997 at Wageningen University and Research. His main research focuses on food fermentation, bioprocess engineering, food nutrition, and pharmacokinetics. The author/coauthor of dozens of scientific publications and books, currently Dr. Yang Zhu is an independent consultant in agro-food and biotechnology in the Netherlands, as well as a guest scientist at Wageningen University, the Netherlands.

Song Liu, PhD, is an associate professor in the School of Biotechnology, Jiangnan University (China). He received his PhD from Jiangnan University (former Wuxi University of Light Industry) in 2011. He mainly engaged in teaching and research on genetic engineering and enzyme engineering. For nearly three years, he undertook three projects, including projects from the Natural Science Foundation of China (31171639), the Foundation Research Project of Jiangsu Province (BK20130132), National High Technology Research and Development Program of China (2015AA021003), and a number of enterprises projects. He had more than 20 papers published in SCI journals, such as *Applied Microbiology and Biotechnology*, *Microbial Cell Factories*, and *FEMS Microbiology Letters*. He was invited to review manuscripts for *PloS ONE* and *Journal of Biotechnology*. He has more than 20 authorized Chinese patents and one U.S. patent.

Contributors

Jian Chen
Key Laboratory of Industrial Biotechnology
Ministry of Education
School of Biotechnology
Jiangnan University
Wuxi, Jiangsu, China

Sheng Chen
State Key Laboratory of Food Science and Technology
Jiangnan University
Wuxi, Jiangsu, China

Guocheng Du
Key Laboratory of Industrial Biotechnology
Ministry of Education
School of Biotechnology
Jiangnan University
Wuxi, Jiangsu, China

Zhen Kang
School of Biotechnology
Jiangnan University
Wuxi, Jiangsu, China

Long Liu
Key Laboratory of Industrial Biotechnology
Ministry of Education
School of Biotechnology
Jiangnan University
Wuxi, Jiangsu, China

Song Liu
Key Laboratory of Industrial Biotechnology
Ministry of Education
School of Biotechnology
Jiangnan University
Wuxi, Jiangsu, China

Yanfeng Liu
Key Laboratory of Industrial Biotechnology
Ministry of Education
School of Biotechnology
Jiangnan University
Wuxi, Jiangsu, China

Lili Lu
State Key Lab of Microbial Technology
National Glycoengineering Research Center
and
Shandong Provincial Key Lab of Carbohydrate Chemistry and Glycobiology
Shandong University
Jinan, China

Qinhong Wang
Key Laboratory of Systems Microbial Biotechnology
Tianjin Institute of Industrial Biotechnology
Chinese Academy of Sciences
Tianjin, China

Zhen-hua Wei
Zhejiang Provincial Key Laboratory of Food Biotechnology
Institute of Bioengineering
Hangzhou Wahaha Group Co., Ltd.
Hangzhou, China

Jing Wu
State Key Laboratory of Food Science and Technology
Jiangnan University
Wuxi, Jiangsu, China

Min Xiao
State Key Lab of Microbial Technology
National Glycoengineering Research Center
and
Shandong Provincial Key Lab of Carbohydrate Chemistry and Glycobiology
Shandong University
Jinan, China

Yong Xu
Jiangsu Co-Innovation Center of Efficient Processing and Utilization of Forest Resources
Jiangsu Province Key Laboratory of Green Biomass-Based Fuels and Chemicals
and
College of Chemical Engineering
Nanjing Forestry University
Nanjing, China

Xiaobei Zhan
Key Laboratory of Carbohydrate Chemistry and Biotechnology
Ministry of Education
School of Biotechnology
Jiangnan University
Wuxi, Jiangsu, China

Guoqiang Zhang
Key Laboratory of Systems Microbial Biotechnology
Tianjin Institute of Industrial Biotechnology
Chinese Academy of Sciences
Tianjin, China

Hongtao Zhang
Key Laboratory of Carbohydrate Chemistry and Biotechnology
Ministry of Education
School of Biotechnology
Jiangnan University
Wuxi, Jiangsu, China

Hongyu Zhang
College of Chemical Engineering
Nanjing Forestry University
and
Jiangsu Province Key Laboratory of Green Biomass-based Fuels and Chemicals
Nanjing, China

Jian-Jiang Zhong
State Key Laboratory of Microbial Metabolism
and
Laboratory of Molecular Biochemical Engineering
School of Life Sciences and Biotechnology
Shanghai Jiao Tong University
Shanghai, China

Xin Zhou
Jiangsu Co-Innovation Center of Efficient Processing and Utilization of Forest Resources
Jiangsu Province Key Laboratory of Green Biomass-Based Fuels and Chemicals
and
College of Chemical Engineering
Nanjing Forestry University
Nanjing, China

Yang Zhu
Department of Biosciences
TNO Quality of Life
Zeist, the Netherlands

1

History and Developments in Functional Carbohydrates

SONG LIU, GUOCHENG DU, JIAN CHEN, AND YANG ZHU

Contents

1.1 Introduction

From the point of nutrition, carbohydrate is one of the most important nutrients in the human diet when daily intake is considered. According to dietary advice, 50–60% of total energy that we consume should come from carbohydrates. Many foods contain large amounts of carbohydrates. Not until the early 1980s did the scientific community begin to focus on the physical diversity of carbohydrates. Since then, the knowledge related to the physiological role of carbohydrates and their effect on health and disease has experienced deep development. "Functional carbohydrates" is the term used to describe those carbohydrates that have more nutritious value than traditional carbohydrates and that play an important role in strengthening immunity, decreasing the level of blood-lipid, regulating the intestinal flora of humans, etc. With the growing diversity of carbohydrates, functional carbohydrates now mainly cover dietary fiber, functional polysaccharides, functional oligosaccharides, sugar alcohols, and other functional monosaccharides.

For achieving good results in application, several functional carbohydrates have recently received considerable attention, including hyaluronic acid, curdlan gum, Ganoderma polysaccharide, cyclodextrin, oligosaccharides from hemicellulose, galactooligosaccharide, glucosamine, N-acetylglucosamine, and various sugar alcohols. In this review, the structures, functions, and sources of these representative functional carbohydrates will be summarized briefly.

1.2 Hyaluronic Acid

Hyaluronic acid is a kind of mucopolysaccharide that is composed of disaccharide repeating units (N-acetyl glucosamine [GlcNAc] and D-glucuronic acid [GlcUA]) by alternate beta-(1-3) and beta-(1-4) glycosidic bonds (Weissmann and Meyer 1954). In 1934, hyaluronic acid was first extracted from bovine vitreous body (Meyer and Palmer 1934). Hyaluronic acid was naturally distributed in various tissues of vertebrate's mesenchymal cells and the capsule of some bacteria (Prehm 1984, 1990; Chong et al. 2005), and it was involved in many important metabolic and physiological processes (Necas et al. 2008). Hyaluronic acid could function as lubricant and buffer zones of joints as well as a filling agent of the colloidal structure of the lens. In addition, hyaluronic acid was also involved in cell immobilization and intracellular signal transduction through those specific or nonspecific interactions with the protein components of cells (Karjalainen et al. 2000). Depending on the sources, the main chain length of hyaluronic acid could be 2000–25,000, or even higher, disaccharides unit (approximately 400 Da for each disaccharide). The spatial structure of hyaluronic acid presented a unique rigid helix column, and directional arrangement of a large number of hydroxyl groups forms strong hydrophilic and hydrophobic regions (Lee and Spicer 2000). Due to the negative charges, the GlcUA carboxyl of hyaluronic acid repel each other, resulting in a huge space that can hold a lot of water (1000 times heavier than hyaluronic acid itself; Scott et al. 1984; Laurent et al. 1996). Due to its unique properties, hyaluronic acid has been widely used in medicine, cosmetics, and food industries (Chong et al. 2005; Kogan et al. 2007).

Hyaluronic acid has long been extracted from animal tissues (e.g., cockscomb). The production of hyaluronic acid was thus limited by the quality of raw materials, extraction costs, and especially the animal epidemic risk (Laurencin and Nair 2008). Hyaluronic acid is currently produced by microbial fermentation using the hemolytic species *Streptococcus* C with a weak pathogenicity (Liu et al. 2011). In contrast to those traditional extraction methods, microbial fermentation has many advantages, including simple production process, abundant raw material, high purity, low cost, and environment friendliness. The research on the production of hyaluronic acid by microbial fermentation was initiated in the 1980s in Japan, and then scientists in Britain and the United States and other countries began to study fermentation production of hyaluronic acid (Triscott and van der Rijn 1986; Johns et al. 1994). The yield and productivity of hyaluronic acid have been greatly improved by using different strategies, such as strain improvement, optimization of metabolic pathway, and culture condition (Huang et al. 2006; Liu et al. 2008). Due to the rapid development of molecular biology and DNA sequencing technology, research on the mechanism of microbial synthesis of hyaluronic acid went deeper. DeAngelis et al. (1993) first cloned the gene of *Streptococcus zooepidemicu* hyaluronic acid synthase and realized the heterologous synthesis of hyaluronic acid. Recently, the production of hyaluronic acid was further improved through the controllable expression of hyaluronidase, together with the optimization of the HA synthetic pathway (Jin et al. 2016).

1.3 Curdlan

Curdlan was an extracellular polysaccharide from *Alcalingenes faecalis* (Harada et al. 1968). It is a homopolysaccharide polymer of glucose connected by the β-1,3-D-glucosidic bond, which was determined by nuclear magnetic resonance spectroscopy (Bluhm and Sarko 1977). Curdlan molecule is composed of 300–500 glucose residues, and the molecular formula was $(C_6H_{10}O_5)_n$ ($n > 250$). The gel strength of curdlan increased with the degree of polymerization, and the curdlan gel is not available when the polymerization degree is less than 50.

Normally, the natural gel obtained from 300–500 glucose residues has strong gel properties. Due to the functional properties, curdlan has been used in the food processing industry to improve the water holding capacity, elasticity, and taste characteristics of vegetarian food, pasta, or quick-frozen foods. At the same time, curdlan-based foods are characterized by being low in calories and hard to absorb due to the absence of enzymes capable of degrading curdlan in the human body (Grandpierre et al. 2008).

Professor Harada and his collaborators in Osaka University have done a lot of pioneering work in curdlan production. In 1964, Harada isolated a strain that can be grown in a medium containing ethylene glycol, named *A. faecalis* var. *myxogenes* 10C3. The strain produced a beta glucan containing approximately 10% succinic acid substituents, and a spontaneous mutant 10C3K that produced only polysaccharide gel was then isolated from the 10C3 (Harada et al. 1968). After that, the Takeda company obtained a mutant strain (*A. faecalis* var. IFO13140) that can achieve stable curdlan production (Harada et al. 1968). In 1968, the Takeda company began to study the fermentation production of curdlan and its application in the food industry. In 1971, Wako Chemical Industry company began to test the large-scale production of curdlan, and the curdlan products were sold as chemical reagents from the middle of 1981. Curdlan has become one of the most important food additives in Japan, South Korea, and the Taiwan Province of China a few years later. At the end of 1996, the U.S. FDA proved curdlan as safe for food processing after long-term safety and food toxicology testing, and it was approved as a food additive. Thus, curdlan became the third food polysaccharide approved by the FDA after xanthan gum and gellan gum.

1.4 Cyclodextrins

Cyclodextrins are cyclic oligomeric compounds formed by more than six glucose units through an alpha-1,4-glycosidic bond, which are obtained by enzymatic hydrolysis and cyclization of starch and related substrates. Based on the amount of glucose units in the ring, cyclodextrins could be divided into α- (6 glucose units), β- (7 glucose units), γ-cyclodextrin (8 glucose units), etc., respectively (Li et al. 2007). These three cyclodextrins are the most commonly used and are called the basic cyclodextrins. In addition to the basic cyclodextrin, those rarely used δ-, ε-, ζ-, and η-cyclodextrins were composed by 9, 10, 11, and 12 glucose units, respectively (Terada 1997). The inner cavity of the cyclodextrins contained many nonpolar hydrogen atoms (in C3 and C5) and oxygen atoms in ether bonds; the secondary hydroxyl groups of C2 and C3 of the glucose units were located at the wide side of the cylinder, while the primary hydroxyl group on the C6 was located at the narrow side of the cylinder. This group layout endowed the cyclodextrin cylinder with a hydrophobic inner cavity and a hydrophilic surface. Due to the structural features, it is possible for cyclodextrin to form inclusion complexes with hydrophobic molecules inside. Because those hydrophobic molecules reside in the cavities, the cyclodextrin inclusion remains stable and is not split as it dissolves in water.

At present, cyclodextrin is mainly used in the food industry (Szente and Szejtli 2004). The major function of cyclodextrin may cover the following aspects: (1) stabilization of certain ingredients in food, including the prevention of evaporation, antioxidation, thermal decomposition, etc.; (2) removal of the food smell and bitter taste of raw fish, salted fish, and other seafood; (3) stabilization of emulsion by the addition of cyclodextrin to foods with high oil content, such as beverages, mayonnaise, sauce, ice cream, and coffee beverages; (4) increasing the foaming ability of foods; (5) transformation of liquid food into solid powder to facilitate packaging, transportation, and storage; (6) improvement of the antiseptic effect of food preservatives; and (7) auxiliary materials of food production, such as tofu coagulant made of lactic acid and cyclodextrins. In the pharmaceutical industry, cyclodextrin is a new type of drug carrier developed in recent years. It can use its hollow cylindrical special structure to swallow the drug molecules and form the nonbonded complexes (Davis and Brewster 2004). Cyclodextrins have been widely used in cardiovascular and cerebrovascular drugs, antituberculosis drugs, high blood pressure drugs, gastrointestinal drugs, and so on. In the field of cosmetics, cyclodextrin was mainly used for solubilizing flavor to prevent volatile loss, increasing transparency of cosmetics, and stabilizing the active ingredients of cosmetics (Holland and Malton 2000). In addition, cyclodextrins have also been applied in biotechnology, agriculture, analytical chemistry, environmental protection, and textile processing (Hedges 1998).

The industrial production of cyclodextrins is currently achieved by using enzymatic synthesis, in which cyclodextrin glucosyl transferase (EC 2.4.1.19, CGT) catalyzed the conversion of starch and related matrix into cyclodextrin by cyclization. The preparation of cyclodextrins generally used two processes: noncontrolled process and controlled process. The noncontrolled process referred to the process without organic compound in the reaction system containing CGT enzyme and starch, but it does not rule out the compound used during selective precipitation of α-, β-, or γ-cyclodextrin. In the controlled process, decyl alcohol, toluene, butanone, and other composite agents, which benefit the formation of inclusion complexes of cyclodextrin, were added into the CGT reaction system, subsequently increasing the specific yield of cyclodextrin through changing the reaction equilibrium. With the continuous expansion of application field, the production and sale of cyclodextrin have shown a large growth in recent years. Japan was the first to realize industrial production of β-cyclodextrin in the world. After that, the United States, Hungary, Germany, the Netherlands, and other countries have done lots of studies relating to cyclodextrin production and its application. To date, Nihon Shokuhin Kako Co., Ltd. (Japan), ENSUIKO Sugar Refining Co., Ltd., (Japan), the Corn Company (United States), Wacker Corporation (Germany), the Chemical Products Company (Hungary), and Cyclolab Ltd. are the main suppliers of cyclodextrin. China is one of the countries importing cyclodextrin. In China, Jiangsu Institute of Food Fermentation completed the industrialization test of β-cyclodextrin with an annual output of 20 tons in 1985, and Beijing Institute of Microbiology (Chinese Academy of Sciences) has also made the pilot production and application of β-cyclodextrin. In 1990, Botai Xie from Shanxi Institute of Microbiology developed a new technology of β-cyclodextrin production that improved the industrial production of cyclodextrin in China. However, the industrial production of γ-cyclodextrin has not been realized in China.

1.5 Galactooligosaccharides

It is generally known that galactooligosaccharides (GOS) are oligosaccharides with superior therapeutic effect and safety among various types of functional oligosaccharides. Moreover, it is also the only nonsynthetic oligosaccharide derived from animal milk. GOS was generally composed of galactose and glucose units linked by beta-1,4-glycosidic bond (Rycroft et al. 2001; Osman et al. 2010), while adjacent galactose units were linked by beta-1,3-, beta-1,4-, and beta-1,6-glycosidic bonds (Alander et al. 2001), resulting in trisaccharides, tetroses, pentasaccharides, and hexose (Campbell et al. 1997). Moreover, some microbial beta-D-galactosidase can catalyze hydrolysis of lactose into glucose and galactose, which were further used to synthesize different oligosaccharides through combining with lactose linked by beta-1,3- and beta-1,6-glycosidic bonds. The physical and chemical properties and biological activity of GOS-disaccharide were similar to other GOS, and it therefore belongs to GOS (Yanahira et al. 1995). Currently, the major sources of beta-galactosidase used for producing GOS include *Aspergillus loryaze*, *Aspergillus niger*, *Bacillus circulans*, *Lactobacillus reuteri*, *Bacillus megaterium*, and *Lactobacillus* (such as *Bifidobacterium bifidum* and *Bifidobacterium longum*) (Balasubramaniam et al. 2005; Klein et al. 2013).

GOS is a water-soluble dietary fiber with small molecular weight, which assumes its fluid state in the intestine. This structure feature endowed the GOS with many characteristics, such as low heat content and viscosity, good water solubility, and cool feeling (Crittenden and Playne 1996). It was found that GOS could not be utilized by *Streptococcus oralis* to produce water-soluble monosaccharide and lactic acid, which could effectively reduce the incidence of dental caries. GOS will not be degraded under pH 3.0 or 180°C, suggesting that GOS could be widely used in baking food and fermented milk products (Adamczak et al. 2009). GOS is an effective proliferative factor of many probiotics in the human intestinal tract, such as *Bifidobacterium infant*, *Bifidobacterium adolescentis*, and *Bifidobacterium longum*, which could regulate the balance of intestinal flora. GOS may benefit the absorption of calcium in the intestinal tract and the accumulation of gray matter in the bone, and it thus has a good therapeutic effect on osteoporosis. Meanwhile, the intake of GOS could improve the absorption of magnesium (Griffin et al. 2002) and regulate levels of cholesterol and lipid in the serum (Campbell et al. 1997; Djouzi and Andrieux 1997). Human trials have also shown that the appropriate amount of GOS could alleviate constipation. Moreover, daily intake of GOS could effectively prevent cancer and stimulate vitamin B, folic acid, niacin, etc. In addition, GOS could also reduce blood pressure (Tomomatsu 1994).

The preparation method of GOS includes extraction from the plant, acid hydrolysis of natural polysaccharide, chemical synthesis, microbial fermentation, and enzymatic synthesis. A small amount of natural GOS exists in human milk and some fruit and vegetables (such as onion, garlic, banana, soybean, and artichoke). However, GOS is colorless and has no electric charge, which makes it hard to extract GOS from the plants. When natural polysaccharide was hydrolyzed by acid, low product yield and product purity of GOS were the main problems. In the chemical synthesis, toxic residues limited the application of GOS in food processing. Production of GOS by microbial fermentation has many byproducts, which make it difficult to separate the target product from the fermentation broth. The enzymatic method used lactose in milk as the substrate to synthesize GOS through microbial beta-galactosidase produced by microbial fermentation. The enzyme-producing strain may include *Aspergillus niger*, *Aspergillus oryzae*, *Kluyveromyces*, and *Kluyveromyces fragilis*. Due to abundant source of substrate and low production cost, the enzymatic method is the main method for mass production of GOS. At present, the domestic and foreign production manufacturers of GOS are as follows: Guangdong Jiangmen Quantum High Tech Biological Engineering Co. Ltd. (China), Guangdong Yunfu Xinjinshan Biological Technology Co. Ltd. (China), Yakult HonshaCo. Ltd. (Japan), Nissin Sugar Manufacturing Co. Ltd. (Japan), Borculo Domo Ingredients Pty Ltd. (Holland), and Kerry Group (USA).

1.6 Functional Hydrolysates of Hemicellulose

Hemicellulose is a heterogeneous polymer composed of xylose, arabinose, galactose, and other monosaccharides. Hemicellulose, when combined on the surface of cellulose microfiber, accounts for 50% of total woody tissue, and it is the second largest biopolymer composition of plant cell wall. The content of hemicellulose varies with the kinds of plants. The hemicellulose content in the sapwood of trees from broad-leaved forests reached approximately 31% while the content in woody leaves and bark was between 10% and 15%. Xylan, a main product of hemicellulose hydrolysis, could be further hydrolyzed into xylooligosaccharides by enzymes. Due to the stable physical and chemical properties, xylooligosaccharides could be added into those foods with low pH, such as lactic acid beverages and vinegar beverages. It can even be used for long-term preservation as it will not decompose and affect its healthcare function. Xylooligosaccharides can also benefit the moisture content of bakery products and the rheological properties of dough. The active site of xylooligosaccharides can adsorb intestinal toxic substances and pathogens, and subsequently improve the body's resistance to disease and activate the immune system. Xylooligosaccharide is also a good feed additive. Xyloglucan is a component of hemicellulose, and its main chain is cellulose while the side chains are oligosaccharides, consisting of a few interconnected xylose (Mirzaeva et al. 1999). Previous studies have shown that it can control the development and metabolism of plants. Xyloglucan is also thought to have the function of regulating the immunoregulation activity of plants, and it may also be one of the main functional components to prevent sunburn. Mannan is a highly branched polymer, among which galactomannan is a polysaccharide containing a mannose backbone and a galactose base. Galactomannan gum has good water solubility and cross-linking properties, and can form stable aqueous solution with high viscosity at low concentration. It is used as a thickener, stabilizer, and binder in many industries, such as petroleum, food, medicine, textile printing and dyeing, chemical, mining and dressing, architectural coatings, wood processing, paper making, and pesticides industries.

1.7 Ganoderma Polysaccharide

Ganoderma lucidum is a rare medicinal fungus in China. Because it could prevent or treat many diseases in ancient times, *G. lucidum* was referred to as "immortality." In recent years, many scholars from China and other Southeast Asian countries, the United States, and Canada have conducted studies of *G. lucidum* (Liu and Zhang 2005). Modern chemical research shows that *G. lucidum* contains many kinds of bioactive components, such as polysaccharides, terpenes, sterols, nucleic acids, proteins, and peptides, among which Ganoderma polysaccharide is one of the key components in *G. lucidum*. There are many kinds of Ganoderma polysaccharides, which can be divided into water-soluble polysaccharides, acidic polysaccharides, and alkaline polysaccharides, according to its solubility. More than 200 kinds of *G. lucidum* polysaccharides have been isolated, most of which are beta-glucan. The glucan polysaccharide linked by beta-1,3-glucose is the backbone of Ganoderma polysaccharides, which were finally assembled by three glucan polysaccharide chains (Bluhm and Sarko 1977). Although Ganoderma polysaccharides showed different chemical structures, the structures of the polysaccharides still shared a lot in common. Generally, the composition of monosaccharide covered D-glucose, D-fructose, D-galactose, D-mannose, D-xylose, L-rhamnose, L-arabinose, etc. The main chain was mainly linked by β-1,3-glycosidic bonds and β-1,4-glycosidic bonds, while the branched chains were linked by β-1,6-glycosidic bonds. The structural characteristic of the three helices of Ganoderma polysaccharide is β-1,3-D-glucan. Because of its special chemical structure and conformation, Ganoderma polysaccharide has various biological activities such as antibacterial activity, antiviral activity, antioxidant, promoting crop growth, inhibiting tumors, lowering blood lipid, and regulating immunity. *G. lucidum* polysaccharide did not directly inhibit or kill tumor cells in human and animals, but it stimulated the activation of the immune system and enhanced the immune function of the host, subsequently inhibiting the activity of tumor cells.

Because of the pharmacological function of Ganoderma polysaccharide, its production technology has become the focus of current research. Based on its high solubility in water and low solubility in alcohol, the method "water extraction and alcohol precipitation" was adopted to isolate the Ganoderma polysaccharide. However, this method was time-consuming and used a large quantity of solvent, and was prone to decomposition of the polysaccharide. Recently, ultrasonic and microwave, complex enzyme extraction technology and other new extraction technologies may realize the rapid and efficient extraction of Ganoderma polysaccharides. To further improve the production of Ganoderma polysaccharide, the fermentation technology has received much attention recently. The production of Ganoderma polysaccharide by fermentation includes solid culture fermentation and submerged fermentation. Solid-state fermentation is an economical and effective way to grow sporocarps. However, the solid-state fermentation needs a long time for cell growth and production, and it is difficult to control the product quality and content. Without the drawbacks of solid-state fermentation, the production of Ganoderma polysaccharide by submerged fermentation is widely concerned. Great progress in the optimization of medium and process has been made for the fermentation of *G. lucidum*. The positive effects of vegetable oil, fatty acid, alcohol, surfactant, and extracts of traditional Chinese medicine on biosynthesis of Ganoderma polysaccharide have been examined (Yang et al. 2000, 2004; Chang et al. 2006; Liu and Zhang 2007). Moreover, two-stage pH control (Sun et al. 2008), fed-batch fermentation (Tang et al. 2011), and multifed batch culture integrated with three-stage light irradiation (Zhu et al. 2010) have been developed for process optimization of *G. lucidum*.

1.8 Glucosamine and N-Acetyl Glucosamine

Glucosamine (GlcN) and its derivatives N-acetyl glucosamine (GlcNAc) have been widely used in nutritional chemicals and pharmaceuticals (Chen et al. 2010; Sitanggang et al. 2012; Liu et al. 2013). As a dietary supplement, GlcN and GlcNAc are used to promote and maintain cartilage and joint health. In addition, clinical trials showed that GlcNAc has a significant effect on the treatment of inflammatory bowel disease. Recent studies have shown that GlcN is able to reduce glucose metabolism through activation of AMP activated protein kinase and promote mitochondrial biogenesis, thereby prolonging cell life (Weimer et al. 2014). GlcNAc could also prolong the life span of cells by using the mechanism of GlcNAc to improve the homeostasis of endoplasmic reticulum proteins, thus playing a role in prolonging cell life (Denzel et al. 2014). GlcN and GlcNAc have been widely used in the treatment of disease, disease prevention, cosmetics production and food production, and in other fields. Thus, GlcN and GlcNAc have a broad market, and the current global market will be expected to total about $2 billion. It is expected that GlcN and GlcNAc global production will reach 50,000 tons in 2017. The major markets of GlcN and GlcNAc are in the Asia Pacific region, North America, and Europe. China is the largest production market and supply market of GlcN and GlcNAc in the world (Liu et al. 2013).

Commercial GlcN and GlcNAc are, however, mainly made from enzymatic hydrolysis of chitin, which has many potential problems in the source of raw materials, environmental protection, and product safety (Sitanggang et al. 2012; Mojarrad et al. 2007). Therefore, GlcN and GlcNAc with their good food safety and non-crustacean levels are urgently required. Microbial fermentation is a sustainable, environmentally friendly, and economical way to produce GlcN and GlcNAc. Since GlcNAc can be converted to GlcN under weak acid conditions, GlcNAc can also be obtained by microbial fermentation use of GlcNAc engineering bacteria, and then further be converted to GlcN (Deng et al. 2005).

At present, the main strains used in the production of GlcN and GlcNAc by the fermentation method are fungi and *Escherichia coli*. GlcN and GlcNAc are the monomers of chitin and chitosan in the fungal cell wall; the hydrolysis of the fungal cell wall can get GlcN. Researchers are paying more attention to fungal fermentation because it is a renewable production method that overcomes the influence of the shortage of raw materials on the production of GlcN. *Aspergillus* sp. BCRC31742 is the strain with the highest yield of GlcN among three wild-type fungi *Rhizopus oligosorus*, *Monascus pilosus*, and *Aspergillus* sp. (Hsieh et al. 2007; Sitanggang et al. 2010; Zhang et al. 2012b). The yield of GlcN produced by *Aspergillus* sp. BCRC31742 can reach 5.48 g/L by medium optimization. The GlcN yield is increased to 7.48 g/L (Sitanggang et al. 2010) after stimulation with methanol. By controlling dissolved oxygen level of two periods, the highest yield of GlcN reached 14.37 g/L (Zhang et al. 2012b). Metabolic engineering methods were applied to the production of GlcN recombinant *E. coli* constructs (Deng et al. 2005, 2006), and the yield of GlcN and GlcNAc reached 17 and 110 g/L, respectively (Deng et al. 2005). However, *E. coli* is not a food-grade strain and is easily polluted by phage in the fermentation process. To solve this problem, we recently used *Bacillus subtilis* 168 as the initial host for metabolic engineering for microbial GlcN and GlcNAc production, which is a wild-type strain of *B. subtilis* and meets the requirement for generally regarded as safe (GRAS) grade for biochemical production. Synthetic biology tools and systems biology methods were combined with metabolic engineering to optimize the cellular properties of engineered *B. subtilis* (Liu et al. 2013, 2014a,b).

1.9 Sugar Alcohols

Sugar alcohols are a kind of polyols containing two or more hydroxyls. However, sugar alcohols can be made of corresponding sugars, in which the aldehyde or ketone is reduced to hydroxyl. For example, sorbitol can be generated by reduction of glucose while reduction of xylose obtains xylitol. This process makes it different from ethylene glycol, propylene glycol, pentaerythritol, and other polyols from petroleum and chemical synthesis. Functional sugar alcohols include sorbitol, mannitol, erythritol, maltitol, lactitol, xylitol, etc. In contrast to sucrose, sugar alcohols have higher thermal stability and acid stability, which make them more strongly adaptable in high-temperature processed foods and low-acid foods. The sweetening properties of sugar alcohols are similar to sucrose, which has a pure, sweet, metallic taste. The sweetness of sugar alcohols is generally lower than sucrose. The sweetness of erythritol, maltitol, sorbitol, and lactitol could be defined as 40%, 75%, 80–90%, and 60% that of sucrose, respectively.

The physiological functions of sugar alcohols can be summarized as four aspects:

1. *Unique metabolism and low calorie.* Erythritol was easily absorbed by the small intestine, and most went into blood circulation. Because the body lacks the metabolism of the erythritol enzyme system, the erythritol cannot be absorbed and is subsequently dispelled in the form of urine. This unique metabolic way determines erythritol's low-calorie characteristics. Thus, sugar alcohol sweeteners are a class of low-calorie sweeteners used for the control of obesity.
2. *Anticaries.* Sugar alcohols (xylitol, erythritol, lactitol, etc.) belong to a unique class of polyhydroxy compounds that can inhibit the growth of *Streptococcal mutants* and acid production, reducing the rate of dental caries infection.
3. *Hypoglycemia reaction.* Sugar alcohols in the human body do not rely on insulin metabolism and can be transformed into glycogen for the body (except erythritol). So sugar alcohols can be used by diabetic patients, giving them only a small amount of energy.
4. *Other health functions.* For example, erythritol is a good antioxidant sweetener with an efficient free radical scavenging effect.

The microorganisms that can produce xylitol in nature include *Enterobacter liquefaciens* (Yoshitake et al. 1973), *Corynebacterium* sp., *Mycobacterium smegmaitis* (Kwon et al. 2006), and *Gluconobacter oxydans* (Suzuki et al. 2002) in bacteria; *Aspergillus*, *Byssochlamys*, *Penicillium*, *Rhizopus*, and *Neurospora* sp. (Suihko 1983), yeast *Candida* (Prakash et al. 2011), and *Hansenula* (Zhang et al. 2012a). The mass production of erythritol by fermentation has been relatively mature, and the microbial fermentation by yeast and bacteria (lactic acid bacteria) that are resistant to high temperature has been widely used in commercial production (Ishizuka et al. 1989; Veiga-Da-Cunha et al. 1992; Seo et al. 2001). Glucose and fructose are used as the substrates to produce sorbitol by biosynthesis. Whole-cell transformation is carried out with *Pseudomonas fluorescens* containing glucose–fructose oxidoreductase. Japan is one of the first countries that produced maltitol on a large scale. To date, the main production method is based on preparation of maltose by using double enzymes, and then the maltose is further transformed into maltitol through chlorination. Mannitol production methods commonly use fructose and glucose as the raw materials, and it covers natural extraction, chemical synthesis, and electrochemical and biological transformation methods. There are many mannitol production strains, including yeast (*Zygosaccharomyces* and *Candida*), lactic bacteria (*Leuconostoc mesenteroides*, *Lactobacillus fermenti*), and filamentous fungi (*Aspergillus* and *Penicillium*). Preparation of lactitol includes hydrogenation with high pressure and nickel catalysis, boron hydride reduction, and electrolytic reduction method.

References

Adamczak, M., Charubin, D. and Bednarski, W. (2009). Influence of reaction medium composition on enzymatic synthesis of galactooligosaccharides and lactulose from lactose concentrates prepared from whey permeate. *Chemical Papers*. **63**(2), 111–116.

Alander, M., Matto, J., Kneifel, W., Johansson, M., Kogler, B., Crittenden, R., Mattila-Sandholm, T. and Saarela, M. (2001). Effect of galacto-oligosaccharide supplementation on human faecal microflora and on survival and persistence of Bifidobacterium lactis Bb-12 in the gastrointestinal tract. *International Dairy Journal*. **11**(10), 817–825.

Balasubramaniam, S., Lee, H. C., Lazan, H., Othman, R. and Ali, Z. M. (2005). Purification and properties of a beta-galactosidase from carambola fruit with significant activity towards cell wall polysaccharides. *Phytochemistry*. **66**(2), 153–163.

Bluhm, T. L. and Sarko, A. (1977). The triple helical structure of lentinan, a linear β-(1-3)-D-glucan. *Canadian Journal of Chemistry*. **55**(2), 293–299.

Campbell, J. M., Fahey, G. C. and Wolf, B. W. (1997). Selected indigestible oligosaccharides affect large bowel mass, cecal and fecal short-chain fatty acids, pH and microflora in rats. *Journal of Nutrition*. **127**(1), 130–136.

Chang, M. Y., Tsai, G. J. and Houng, J. Y. (2006). Optimization of the medium composition for the submerged culture of *Ganoderma lucidum* by Taguchi array design and steepest ascent method. *Enzyme and Microbial Technology*. **38**(3–4), 407–414.

Chen, J. K., Shen, C. R. and Liu, C. L. (2010). N-acetylglucosamine: Production and application. *Marine Drugs*. **8**(9), 2493–2516.

Chong, B. F., Blank, L. M., Mclaughlin, R. and Nielsen, L. K. (2005). Microbial hyaluronic acid production. *Applied Microbiology and Biotechnology*. **66**(4), 341–351.

Crittenden, R. G. and Playne, M. J. (1996). Production, properties and applications of food-grade oligosaccharides. *Trends in Food Science & Technology*. **7**(11), 353–361.

Davis, M. E. and Brewster, M. E. (2004). Cyclodextrin-based pharmaceutics: Past, present and future. *Nature Reviews Drug Discovery*. **3**(12), 1023–1035.

DeAngelis, P. L., Papaconstantinou, J. and Weigel, P. H. (1993). Isolation of a *Streptococcus pyogenes* gene locus that directs hyaluronan biosynthesis in acapsular mutants and in heterologous bacteria. *Journal of Biological Chemistry*. **268**(20), 14568–14571.

Deng, M. D., Severson, D. K., Grund, A. D. et al. (2005). Metabolic engineering of *Escherichia coli* for industrial production of glucosamine and N-acetylglucosamine. *Metabolic Engineering*. **7**(3), 201–214.

Deng, M. D., Wassink, S. L. and Grund, A. D. (2006). Engineering a new pathway for N-acetylglucosamine production: Coupling a catabolic enzyme, glucosamine-6-phosphate deaminase, with a biosynthetic enzyme, glucosamine-6-phosphate N-acetyltransferase. *Enzyme and Microbial Technology*. **39**(4), 828–834.

Denzel, M. S., Storm, N. J., Gutschmidt, A. et al. (2014). Hexosamine pathway metabolites enhance protein quality control and prolong life. *Cell*. **156**(6), 1167–1178.

Djouzi, Z. and Andrieux, C. (1997). Compared effects of three oligosaccharides on metabolism of intestinal microflora in rats inoculated with a human faecal flora. *British Journal of Nutrition*. **78**(2), 313–324.

Grandpierre, C., Janssen, H. G., Laroche, C., Michaud, P. and Warrand, J. (2008). Enzymatic and chemical degradation of curdlan targeting the production of beta-(1 -> 3) oligoglucans. *Carbohydrate Polymers*. **71**(2), 277–286.

Griffin, I. J., Davila, P. M. and Abrams, S. A. (2002). Non-digestible oligosaccharides and calcium absorption in girls with adequate calcium intakes. *British Journal of Nutrition*. **87**, S187–S191.

Harada, T., Misaki, A. and Saito, H. (1968). Curdlan: A bacterial gel-forming beta-1,3-glucan. *Archives of Biochemistry and Biophysics*. **124**(1), 292–298.

Hedges, A. R. (1998). Industrial applications of cyclodextrins. *Chemical Reviews*. **98**(5), 2035–2044.

Holland L. R. G. and Malton, P. (2000). Cosmetic compositions comprising cyclic oligosaccharides and fragrance. WO Pat. 2000/067716.

Hsieh, J. W., Wu, H. S., Wei, Y. H. and Wang, S. S. (2007). Determination and kinetics of producing glucosamine using fungi. *Biotechnology Progress*. **23**(5), 1009–1016.

Huang, W.-C., Chen, S.-J. and Chen, T.-L. (2006). The role of dissolved oxygen and function of agitation in hyaluronic acid fermentation. *Biochemical Engineering Journal*. **32**(3), 239–243.

Ishizuka, H., Wako, K., Kasumi, T. and Sasaki, T. (1989). Breeding of a mutant of *Aureobasidium sp.* with high erythritol production. *Journal of Fermentation and Bioengineering*. **68**(5), 310–314.

Jin, P., Kang, Z., Yuan, P., Du, G. and Chen, J. (2016). Production of specific-molecular-weight hyaluronan by metabolically engineered *Bacillus subtilis* 168. *Metabolic Engineering*. **35**, 21–30.

Johns, M. R., Goh, L. T. and Oeggerli, A. (1994). Effect of pH, agitation and aeration on hyaluronic-acid production by *Streptococcus–Zooepidemicus*. *Biotechnology Letters*. **16**(5), 507–512.

Karjalainen, J. M., Tammi, R. H., Tammi, M. I., Eskelinen, M. J., Agren, U. M., Parkkinen, J. J., Alhava, E. M. and Kosma, V. M. (2000). Reduced level of CD44 and hyaluronan associated with unfavorable prognosis in clinical stage I cutaneous melanoma. *American Journal of Pathology*. **157**(3), 957–965.

Klein, M. P., Fallavena, L. P., Schoeffer, J. d. N., Ayub, M. A. Z., Rodrigues, R. C., Ninow, J. L. and Hertz, P. F. (2013). High stability of immobilized beta-D-galactosidase for lactose hydrolysis and galactooligosaccharides synthesis. *Carbohydrate Polymers*. **95**(1), 465–470.

Kogan, G., Soltes, L., Stern, R. and Gemeiner, P. (2007). Hyaluronic acid: A natural biopolymer with a broad range of biomedical and industrial applications. *Biotechnology Letters*. **29**(1), 17–25.

Kwon, S.-G., Park, S.-W. and Oh, D.-K. (2006). Increase of xylitol productivity by cell-recycle fermentation of *Candida tropicalis* using submerged membrane bioreactor. *Journal of Bioscience and Bioengineering*. **101**(1), 13–18.

Laurencin, C. T. and Nair, L. (2008). The FDA and safety—Beyond the heparin crisis. *Nature Biotechnology*. **26**(6), 621–623.

Laurent, T. C., Laurent, U. B. and Fraser, J. R. (1996). The structure and function of hyaluronan: An overview. *Immunology and Cell Biology*. **74**(2), A1–7.

Lee, J. Y. and Spicer, A. P. (2000). Hyaluronan: A multifunctional, megaDalton, stealth molecule. *Current Opinion in Cell Biology*. **12**(5), 581–586.

Li, Z. F., Wang, M., Wang, F., Gu, Z. B., Du, G. C., Wu, J. and Chen, J. (2007). gamma-Cyclodextrin: A review on enzymatic production and applications. *Applied Microbiology and Biotechnology*. **77**(2), 245–255.

Liu, G. Q. and Zhang, K. C. (2005). Mechanisms of the anticancer action of *Ganoderma lucidum* (Leyss. ex Fr.) Karst.: A new understanding. *Journal of Integrative Plant Biology*. **47**(2), 129–135.

Liu, G. Q. and Zhang, K. C. (2007). Enhancement of polysaccharides production in *Ganoderma lucidum* by the addition of ethyl acetate extracts from *Eupolyphaga sinensis* and *Catharsius molossus*. *Applied Microbiology and Biotechnology*. **74**(3), 572–577.

Liu, L., Du, G., Chen, J., Wang, M. and Sun, J. (2008). Enhanced hyaluronic acid production by a two-stage culture strategy based on the modeling of batch and fed-batch cultivation of *Streptococcus zooepidemicus*. *Bioresource Technology*. **99**(17), 8532–8536.

Liu, L., Liu, Y. F., Li, J. H., Du, G. C. and Chen, J. (2011). Microbial production of hyaluronic acid: Current state, challenges, and perspectives. *Microbial Cell Factories*. **10**, 99–107.

Liu, L., Liu, Y. F., Shin, H. D., Chen, R., Li, J. H., Du, G. C. and Chen, J. (2013). Microbial production of glucosamine and N-acetylglucosamine: Advances and perspectives. *Applied Microbiology and Biotechnology*. **97**(14), 6149–6158.

Liu, Y., Liu, L., Shin, H. D., Chen, R. R., Li, J., Du, G. and Chen, J. (2013). Pathway engineering of *Bacillus subtilis* for microbial production of N-acetylglucosamine. *Metabolic Engineering*. **19**, 107–115.

Liu, Y., Zhu, Y., Li, J., Shin, H. D., Chen, R. R., Du, G., Liu, L. and Chen, J. (2014a). Modular pathway engineering of *Bacillus subtilis* for improved N-acetylglucosamine production. *Metabolic Engineering*. **23**, 42–52.

Liu, Y., Zhu, Y., Ma, W., Shin, H. D., Li, J., Liu, L., Du, G. and Chen, J. (2014b). Spatial modulation of key pathway enzymes by DNA-guided scaffold system and respiration chain engineering for improved N-acetylglucosamine production by *Bacillus subtilis*. *Metabolic Engineering*. **24**, 61–69.

Meyer, K. and Palmer, J. W. (1934). The polysaccharide of the vitreous humor. *Journal of Biological Chemistry*. **107**(3), 629–634.

Mirzaeva, M. R., Rakhmanberdyeva, R. K. and Rakhimov, D. A. (1999). Structure of galactomannan from seeds of *Gleditsia macracantha*. *Chemistry of Natural Compounds*. **35**(5), 505–507.

Mojarrad, J. S., Nemati, M., Valizadeh, H., Ansarin, M. and Bourbour, S. (2007). Preparation of glucosamine from exoskeleton of shrimp and predicting production yield by response surface methodology. *Journal of Agricultural and Food Chemistry*. **55**(6), 2246–2250.

Necas, J., Bartosikova, L., Brauner, P. and Kolar, J. (2008). Hyaluronic acid (hyaluronan): A review. *Veterinarni Medicina*. **53**(8), 397–411.

Osman, A., Tzortzis, G., Rastall, R. A. and Charalampopoulos, D. (2010). A comprehensive investigation of the synthesis of prebiotic galactooligosaccharides by whole cells of *Bifidobacterium bifidum* NCIMB 41171. *Journal of Biotechnology*. **150**(1), 140–148.

Prakash, G., Varma, A., Prabhune, A., Shouche, Y. and Rao, M. (2011). Microbial production of xylitol from D-xylose and sugarcane bagasse hemicellulose using newly isolated thermotolerant yeast *Debaryomyces hansenii*. *Bioresource Technology*. **102**(3), 3304–3308.

Prehm, P. (1984). Hyaluronate is synthesized at plasma membranes. *Biochemical Journal*. **220**(2), 597–600.

Prehm, P. (1990). Release of hyaluronate from eukaryotic cells. *Biochemical Journal*. **267**(1), 185–189.

Rycroft, C. E., Jones, M. R., Gibson, G. R. and Rastall, R. A. (2001). A comparative in vitro evaluation of the fermentation properties of prebiotic oligosaccharides. *Journal of Applied Microbiology*. **91**(5), 878–887.

Scott, J. E., Heatley, F. and Hull, W. E. (1984). Secondary structure of hyaluronate in solution. A 1H-n.m.r. investigation at 300 and 500 MHz in [2H6]dimethyl sulphoxide solution. *Biochemical Journal*. **220**(1), 197–205.

Seo, J. H., Ryu, Y. W., Jung, S. R. and Kim, S. Y. (2001). Fermentation process for preparing erythritol by a high salt tolerant mutant of *Candida sp*. Google Patents.

Sitanggang, A. B., Sophia, L. and Wu, H.-S. (2012). Aspects of glucosamine production using microorganisms. *International Food Research Journal*. **19**(2), 393–404.

Sitanggang, A. B., Wu, H.-S., Wang, S. S. and Ho, Y. C. (2010). Effect of pellet size and stimulating factor on the glucosamine production using *Aspergillus sp* BCRC 31742. *Bioresource Technology*. **101**(10), 3595–3601.

Suihko, M.-L. (1983). The fermentation of different carbon sources by *Fusarium oxysporum*. *Biotechnology Letters*. **5**(11), 721–724.

Sun, J. X., Zhang, Q. and Zhu, H. X. (2008). Optimization on fermentation conditions for *Ganoderma fungus*. *China Brewing*. (7), 27–29.

Suzuki, S.-I., Sugiyama, M., Mihara, Y., Hashiguchi, K.-I. and Yokozeki, K. (2002). Novel enzymatic method for the production of xylitol from D-arabitol by *Gluconobacter oxydans*. *Bioscience, Biotechnology, and Biochemistry*. **66**(12), 2614–2620.

Szente, L. and Szejtli, J. (2004). Cyclodextrins as food ingredients. *Trends in Food Science & Technology*. **15**(3–4), 137–142.

Tang, Y. J., Zhang, W., Liu, R. S., Zhu, L. W. and Zhong, J. J. (2011). Scale-up study on the fed-batch fermentation of *Ganoderma lucidum* for the hyperproduction of ganoderic acid and *Ganoderma* polysaccharides. *Process Biochemistry*. **46**(1), 404–408.

Terada, Y., Yanase, M., Takata, H., Takaha, T. and Okada, S. (1997). Cyclodextrins are not the major cyclic α-1,4-glucans produced by the initial action of cyclodextrin glucanotransferase on amylose. *Journal of Biological Chemistry*. **272**(25), 15729–15733.

Tomomatsu, H. (1994). Health-effects of oligosaccharides. *Food Technology*. **48**(10), 61–65.

Triscott, M. X. and van der Rijn, I. (1986). Solubilization of hyaluronic acid synthase activity from *Streptococci* and its activation with phospholipids. *Journal of Biological Chemistry*. **261**(13), 6004–6009.

Veiga-Da-Cunha, M., Firme, P., San Romão, M. V. and Santos, H. (1992). Application of 13C nuclear magnetic resonance to elucidate the unexpected biosynthesis of erythritol by *Leuconostoc oenos*. *Applied and Environmental Microbiology*. **58**(7), 2271–2279.

Weimer, S., Priebs, J., Kuhlow, D. et al. (2014). D-Glucosamine supplementation extends life span of nematodes and of ageing mice. *Nature Communications*. **5**(4), 3563.

Weissmann, B. and Meyer, K. (1954). The structure of hyalobiuronic acid and of hyaluronic acid from umbilical cord. *Journal of the American Chemical Society*. **76**(7), 1753–1757.

Yanahira, S., Kobayashi, T., Suguri, T., Nakakoshi, M., Miura, S., Ishikawa, H. and Nakajima, I. (1995). Formation of oligosaccharides from lactose by *Bacillus circulans* beta-galactosidase. *Bioscience Biotechnology and Biochemistry*. **59**(6), 1021–1026.

Yang, F. C., Ke, Y. F. and Kuo, S. S. (2000). Effect of fatty acids on the mycelial growth and polysaccharide formation by *Ganoderma lucidum* in shake flask cultures. *Enzyme and Microbial Technology*. **27**(3–5), 295–301.

Yang, H. L., Wu, T. X. and Zhang, K. C. (2004). Enhancement of mycelial growth and polysaccharide production in *Ganoderma lucidum* (the Chinese medicinal fungus, 'Lingzhi') by the addition of ethanol. *Biotechnology Letters*. **26**(10), 841–844.

Yoshitake, J., Ishizaki, H., Shimamura, M. and Imai, T. (1973). Xylitol production by an Enterobacter species. *Agricultural and Biological Chemistry*. **37**(10), 2261–2267.

Zhang, J., Geng, A., Yao, C., Lu, Y. and Li, Q. (2012a). Xylitol production from D-xylose and horticultural waste hemicellulosic hydrolysate by a new isolate of *Candida athensensis* SB18. *Bioresource Technology*. **105**, 134–141.

Zhang, J., Liu, L., Li, J., Du, G. and Chen, J. (2012b). Enhanced glucosamine production by *Aspergillus sp.* BCRC 31742 based on the time-variant kinetics analysis of dissolved oxygen level. *Bioresource Technology*. **111**, 507–511.

Zhu, L. W., Zhong, J. J. and Tang, Y. J. (2010). Multi-fed batch culture integrated with three-stage light irradiation and multiple additions of copper ions for the hyperproduction of ganoderic acid and *Ganoderma* polysaccharides by the medicinal mushroom *Ganoderma lucidum*. *Process Biochemistry*. **45**(12), 1904–1911.

2

Microbial Production of Hyaluronic Acid

Current State, Challenges, and Perspectives

ZHEN KANG, LONG LIU, AND SONG LIU

Contents

2.1 Introduction

Hyaluronic acid (HA) is a linear and unbranched high molecular weight polysaccharide. It is composed of disaccharide repeats of D-glucuronic acid (GlcUA) and N-acetylglucosamine (GlcNAc) joined alternately by β-1,3 and β-1,4 glycosidic bonds (Figure 2.1). The molecular weight of HA from different sources is highly variable, ranging from 10^4 to 10^7 Da. In the human body, HA occurs in the salt hyaluronate form and is found in high concentrations in the skin, umbilical cord, and vitreous humor (Kogan et al. 2007). In addition, HA is also discovered in the capsules of many microbial strains (e.g., *streptococci* sp). Due to its distinctive moisturizing retention ability and viscoelasticity, coupled with its lack of immunogenicity and toxicity,

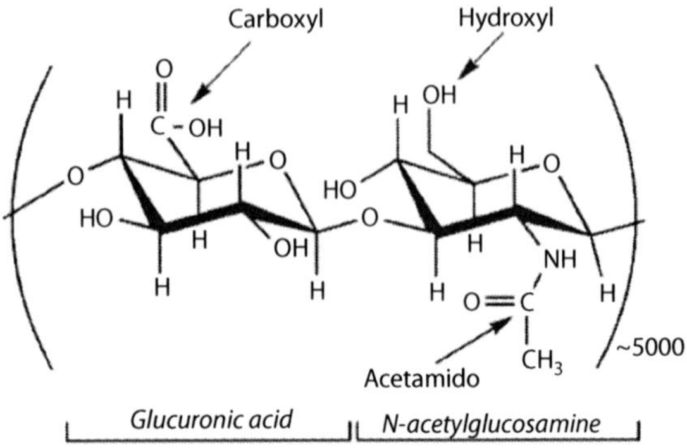

Figure 2.1 Structure of disaccharide repeating unit of HA.

and its significant structural, rheological, physiological, and active biological functions, HA has various applications in the cosmetic, biomedical, and food industries (Chong et al. 2005). Traditionally, HA was extracted from rooster combs while now it is mainly produced via microbial fermentation with lower production costs and less environmental pollution (Chong and Nielsen 2003a,b; Chien and Lee 2007a,b; Liu, Wang et al. 2008). HA has been successfully produced at an industrial scale with *Streptococcus* sp. as the main producer. Nevertheless, the production of HA from *Streptococcus* sp. is facing a growing concern due to the fact that *Streptococci* are pathogenic (Widner et al. 2005). In this background, construction of recombinant strains for producing HA has attracted increasing interest; for instance, Novozymes has realized production of HA with recombinant *Bacillus subtilis* on an industrial scale (Widner et al. 2005).

2.2 History and HA Market

2.2.1 History

In 1934, Karl Meyer and John Palmer described a new polysaccharide isolated from bovine vitreous humor. They found that the substance contained uronic acid and aminosugar, and accordingly named the polysaccharide as "hyaluronic acid" (Meyer and Palmer 1934). The term "hyaluronan" was introduced in 1986 to conform to polysaccharide nomenclature. During the 1930s and 1940s, HA was isolated from many sources including the vitreous body, umbilical cord, rooster comb, and *streptococci* (Kendall et al. 1937). The chemical structure of HA was essentially solved by Karl Meyer and his coworkers, who found that HA consists of disaccharide repeats of D-glucuronic acid (GlcUA) and N-acetylglucosamine (GlcNAc) joined alternately by β-1,3 and β-1,4 glycosidic bonds (Figure 2.1). The physicochemical characterization of HA was conducted during the 1950s and 1960s. At a concentration as low as 0.1%, the HA chains were entangled, resulting in an extremely high and shear-dependent viscosity (Laurent et al. 1996). These properties enabled HA to regulate water balance and flow resistance, to act as a lubricant, and to stabilize structures (Chong et al. 2005). The original development of HA as a product used in clinical medicine is thanks to Endre Balazs, who developed the first noninflammatory, highly purified high molecular weight HA from umbilical cords and rooster combs (Balazs and Sweeney 1968). In the early 1980s, HA had been used as a major material in ophthalmic surgery.

The rooster comb-based extraction process is facing a growing concern over the use of animal-derived components in biomedical and pharmaceutical applications. Hence, microbial production of HA has emerged as a new alternative. The first commercially fermented HA was produced from *Streptococcus zooepidemicus*, which remains the current common strain in the industrial production of HA (Chong and Nielsen 2003a,b; Krahulec and Krahulcova 2006). Nevertheless, the presence of bacterial endotoxins in HA from streptococcal fermentation limits the application of HA in biomedical field (Widner et al. 2005; Chien and Lee 2007a). Therefore, recombinant HA production has emerged as an attractive alternative. Both Gram-positive and Gram-negative bacteria were used as hosts, including *Bacillus* sp. (Widner et al. 2005; Chien and Lee 2007), *Lactococcus lactis* (Chien and Lee 2007b), *Agrobacterium* sp. (Mao and Chen 2007), and *Escherichia coli* (Yu and Stephanopoulos 2008). The applications of HA depend on its molecular weight, which is an important quality parameter for characterizing commercial HA products. Yet, the fermentation product is a mixture of HAs with different molecular weights. Obtaining HA with a uniform molecular weight represents a challenge and much work has been conducted to elucidate the molecular weight control mechanism, which is a current research focus in the field of microbial HA production (Chen, Marcellin et al. 2009; Sheng et al. 2009; Jagannath and Ramachandran 2010).

2.2.2 *HA Market*

The current worldwide market for HA is estimated to be worth over $1 billion (Chong et al. 2005). With the knee osteoarthritis patient population increasing by 26% from 15 million in 2000 to 19 million in 2010, the demand for viscosupplements is expected to escalate. In the United States, the first single-injection HA viscosupplementation product, Synvisc-One, was approved in February 2009, and the product gained rapid acceptance by patients and physicians because of its convenience. The European HA viscosupplementation market is shifting toward shorter treatment regimens and the convenience of having to undergo the procedure only once has continously attracted more patients through 2013. In the Asia Pacific, the HA viscosupplementation market will be favorably affected by both the aging and physically active demographics, as well as rising awareness of the treatment's benefits among physicians and patients.

The global market for dermal fillers is booming, at approximately $759 million during 2009, according to Medical Insight Inc. Nowadays, there are almost 100 different dermal fillers on the market, and about half of them are based on HA. The American Society for Aesthetic Plastic Surgery reports that about 23,000 dermatologists, plastic surgeons, and cosmetic surgeons in the United States performed more than 11.8 million surgical and nonsurgical cosmetic surgery procedures in 2004, generating $12.5 billion in fees (Liu et al. 2011). The dermal filler market is expanding at an annual rate of more than 25% through 2011 in the United States, and 20% throughout the rest of the world, reaching $1.5 billion in global sales. The launch of QMed's Restylane, with NASHA technology (nonanimal stabilized HA), has ushered in a new era in dermal enhancement. This filler addresses many of the issues with traditional bovine collagen fillers, namely shelf life, skin testing, and its animal origin.

2.3 Microbial Production of HA with *Streptococcus zooepidemicus*

Microbial HA production on an industrial scale was first achieved in the 1980s by Shiseido. The commonly used strain in HA production is *Streptococcus zooepidemicus*, which produces 6–7 g/L HA under suitable culture conditions. Figure 2.2 shows the synthesis pathway of

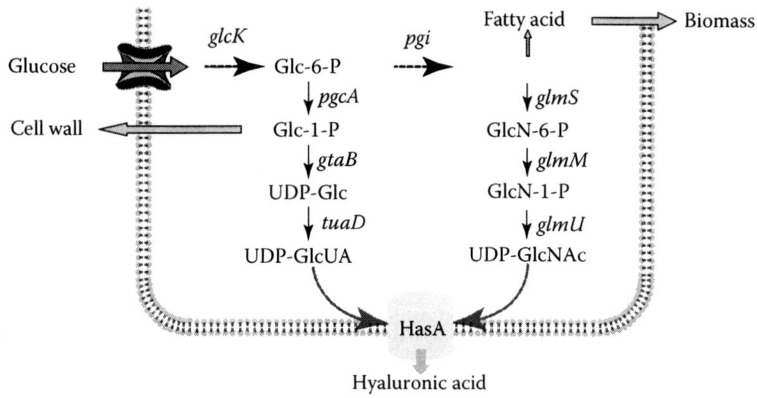

Figure 2.2 Biological synthesis pathway of HA in *Streptococcus zooepidemicus.*

HA in *S. zooepidemicus*. However, some challenges still exist for HA production with *S. zooepidemicus*. First, the broth viscosity reached as high as 400–500 mPas at 4–5 g/L HA, which caused poor mixing and low oxygen mass transfer rate, and thus HA production was severely limited. Second, there exists strong competition between HA synthesis and cell growth for the common precursors such as uridine diphosphate (UDP)-N-acetyl-glucosamine and UDP-glucuronic acid. Third, lactic acid is a main byproduct of HA fermentation, and the accumulation of lactic acid results in a strong inhibition of cell growth and HA synthesis. Extensive studies have been conducted to improve HA production by *S. zooepidemicus*, and the recent advances are summarized next.

2.3.1 Biosynthesis Pathway of HA

The sugar backbone of HA is derived from glucose-6-phosphate and fructose-6-phosphate. The HA synthesis pathway can be divided into two sets. In the first set of reactions, glucose-6-phosphate is converted to glucose-1-phosphate by a-phosphoglucomutase. UDP-glucose pyrophosphorylase adds UTP to glucose-1-phosphate to produce UDP-glucose. Finally, oxidation of the primary alcohol in UDP-glucose by UDP-glucose dehydrogenase yields the first HA precursor, UDP-glucuronic acid. In the second set of reactions, glutamine fructose-6-phosphate amidotransferase transfers the amido group from glutamine to fructose-6-phosphate to yield glucosamine-6-phosphate.

Phosphate group rearrangement by phosphoglucosamine mutase produces glucosamine-1-phosphate. The acetylated form of this compound is produced in the next step by phosphoglucosamine acetyltransferase. Finally, N-acetylglucosamine-1-phosphate pyrophosphorylase activates the intermediate by the addition of UTP, thus yielding the second HA precursor, UDP-N-acetylglucosamine.

Figure 2.2 also shows that HA synthesis and cell growth share precursors such as glucose-1-phosphate, UDP-glucose, and UDP-N-acetylglucosamine. Thus, there is competition between HA synthesis and cell growth to consume the same precursors, and a high specific growth rate is not favorable for HA synthesis (Armstrong et al. 1997). In addition, it can be seen that the glycolysis and HA synthesis compete for the carbon flux. Therefore, weakening the glycolytic process and reducing the rate of biomass formation are effective for the enhancement of HA titer and molecular weight. For example, HA titer was improved from 5.0 to 6.5 g/L by reducing the biomass formation rate via an intermittent alkaline stress strategy, where the cyclical pH switch from 7.0 (optimal for cell growth) to 8.5 (sub-optimal for cell growth) was done every 2 h for 6–16 h (Liu, Wang et al. 2008). Both the cyclical temperature switch from 37°C to 30°C (reducing cell growth rate) and the addition of pyruvate (weakening the glycolytic process) can significantly increase the HA titer (Jagannath and Ramachandran 2010).

2.3.2 HA Fermentation Optimization

Streptococci spp. are nutritionally fastidious lactic acid bacteria and cannot synthesize some amino acids (Armstrong et al. 1997). Supplementing some amino acids such as arginine and lysine in the medium was favorable for cell growth and HA production (Liu, Sun et al. 2009). The chemically defined medium (CDM), containing some nutritional factors essential to growth, can also be used for the culture of *S. zooepidemicus* with the same HA concentration and specific HA production rate relative to complex medium (Armstrong et al. 1997). Zhang et al. (2006) developed a serum-free medium with starch as the exclusive carbon source, and HA concentration reached 6.7 g/L. The mineral ions and initial glucose concentration also had significant effects on microbial HA production (Pires, Eguchi et al.

2010; Pires and Santana 2010). The absence of glucose resulted in a mixed acid metabolism independent of the oxygen supply, while, for the initial glucose concentrations ranging from 5 to 90 g/L, the homolactic metabolism was prevalent (Pires and Santana 2010).

In addition to the culture medium, culture conditions (pH, temperature, agitation speed, aeration rate, shear stress, dissolved oxygen, and bioreactor type) also significantly; influence the microbial HA production. The pH and temperature for HA production by *S. zooepidemicus* were usually at 7.0°C and 37°C, respectively (Johns et al. 1994; Kim et al. 1996). The microbial HA production by *S. zooepidemicus* is a typically viscous process, and thus, mixing performance and oxygen mass transfer rate significantly influence HA production. The effects of agitation speed, aeration rate, shear stress, and dissolved oxygen on microbial HA production have been extensively studied (Johns et al. 1994; Kim et al. 1996; Huang et al. 2006; Duan et al. 2008, 2009; Liu, Du et al. 2009; Liu, Sun et al. 2009; Wu et al. 2009). Compared with an anaerobic culture, an aerobic culture gives higher HA titer and molecular weight (Johns et al. 1994; Armstrong et al. 1997). For example, Armstrong and Johns observed a 20% increase in HA titer when *S. zooepidemicus* were grown under aerobic conditions (Armstrong et al. 1997). Johns et al. also reported that the aerated culture gave higher HA concentration and yield than the equivalent anaerobic fermentation (Johns et al. 1994). The stimulation effects of aeration on HA production can be explained by the following. (1) Oxygen may stimulate the HA synthesis as the aggregation of streptococcal cells mediated by their HA capsule shielded them from oxygen metabolites. (2) Dissolved oxygen in the medium can redirect the carbon flux toward lactic acid to acetic acid and thus more ATP can be generated (YATP/glucose is 3 mol/mol with acetate production against 2 mol/mol with lactate production). The extra ATP generated during the formation of acetate is favorable for the attainment of higher HA titer. (3) The aeration can enhance microbial production of low molecular weight HA acetyl-CoA accumulation as such that more acetyl-CoA can be diverged from the central carbon metabolism to replenish acetyl-CoA for the synthesis of HA (Wu et al. 2009). Yet, there is a considerable divergence on the impacts of agitation speed and aeration on the microbial HA production. It was observed that HA production was not affected by aeration rate, whereas it decreased with the increase in agitation speed (Kim et al. 1996). Hasegawa et al. (1999) reported that HA production increased with the increase in aeration rate and agitation speed; nevertheless, too high agitation speed caused cell damage and led to a drop in HA concentration. The energy status was improved via the overexpression of NADH oxidase in *S. zooepidemicus*; however, little impact on HA titer was observed (Chong and Nielsen 2003a). This divergence may be explained by a recent study that indicated that there existed a critical dissolved oxygen level of 5% air saturation for HA synthesis (Huang et al. 2006). That is to say, when dissolved oxygen level was lower than 5% of air saturation, the increase in aeration rate and agitation speed was favorable for microbial HA production; and when dissolved oxygen level was higher than 5% of air saturation, there was little impact of agitation speed and aeration rate on HA production.

In addition, various fermentation modes, such as batch, repeated batch, fed-batch, and continuous culture have been used for HA production (Cooney et al. 1999; Blank et al. 2005; Huang et al. 2008; Liu, Du et al. 2008; Chen, Chen et al. 2009; Im et al. 2009; Don and Shoparwe 2010; Vazquez et al. 2010) (Table 2.1). Batch culture is the dominant operation mode for HA production. Compared to batch culture, fed-batch culture can shorten fermentation time, and thus, increase HA productivity (Vazquez et al. 2010). The combination of fed-batch and batch was found to be effective for HA production. Namely, *S. zooepidemicus* were cultured in a fed-batch mode with sucrose concentration at 1.0 g/L for 0–8 h, and then batch culture was performed for 8–20 h with an initial sucrose concentration of 15 g/L. With this two-stage culture strategy, HA production was increased by 32% compared to the batch culture (Liu, Du et al. 2008). Recently, the repeated batch culture has also been employed for HA production, and HA productivity was significantly enhanced (Huang et al. 2008; Chen, Chen et al. 2009). In an operation that seeded 31% cell, the volumetric production rate of the repeated batch culture (0.59 g HA/[L·h]) was found to be 2.5-fold of the batch culture (0.24 g HA/[L·h]). Compared with batch operation, continuous operation can extend culture period, reduce the time spent on reactor turnover, and decrease the polydispersity of molecular weight (Armstrong et al. 1997; Blank et al. 2005). HA production in a chemostat was, however,

Table 2.1 Overview of the Production of HA and Its Molecular Weight under Different Culture Conditions

MICROORGANISM	VOLUME	CULTURE MEDIUM	HA TITER AND MOLECULAR WEIGHT (MW, Da)	REFERENCES
S. zooepidemicus (ATCC 35246)	2.5 L	Maltose 20 g/L, CDM	2.14 g/L; 2.1×10^6	(Chong and Nielsen 2003a)
S. zooepidemicus (ATCC 39920)	3.7 L	CDM	3.66 g/L; 3.8×10^6	(Jagannath and Ramachandran 2010)
S. zooepidemicus (ATCC 35246)	2 L	Glucose 60 g/L, CDM	4.2 g/L; 3.2×10^6	(Armstrong et al. 1997)
S. zooepidemicus WSH-24	7 L	Yeast extract 25 g/L, sucrose 70 g/L	6.7 g/L; n.d.	(Liu, Sun et al. 2009)
S. zooepidemicus NJUST01	0.5 L	Starch 50 g/L, glucose 3 g/L, peptone 5 g/L	6.7 g/L; n.d.	(Zhang et al. 2006)
S. zooepidemicus G1 (mutant of ATCC39920)	5 L	Glucose 40 g/L, polypeptone 20 g/L, yeast extract 10 g/L	3.5 g/L; 2.19×10^6	(Duan et al. 2008)
S. zooepidemicus (ATCC 35246)	2 L	Glucose 20 g/L, yeast extract 10 g/L	2.1 g/L; n.d.	(Johns et al. 1994)
S. zooepidemicus (ATCC 39920)	3 L	Glucose 20 g/L, yeast extract 10 g/L	2.3 g/L; n.d.	(Huang et al. 2006)
S. zooepidemicus WSH-24	7 L	Yeast extract 25 g/L, sucrose 70 g/L	6.6 g/L; n.d.	(Liu, Du et al. 2009)
S. zooepidemicus (ATCC 39920)	2.5 L	glucose 5 g/L, yeast extract 2.5 g/L	3.5 g/L; n.d	(Chen, Chen et al. 2009)
Streptococcus sp. (KCTC1139BP)	75 L	Glucose 40 g/L, yeast extract 7.5 g/L, casein peptone 10 g/L	6.94 g/L; 5.9×10^6	(Im et al. 2009)
S. zooepidemicus WSH-24	7 L	Yeast extract 25 g/L, sucrose 70 g/L	6.6 g/L; n.d	(Liu, Du et al. 2008)
S. zooepidemicus (ATCC 35246)	2 L	Mussel processing wastewater 50 g/L, tuna peptone 8 g/L	2.46 g/L; 2.5×10^6	(Vazquez et al. 2010)

(*Continued*)

Table 2.1 (Continued) Overview of the Production of HA and Its Molecular Weight under Different Culture Conditions

MICROORGANISM	VOLUME	CULTURE MEDIUM	HA TITER AND MOLECULAR WEIGHT (MW, Da)	REFERENCES
S. zooepidemicus (ATCC 35246)	3 L	Yeast extract 10 g/L and a mixture of inorganic salts	2.2 g/L; n.d	(Cooney et al. 1999)
S. zooepidemicus (ATCC 39920)	2 L	Glucose (10-60) g/L, yeast extract 10 g/L	1.8 g/L; 2.52×10^6	(Don and Shoparwe 2010)
S. zooepidemicus mutant	100 L	CDM	6–7 g/L; 3.2×10^6	(Kim et al. 1996)
Recombinant *Bacillus subtilis*	0.25 L	Modified minimal medium, 10 g/L glucose	1.8 g/L; n.d.	(Chien and Lee 2007b)
Recombinant *Lactococcus lactis*	2.4 L	M17 medium, 20 g/L glucose	1.8 g/L; n.d	(Prasad et al. 2010)

not easily achieved due to the instability of the HA-producing phenotype of highly encapsulated *Streptococci* strains at high dilution rate (Blank et al. 2005). The highest dilution rate for stable HA production in a chemostat culture was 0.4 h^{-1} (Blank et al. 2005). Therefore, the industrial production of HA cannot be achieved in continuous cultivation (Chong et al. 2005).

2.4 Microbial Production of HA with Other Systems

Recently, the recombinant HA production has emerged as an attractive alternative that could alleviate safety concerns stemming from pathogenic *S. zooepidemicus* and avian products. Host bacteria, both Gram-positive and Gram-negative, include *Bacillus* sp. (Widner et al. 2005; Chien and Lee 2007a), *L. lactis* (Chien and Lee 2007b), *Agrobacterium* sp. (Mao and Chen 2007), and *E. coli* (Jongsareejit et al. 2007; Yu and Stephanopoulos 2008; Mao et al. 2009). An *E. coli* strain (JM109) was engineered into an efficient HA producer by coexpressing the HA synthase from *Pasteurella multocida* and UDP-glucose dehydrogenase from *E. coli* K5 strain (Mao et al. 2009). The engineered strain produced 0.5 g/L HA in shaker flask and 2.0–3.8 g/L HA in a fed-batch culture process in a 1-L bioreactor (Mao et al. 2009). *L. lactis* was engineered by introducing the HA synthetic machinery from the operon of *S. zooepidemicus*, and it was found that the insertion of UDP-glucose pyrophosphorylase (encoded by *hasC*) gene in addition to the HA synthase (hasA) and UDP-glucose dehydrogenase (encoded by *hasB*) genes can significantly increase HA production (Prasad et al. 2010). The recombinant *L. lactis* NZ9000 strain transformed with the plasmid pSJR3 (coexpressing *hasA*, *hasB*, and *hasC* genes) gave a maximum of 1.8 g/L HA in a 2.4-L batch bioreactor (Prasad et al. 2010). The *hasA* gene from *S. zooepidemicus* was expressed in *B. subtilis* for the production of HA, and it was found that the production of UDP-glucuronic acid is limiting in *B. subtilis* and that overexpressing the *hasA* gene along with the endogenous *tuaD* gene is sufficient for high-level production of HA in *B. subtilis* (Widner et al. 2005). *Agrobacterium* sp. ATCC 31749 was engineered by coexpressing HA synthase gene from *P. multocida*, along with a *kfiD* gene encoding UDP-glucose dehydrogenase from *E. coli* K5 strain (Mao and Chen 2007). Coexpression of these two heterologous enzymes enables *Agrobacterium* to produce 0.3 g/L HA in shaker flask cultivation (Mao and Chen 2007). Table 2.1 shows the HA production with different strains under different culture conditions. Though HA from *Bacillus* is commercially available, in general, the recombinant strains produced a lower HA titer than *streptococci* did, and the forthcoming research should focus on the construction of an efficient HA producer with metabolic and genetic tools.

2.5 Microbial Production of Low Molecular Weight HA

Molecular weight is an important quality parameter for a commercial HA product, as it determines the HA's rheological properties, affects physiological response, and defines suitable applications (Armstrong and Johns 1997; Blank et al. 2008). HA with a high molecular weight (greater than 10 kDa) has good viscoelasticity, moisture retention, and mucoadhesion qualities desirable in the areas of ophthalmology, orthopedics, wound healing, and cosmetics. Whereas HA with a relatively low molecular weight (2–3.5 kDa) or HA oligosaccharides (10–20 sugars in length) have shown to promote angiogenesis, induce expression of inflammatory mediators, and inhibit tumor growth

(Sheng et al. 2009). Compared with anaerobic condition, aeration can increase the molecular weight of HA due to the fact that more energy can be produced under aerobic conditions (Armstrong and Johns 1997).

Moreover, a high dissolved oxygen level favored a high molecular weight, while a high shear stress led to a lower molecular weight (Duan et al. 2008). The decrease in HA molecular weight at high shear stress was caused by the reactive oxygen species generated by NADH oxidase. Thus, a combination of high dissolved oxygen level and mild shear stress may be an effective strategy to enhance HA molecular weight. Besides the culture conditions, the balance between the synthesis rate of HA and the providing rate of precursor sugars was also important for the molecular weight. A high ratio of HA synthase gene (HasA) to UDP-glucose-6-dehydrogenase gene (HasB) resulted in a lower HA molecular weight (Sheng et al. 2009). Altering this ratio affected the concentration of precursor sugars and ultimately affected the HA size, and it was an effective approach to control HA molecular weight (Sheng et al. 2009). Of the two sugar precursors, UDP-glucuronic acid and UDP-N-acetylglucosamine, the latter exerted a dominant effect on molecular weight (Sheng et al. 2009). An overexpression of the genes involved in UDP-glucuronic acid biosynthesis decreased molecular weight; whereas an overexpression of the genes involved in UDP-N-acetylglucosamine biosynthesis increased molecular weight (Chen, Marcellin et al. 2009). Thus, manipulating an appropriate balance of UDP-N-acetylglucosamine and UDP-glucuronic acid was necessary to obtain HA with high molecular weight. In addition, the balance of glycolytic rate and HA synthesis rate was also important for the molecular weight of HA (Jagannath and Ramachandran 2010).

2.6 Enzymatic Production of Oligosaccharides

To date, many studies have shown that low molecular weight HA oligosaccharides have unique biological activities (Lesley et al. 2000). Smaller HA oligosaccharides can stimulate fibroblast proliferation and collagen synthesis (Rooney et al. 1993) and can selectively kill many types of cancer cells via disruption of the receptor–hyaluronan interaction (Zeng et al. 1998; Seeberger and Werz 2007; Toole et al. 2008). HA octasaccharides (HA8) and decasaccharides (HA10) have significant suppressive effects on cancer cells (Urakawa et al. 2012), and HA tetrasaccharides (HA4) and hexasaccharides (HA6) can induce dendritic cell maturation via Toll-like receptor (TLR)-4 associated with the antigen-presenting cells of the innate immune system (Termeer et al. 2002). In addition, lower molecular weight HA oligosaccharides are easily absorbed by the body and serve as precursors for the synthesis of higher molecular weight HA molecules and other substances. Thus, a specific narrow spectrum of HA oligosaccharides could have broad applications in medicine, food, and cosmetics. Low molecular weight HA is mainly produced by the degradation of high molecular weight HA by physical and chemical methods (Stern et al. 2007). However, the products of these methods have a broad range of molecular weights (>3,000 Da), making it difficult to obtain HA oligosaccharides with specific molecular weights. Many chemical approaches have been developed for the de novo synthesis of HA oligosaccharides (Seeberger and Werz 2007). However, these complex processes are time-consuming, carbohydrate oligosaccharide backbones are rare, and the substrate UDP-sugars are expensive, limiting the applications of these synthetic methods in large-scale production (Boltje et al. 2009). In contrast, the enzymatic production of HA oligosaccharides with a well-characterized hyaluronidase (HAase) is promising and attractive because of its unique advantages, such as mild operation conditions, high degradation rates, and high product uniformity (Weijers et al. 2008).

HAases (also previously referred to as "spreading factors") are a large class of glycosidases that predominantly degrade HA. Based on substrate specificities and hydrolysis products, HAases are commonly grouped into three families (Meyer and Rapport 1952): hyaluronate lyases (EC 4.2.2.1, *Streptococcus* hyaluronate lyase), hyaluronate 4-glycanohydrolases (EC 3.2.1.35, Bovine testicular hyaluronidase, BTH), and hyaluronate 3-glycanohydrolases (EC 3.2.1.36, Leech HAase; Figure 2.3).

Commercial BTH has been widely used in clinical medicine, and its hydrolysis mechanism has been studied extensively (El-Safory et al. 2010). The disadvantages of the enzymatic production of specific or narrow-spectrum HA oligosaccharides by BTH include the limited source material (bovine testes), its considerably high price, and the broad range of degradation products (Tawada et al. 2002). Compared with BTH and *Streptococcus* hyaluronate lyase, leech HAase has higher substrate specificity and a narrow spectrum of enzymatic products (Linker et al. 1957, 1960). In addition, the use of recombinant leech HAase does not pose any risk of animal cross-infection. Therefore,

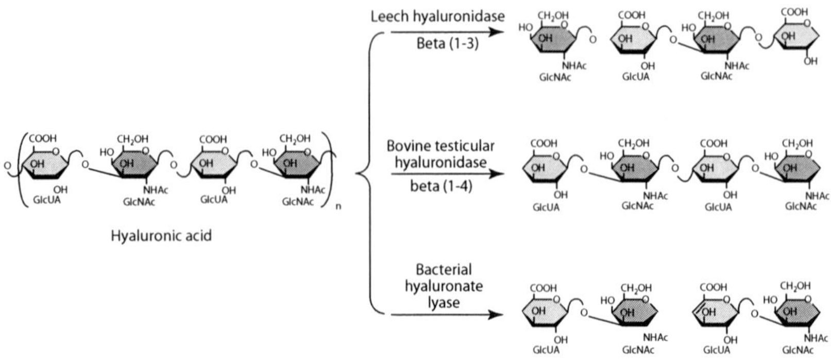

Figure 2.3 Categories of hyaluronidase.

high-level production of recombinant leech HAase would be of great significance for both clinical medical treatment (such as surgery, ophthalmology, and internal medicine) and producing narrow-spectrum HA oligosaccharides at the industrial scale.

More recently, Jin et al. cloned and characterized the first leech HAase-encoding gene, LHyal. Through a combination of N-terminal engineering and fed-batch fermentation strategies, they achieved high-level production of leech HAases in recombinant *Pichia pastoris*. After optimization, the production level was increased to 8.42×10^5 U/mL. More importantly, by exploring the hydrolytic process and mechanism, they achieved enzymatic production of narrowband HA oligosaccharides (Figure 2.4). The construction of robust microbial factories for the large-scale production of recombinant HAase and the development of the enzymatic approach for the preparation of specific HA oligosaccharides would greatly promote progress in related research areas, such as the chemical synthesis of oligosaccharides and cancer therapy.

2.7 Perspectives: Challenges and Opportunities

Though great progress has been achieved on the microbial production of HA with *S. zooepidemicus* and the recombinant production systems, several challenges remain.

First, the continuous rise in the cost of raw materials weakens the commercial competiveness of microbial HA production, and thus it is necessary to find a cheaper substrate replacement to reduce production cost. Furthermore, the needs of a sustainable society point to the

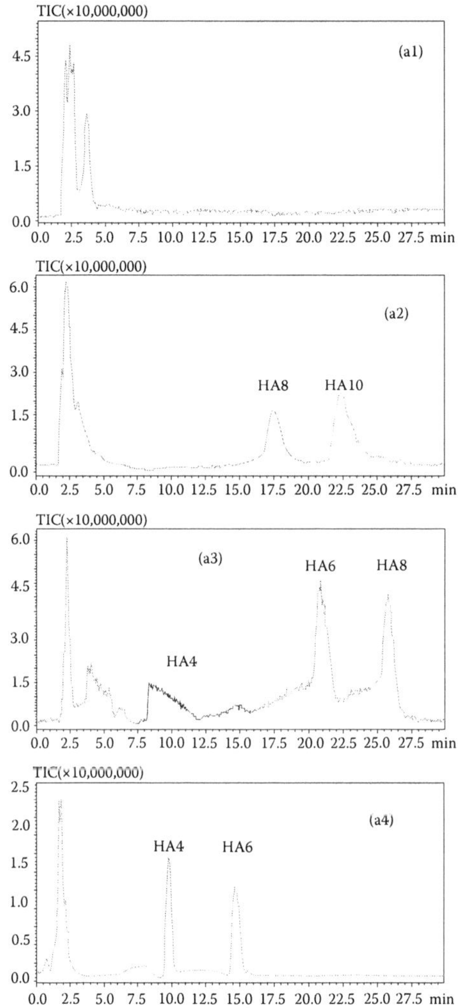

Figure 2.4 Enzymatic production of hyaluroan oligosaccharides. (*Continued*)

conversion of renewable resources, such as agricultural derivatives, into valuable bioproducts. Thus, exploring the feasibility of producing HA with cheap crude materials or waste from other industrial processes is worth investigating. Mussel processing wastewater and tuna peptone from viscera residue are used for HA production by *S. zooepidemicus*, and the economic analysis indicates that the production cost can be reduced by more than 30% with the byproducts as the culture medium (Vazquez et al. 2010). Agricultural resource derivatives, such as cashew apple juice, are a promising medium for the microbial HA production

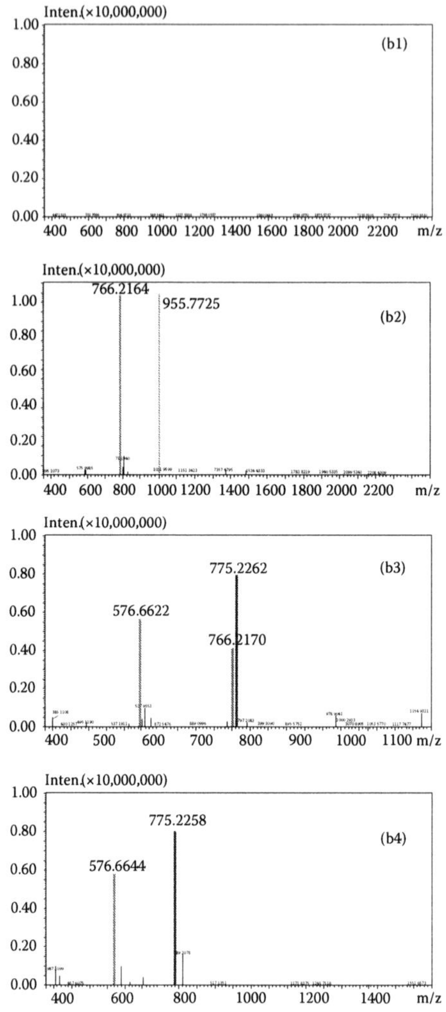

Figure 2.4 (Continued) Enzymatic production of hyaluroan oligosaccharides. (*Continued*)

(Pires et al. 2010). For another example, the large amounts of crude glycerol produced in the biodiesel industry, if not properly treated, pose a significant environmental concern. Therefore, we can explore the potential of microbial HA production with crude glycerol as a substrate. Of course, process engineering for the efficient treatment of crude material and metabolic engineering of microbes for the efficient utilization of raw substrates should be considered to achieve this objective.

Second, whether for *S. zooepidemicus* or the recombinant systems like *E. coli*, *B. subtilis*, and *L. lactis*, the key factors limiting

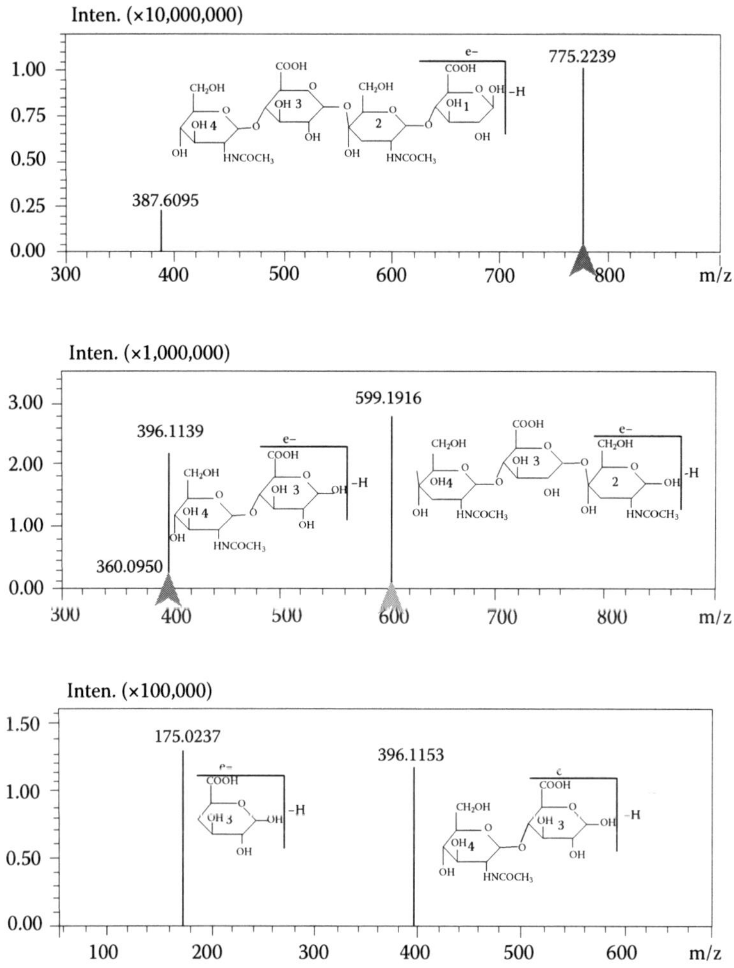

Figure 2.4 (Continued) Enzymatic production of hyaluroan oligosaccharides.

HA synthesis need to be further clarified. The tools of metabolic engineering, such as metabolic flux analysis (MFA) and metabolic control analysis (MCA), can be employed to develop a rational strategy to improve HA yield and molecular weight. MFA is an analysis technique used to calculate and analyze the flux distribution of the entire biochemical reaction network during a process. MCA quantifies the relation between genetic modifications or environmental changes and cellular process responses (Wang and Hatzimanikatis 2006). MCA introduces the control coefficients to quantify the fractional change of cellular output, such as metabolite concentrations and metabolic fluxes, in response to fractional change of system parameters, such as enzyme activities and growth conditions (de Groot et al. 2005). The combination of MFA and MCA can be used to investigate the metabolic responses of HA producer to the environmental changes or the expression of key genes related with HA synthesis. With the information gathered from MFA and MCA, the optimal strategies (both process control and key genes expression) can be determined to improve HA titer and molecular weight.

Third, it is necessary to obtain specially designated molecular weight or uniform size-defined HA to extend the applications of HA and make better HA containing biomedical products. To achieve low polydispersity, we must know the regulatory mechanisms of initiation and elongation during the HA polymer synthesis process. Despite an HA polymerization model having been put forward, and some key intracellular metabolites influencing molecular weight having been clarified, much work needs to be performed to understand the mechanism of molecular weight control.

References

Armstrong, D. C., Cooney, M. J. and Johns, M. R. (1997). Growth and amino acid requirements of hyaluronic-acid-producing *Streptococcus zooepidemicus*. *Applied Microbiology and Biotechnology*. **47**(3), 309–312.

Armstrong, D. C. and Johns, M. R. (1997). Culture conditions affect the molecular weight properties of hyaluronic acid produced by *Streptococcus zooepidemicus*. *Applied and Environmental Microbiology*. **63**(7), 2759–2764.

Balazs, E. and Sweeney, D. (1968). The injection of hyaluronic acid and reconstituted vitreous into the vitreous cavity. In A. McPherson (ed.), *New and Controversial Aspects of Retinal Detachment*. New York: Harper and Row, pp. 371–376.

Blank, L. M., Hugenholtz, P. and Nielsen, L. K. (2008). Evolution of the hyaluronic acid synthesis (has) operon in *Streptococcus zooepidemicus* and other pathogenic streptococci. *Journal of Molecular Evolution*. **67**(1), 13–22.

Blank, L. M., McLaughlin, R. L. and Nielsen, L. K. (2005). Stable production of hyaluronic acid in *Streptococcus zooepidemicus* chemostats operated at high dilution rate. *Biotechnology and Bioengineering*. **90**(6), 685–693.

Boltje, T. J., Buskas, T. and Boons, G.-J. (2009). Opportunities and challenges in synthetic oligosaccharide and glycoconjugate research. *Nature Chemistry*. **1**(8), 611–622.

Chen, S.-J., Chen, J.-L., Huang, W.-C. and Chen, H.-L. (2009). Fermentation process development for hyaluronic acid production by *Streptococcus zooepidemicus* ATCC 39920. *Korean Journal of Chemical Engineering.* **26**(2), 428–432.

Chen, W. Y., Marcellin, E., Hung, J. and Nielsen, L. K. (2009). Hyaluronan molecular weight is controlled by UDP-N-acetylglucosamine concentration in *Streptococcus zooepidemicus. Journal of Biological Chemistry.* **284**(27), 18007–18014.

Chien, L. J. and Lee, C. K. (2007a). Enhanced hyaluronic acid production in *Bacillus subtilis* by coexpressing bacterial hemoglobin. *Biotechnology Progress.* **23**(5), 1017–1022.

Chien, L. J. and Lee, C. K. (2007b). Hyaluronic acid production by recombinant *Lactococcus lactis. Applied Microbiology and Biotechnology.* **77**(2), 339–346.

Chong, B. F., Blank, L. M., McLaughlin, R. and Nielsen, L. K. (2005). Microbial hyaluronic acid production. *Applied Microbiology and Biotechnology.* **66**(4), 341–351.

Chong, B. F. and Nielsen, L. K. (2003a). Aerobic cultivation of *Streptococcus zooepidemicus* and the role of NADH oxidase. *Biochemical Engineering Journal.* **16**(2), 153–162.

Chong, B. F. and Nielsen, L. K. (2003b). Amplifying the cellular reduction potential of *Streptococcus zooepidemicus. Journal of Biotechnology.* **100**(1), 33–41.

Cooney, M. J., Goh, L. T., Lee, P. L. and Johns, M. R. (1999). Structured model-based analysis and control of the hyaluronic acid fermentation by *Streptococcus zooepidemicus*: Physiological implications of glucose and complex nitrogen-limited growth. *Biotechnology Progress.* **15**(5), 898–910.

de Groot, M. J. L., Prathumpai, W., Visser, J. and Ruijter, G. J. G. (2005). Metabolic control analysis of *Aspergillus niger* L-arabinose catabolism. *Biotechnology Progress.* **21**(6), 1610–1616.

Don, M. M. and Shoparwe, N. F. (2010). Kinetics of hyaluronic acid production by *Streptococcus zooepidemicus* considering the effect of glucose. *Biochemical Engineering Journal.* **49**(1), 95–103.

Duan, X. J., Niu, H. X., Tan, W. S. and Zhang, X. (2009). Mechanism analysis of effect of oxygen on molecular weight of hyaluronic acid produced by *Streptococcus zooepidemicus. Journal of Microbiology and Biotechnology.* **19**(3), 299–306.

Duan, X. J., Yang, L., Zhang, X. and Tan, W. S. (2008). Effect of oxygen and shear stress on molecular weight of hyaluronic acid. *Journal of Microbiology and Biotechnology.* **18**(4), 718–724.

El-Safory, N. S., Fazary, A. E. and Lee, C.-K. (2010). Hyaluronidases, a group of glycosidases: Current and future perspectives. *Carbohydrate Polymers.* **81**(2), 165–181.

Hasegawa, S., Nagatsuru, M., Shibutani, M., Yamamoto, S. and Hasebe, S. (1999). Productivity of concentrated hyaluronic acid using a Maxblend fermentor. *Journal of Bioscience and Bioengineering.* **88**(1), 68–71.

Huang, W.-C., Chen, S.-J. and Chen, T.-L. (2006). The role of dissolved oxygen and function of agitation in hyaluronic acid fermentation. *Biochemical Engineering Journal.* **32**(3), 239–243.

Huang, W.-C., Chen, S.-J. and Chen, T.-L. (2008). Production of hyaluronic acid by repeated batch fermentation. *Biochemical Engineering Journal.* **40**(3), 460–464.

Im, J. H., Song, J. M., Kang, J. H. and Kang, D. J. (2009). Optimization of medium components for high-molecular-weight hyaluronic acid production by *Streptococcus sp.* ID9102 via a statistical approach. *Journal of Industrial Microbiology & Biotechnology.* **36**(11), 1337–1344.

Jagannath, S. and Ramachandran, K. B. (2010). Influence of competing metabolic processes on the molecular weight of hyaluronic acid synthesized by *Streptococcus zooepidemicus. Biochemical Engineering Journal.* **48**(2), 148–158.

Jin, P., Kang, Z., Zhang, N., Du, G. and Chen, J. (2014). High-yield novel leech hyaluronidase to expedite the preparation of specific hyaluronan oligomers. *Science Report.* **4**, 4471.

Johns, M. R., Goh, L. T. and Oeggerli, A. (1994). Effect of pH, agitation and aeration on hyaluronic acid production by *Streptococcus zooepidemicus. Biotechnology Letters.* **16**(5), 507–512.

Jongsareejit, B., Bhumiratana, A., Morikawa, M. and Kanaya, S. (2007). Cloning of hyaluronan synthase (sz-has) gene from *Streptococcus zooepidemicus* in *Escherichia coli. Education and Training of the Mentally Retarded.* **33**(4), 389–395.

Kendall, F. E., Heidelberger, M. and Dawson, M. H. (1937). A serologically inactive polysaccharide elaborated by mucoid strains of group A hemolytic streptococcus. *Journal of Biological Chemistry.* **118**(1), 61–69.

Kim, J. H., Yoo, S. J., Oh, D. K., Kweon, Y. G., Park, D. W., Lee, C. H. and Gil, G. H. (1996). Selection of a *Streptococcus equi* mutant and optimization of culture conditions for the production of high molecular weight hyaluronic acid. *Enzyme and Microbial Technology.* **19**(6), 440–445.

Kogan, G., Soltes, L., Stern, R. and Gemeiner, P. (2007). Hyaluronic acid: A natural biopolymer with a broad range of biomedical and industrial applications. *Biotechnology Letters.* **29**(1), 17–25.

Krahulec, J. and Krahulcova, J. (2006). Increase in hyaluronic acid production by *Streptococcus equi subsp. zooepidemicus* strain deficient in beta-glucuronidase in laboratory conditions. A*pplied Microbiology and Biotechnology.* **71**(4), 415–422.

Laurent, T. C., Laurent, U. B. and Fraser, J. R. (1996). The structure and function of hyaluronan: An overview. *Immunology and Cell Biology.* **74**(2), A1–7.

Lesley, J., Hascall, V. C., Tammi, M. and Hyman, R. (2000). Hyaluronan binding by cell surface CD44. *Journal of Biological Chemistry.* **275**(35), 26967–26975.

Linker, A., Hoffman, P. and Meyer, K. (1957). The hyaluronidase of the leech: An endoglucuronidase. *Nature.* **180**(4590), 810–811.

Linker, A., Meyer, K. and Hoffman, P. (1960). The production of hyaluronate oligosaccharides by leech hyaluronidase and alkali. *Journal of Biological Chemistry.* **235**(4), 924–927.

Liu, L., Du, G., Chen, J., Wang, M. and Sun, J. (2008). Enhanced hyaluronic acid production by a two-stage culture strategy based on the modeling of batch and fed-batch cultivation of *Streptococcus zooepidemicus*. *Bioresource Technology*. **99**(17), 8532–8536.

Liu, L., Du, G., Chen, J., Wang, M. and Sun, J. (2009). Comparative study on the influence of dissolved oxygen control approaches on the microbial hyaluronic acid production of *Streptococcus zooepidemicus*. *Bioprocess and Biosystems Engineering*. **32**(6), 755–763.

Liu, L., Liu, Y., Li, J., Du, G. and Chen, J. (2011). Microbial production of hyaluronic acid: Current state, challenges, and perspectives. *Microbial Cell Factories*. **10**(1), 1–9.

Liu, L., Sun, J., Xu, W., Du, G. and Chen, J. (2009). Modeling and optimization of microbial hyaluronic acid production by *Streptococcus zooepidemicus* using radial basis function neural network coupling quantum-behaved particle swarm optimization algorithm. *Biotechnology Progress*. **25**(6), 1819–1825.

Liu, L., Wang, M., Du, G. and Chen, J. (2008). Enhanced hyaluronic acid production of *Streptococcus zooepidemicus* by an intermittent alkaline-stress strategy. *Letters in Applied Microbiology*. **46**(3), 383–388.

Mao, Z. and Chen, R. R. (2007). Recombinant synthesis of hyaluronan by *Agrobacterium sp*. *Biotechnology Progress*. **23**(5), 1038–1042.

Mao, Z., Shin, H. D. and Chen, R. (2009). A recombinant *E. coli* bioprocess for hyaluronan synthesis. *Applied Microbiology and Biotechnology*. **84**(1), 63–69.

Meyer, K. and Palmer, J. W. (1934). The polysaccharide of the vitreous humor. *Journal of Biological Chemistry*. **107**(3), 629–634.

Meyer, K. and Rapport, M. M. (1952). Hyaluronidases. *Advances in Enzymology and Related Subjects of Biochemistry*. **13**, 199–236.

Pires, A. M., Eguchi, S. Y. and Santana, M. H. (2010). The influence of mineral ions on the microbial production and molecular weight of hyaluronic acid. *Applied Biochemistry and Biotechnology*. **162**(8), 2125–2135.

Pires, A. M., Macedo, A. C., Eguchi, S. Y. and Santana, M. H. A. (2010). Microbial production of hyaluronic acid from agricultural resource derivatives. *Bioresource Technology*. **101**(16), 6506–6509.

Pires, A. M. and Santana, M. H. (2010). Metabolic effects of the initial glucose concentration on microbial production of hyaluronic acid. *Applied Biochemistry and Biotechnology*. **162**(6), 1751–1761.

Prasad, S. B., Jayaraman, G. and Ramachandran, K. B. (2010). Hyaluronic acid production is enhanced by the additional co-expression of UDP-glucose pyrophosphorylase in *Lactococcus lactis*. *Applied Microbiology and Biotechnology*. **86**(1), 273–283.

Rooney, P., Wang, M., Kumar, P. and Kumar, S. (1993). Angiogenic oligosaccharides of hyaluronan enhance the production of collagens by endothelialcells. *Journal of Cell Science*. **105**, 213–218.

Seeberger, P. H. and Werz, D. B. (2007). Synthesis and medical applications of oligosaccharides. *Nature*. **446**(7139), 1046–1051.

Sheng, J. Z., Ling, P. X., Zhu, X. Q., Guo, X. P., Zhang, T. M., He, Y. L. and Wang, F. S. (2009). Use of induction promoters to regulate hyaluronan synthase and UDP-glucose-6-dehydrogenase of *Streptococcus zooepidemicus* expression in *Lactococcus lactis*: A case study of the regulation mechanism of hyaluronic acid polymer. *Journal of Applied Microbiology*. **107**(1), 136–144.

Stern, R., Kogan, G., Jedrzejas, M. J. and Soltes, L. (2007). The many ways to cleave hyaluronan. *Biotechnology Advances*. **25**(6), 537–557.

Tawada, A., Masa, T., Oonuki, Y., Watanabe, A., Matsuzaki, Y. and Asari, A. (2002). Large-scale preparation, purification, and characterization of hyaluronan oligosaccharides from 4-mers to 52-mers. *Glycobiology*. **12**(7), 421–426.

Termeer, C., Benedix, F., Sleeman, J., Fieber, C., Voith, U., Ahrens, T., Miyake, K., Freudenberg, M., Galanos, C. and Simon, J. C. (2002). Oligosaccharides of hyaluronan activate dendritic cells via toll-like receptor 4. *Journal of Experimental Medicine*. **195**(1), 99–111.

Toole, B. P., Ghatak, S. and Misra, S. (2008). Hyaluronan oligosaccharides as a potential anticancer therapeutic. *Current Pharmaceutical Biotechnology*. **9**(4), 249–252.

Urakawa, H., Nishida, Y., Knudson, W., Knudson, C. B., Arai, E., Kozawa, E., Futamura, N., Wasa, J. and Ishiguro, N. (2012). Therapeutic potential of hyaluronan oligosaccharides for bone metastasis of breast cancer. *Journal of Orthopaedic Research*. **30**(4), 662–672.

Vazquez, J. A., Montemayor, M. I., Fraguas, J. and Murado, M. A. (2010). Hyaluronic acid production by *Streptococcus zooepidemicus* in marine by-products media from mussel processing wastewaters and tuna peptone viscera. *Microbial Cell Factories*. **9**, 46.

Wang, L. Q. and Hatzimanikatis, V. (2006). Metabolic engineering under uncertainty. I: Framework development. *Metabolic Engineering*. **8**(2), 133–141.

Weijers, C. A., Franssen, M. C. and Visser, G. M. (2008). Glycosyltransferase-catalyzed synthesis of bioactive oligosaccharides. *Biotechnology Advances*. **26**(5), 436–456.

Widner, B., Behr, R., Von Dollen, S., Tang, M., Heu, T., Sloma, A., Sternberg, D., Deangelis, P. L., Weigel, P. H. and Brown, S. (2005). Hyaluronic acid production in *Bacillus subtilis*. *Applied and Environmental Microbiology*. **71**(7), 3747–3752.

Wu, T. F., Huang, W. C., Chen, Y. C., Tsay, Y. G. and Chang, C. S. (2009). Proteomic investigation of the impact of oxygen on the protein profiles of hyaluronic acid-producing *Streptococcus zooepidemicus*. *Proteomics*. **9**(19), 4507–4518.

Yu, H. and Stephanopoulos, G. (2008). Metabolic engineering of *Escherichia coli* for biosynthesis of hyaluronic acid. *Metabolic Engineering*. **10**(1), 24–32.

Zeng, C., Toole, B. P., Kinney, S. D., Kuo, J. W. and Stamenkovic, I. (1998). Inhibition of tumor growth in vivo by hyaluronan oligomers. *International Journal of Cancer*. **77**(3), 396–401.

Zhang, J., Ding, X., Yang, L. and Kong, Z. (2006). A serum-free medium for colony growth and hyaluronic acid production by *Streptococcus zooepidemicus* NJUST01. *Applied Microbiology and Biotechnology*. **72**(1), 168–172.

3

Curdlan Oligosaccharides

Production and Application

XIAOBEI ZHAN AND HONGTAO ZHANG

Contents

3.1 Introduction

Curdlan is a microbial extrapolysaccharide produced by *Agrobacterium* sp. It consists of β1,3-linked glucose with average degree of polymerization (DP) values between 135 and 455 (Figure 3.1a), and has been widely used as a food additive in noodles and medicines (Na et al. 2000; Verma and Gu 2012). However, curdlan is insoluble in water; thus, it cannot be readily absorbed by the intestines. In addition, the human body does not have enzymes that can degrade curdlan. Interestingly,

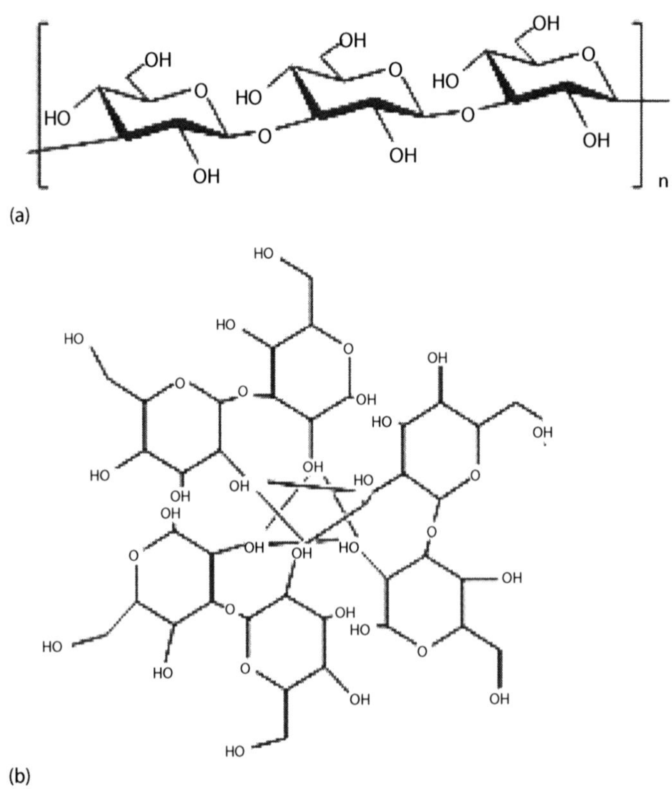

Figure 3.1 Structure of curdlan (a) the structure of curdlan and (b) the top-down view of curdlan.

β1,3-linked glucose widely exists in the cell wall of mushroom alga and some pathogenic fungus, and it can recognize pattern recognition receptors (PRRs) of humans and plants and then implement biological activities (Brown and Gordon 2005; Novak and Vetvicka 2008). For example, low molecular weight β1,3-glucooligosaccharide can kill cancer cells via the CR3 receptor (Thornton et al. 1996; Saraswat-Ohri et al. 2011) and can also activate the immune system via Dectin-1 (Thornton et al. 1996), especially since the potential application of β1,3-glucooligosaccharides as a functional food has received considerable attention (Unno et al. 2005; Zhang 2011). Curdlan is an ideal and cheap raw material for producing β1,3-glucooligosaccharides given that it is a homogenous microbial β1,3-glucan. However, the water-insoluble property of curdlan is the predominant barrier for the degradation of curdlan to acquire β1,3-glucooligosaccharides (also called curdlan oligosaccharides [CurdOs]). Therefore, a systematic review of the production and application of CurdOs is very useful for further research. In this chapter, we focus on introducing different methods for the production of oligosaccharides from the water-insoluble curdlan, which is the main low-priced material for oligosaccharide production. We also update the curdlan production as well as review the application progress of CurdOs.

3.2 Curdlan

3.2.1 Basic Chemical and Physical Properties of Curdlan

Curdlan is a neutral polysaccharide that consists of linearly linked β-1, 3-glucose backbone, and it also exists as a triple-helical conformation in solid and liquid states (Figure 3.1b). Although curdlan is insoluble in water (McIntosh et al. 2005), it is soluble in formic acid and some aprotic solvents, such as N-methylmorpholine-N-oxide (NMMO), dimethyl sulfoxide (DMSO; Tada et al. 2001), and dimethylacetamide containing lithium chloride (DMAc/LiCl). The stiff triple helix can dissociate into single-strand coils by increasing alkaline or DMSO concentrations, resulting from the destruction of the intra- and intermolecular hydrogen bonding. Interestingly, the dissociated coils can self-assemble into triple-helical structures once the hydrogen bonds form after a long time dialysis against water, thus indicating that curdlan can show strong self-assembling abilities.

Curdlan associates at 25°C and curdlan association is enhanced at an NaOH concentration lower than 0.05N (Puertas et al. 2014). When the concentration of NaOH is lower than 0.02N and 0.01N, there is no dissociation of single molecules by heating (Toshifumi and Kozo 1999). Curdlan can form gel easily and curdlan gels are thermally irreversible, following a typical procedure: curdlan was dissolved in 0.01N NaOH aqueous solution using 10 wt% curdlan and was then heated above 60°C. Two types of gels can form depending on the heating temperature. One of which is a high-set thermal nonreversible gel (–80°C), whereas the other is a low-set thermal reversible gel (–55°C) (Konno and Harada 1992; Kwak and Lafleur 2003). The remarkable rheological and thermal behaviors of curdlan have led to its application as a thickening agent and a replacement for fat in the food industry.

3.2.2 Production of Curdlan as Essential Raw Material for CurdOs Production

3.2.2.1 Typical Microorganism for Curdlan Production and the Biosynthesis Way Curdlan and curdlan-like polysaccharides have been widely present in bacteria, alga, fungi, and so on. *Agrobacterium* sp. 10C3 (IFO 13714) (formerly called *Alcaligenes faecalis* var. *myxogenes* 10C3) was first reported to produce curdlan by Harada et al. in 1966. They also reported the chemical structure of curdlan. Three pathways of curdlan synthesis in *Agrobacterium* sp. ATCC 31749 were proposed by our group in detail (Zhan et al. 2012): (1) glucose is used to produce UDP-glucose, which is the precursor of curdlan synthesis; (2) glucose is used via TCA cycle to produce $NADH_2$, $NADH_2$ via electron transport chain (ETC) producing ATP; and (3) curdlan is synthesized using UDP-glucose and ATP by bacteria (Figure 3.2).

Curdlan is widely used in the food industry because of its special solubility and rheological properties. The market demand for curdlan has increased in Korea, Taiwan, and Japan since 1989. Research work for curdlan production is focused on the selection of high-producing bacteria strains (Table 3.1), optimization of the fermentation medium (Thornton et al. 1996; Fang et al. 2013; Gao et al. 2010; Rafigh et al. 2014), and fermentation process (Zheng et al. 2014; Zhang et al. 2015).

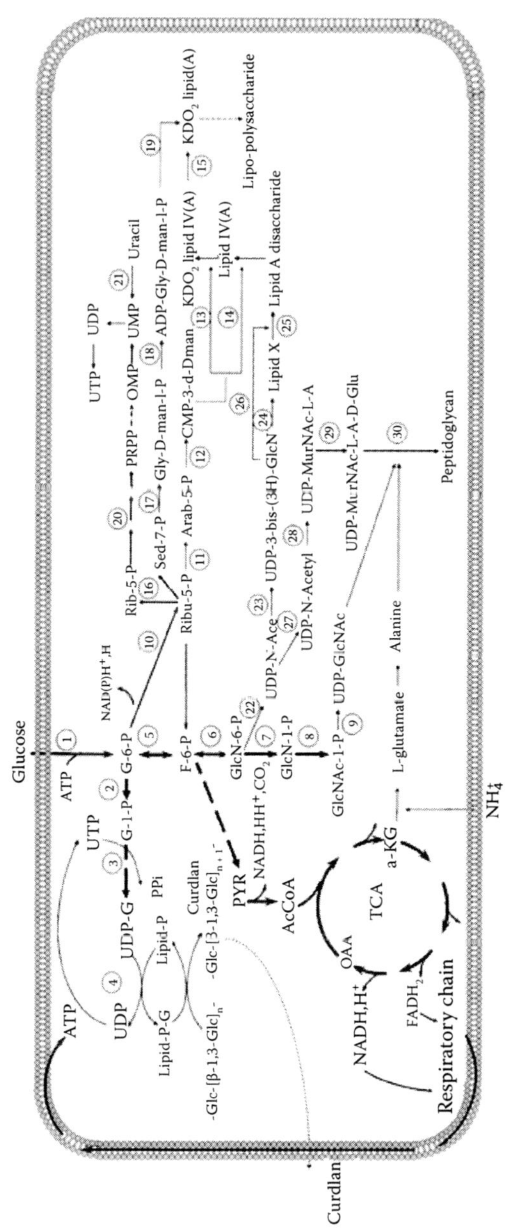

Figure 3.2 Metabolic pathways of curdlan biosynthesis in *Agrobacterium* sp. ATCC 31749.

Table 3.1 Bacteria for Producing Bacterial Curdlan

BACTERIAL SOURCE	REFERENCE
Agrobacterium sp. 10C3	(Nakanishi et al. 1976)
Agrobacterium sp. ATCC 31749,31750	(Phillips et al. 1983; Kalyanasundaram et al. 2012; Zhang et al. 2012)
Agrobacterium sp. Biovar I GA-27 and GA-33	(Kanegae et al. 1996)
A. radiobacter IFO 12607, 12665, 13127, 13256A, and *A. rhizogenes* IFO 13259	(Nakanishi et al. 1976)
Agrobacterium sp. IFO 13140	(Martinez et al. 2015)
Rhizobium trifolii J60	(Ghai et al. 1981)
Rhizobium sp. TISTR 64B	(Footrakul et al. 1981)
Rhizobium niftal 600	(Harada et al. 1966)
Paenibacillus polymyxa ATCC 21830	(Rafigh et al. 2014)
Bacillus sp. SNC07	(Gummadi and Kumar 2005)
Rhizobium radiobacter ATCC 6466™	(Salah et al. 2011)

3.2.2.2 Optimization of the Curdlan Fermentation Medium and Production Processes Fermentation is an efficient method for producing target products using microorganisms. Usually, temperature, pH, and dissolved oxygen are the key parameters from the fermentation process optimization. During curdlan biosynthesis, UTP plays an important role in UDP-glucose generation, which is one of the key precursors for curdlan production (Figure 3.2). Higher intracellular levels of uridine monophosphate under nitrogen-limited conditions can enhance metabolic flux for curdlan synthesis in *Agrobacterium* species (Kim et al. 1999). To supply enough UTP for curdlan synthesis, low-polyphosphates $(NaPO_3)_6$ with higher energy phosphate bonds were employed to substitute for KH_2PO_4–K_2HPO_4 in curdlan fermentation medium. The curdlan yields were enhanced 134% when phosphate concentrations of 0.048 mol/L of $(NaPO_3)_6$ were added in the medium. The maximum curdlan yield of 30 ± 1.02 g/L was obtained with the addition of 0.048 mol/L of $(NaPO_3)_6$ with 5 g/L $CaCO_3$ in the medium. This revealed that enough adenosine triphosphate with high-energy phosphate bonds is required for uridine triphosphate regeneration in *Agrobacterium* sp. ATCC 31749 for curdlan biosynthesis (Yu et al. 2011).

Uridine diphosphate (UDP)-glucose is a precursor for curdlan biosynthesis and is synthesized from glucose-1-phosphate and uridine triphosphate (UTP). Uracil was the precursor for UTP generation; therefore, uracil addition was explored to improve curdlan fermentation. Uracil was used as a nitrogen source for cell growth when it was added after the ammonium was exhausted, causing a decrease in curdlan production. However, with the addition of both sucrose (200 g) and uracil (1.5 g) at the same time, the curdlan production was increased up to 93 g/L after 160 h fermentation (Lee and Lee 2000). In addition, supplementation of pyrimidine base into the culture medium can also increase curdlan production by *Agrobacterium* sp. ATCC 31749 (West 2006), where the enhancement of curdlan is closely related to the increase in cell concentration.

UTP, mainly produced via ETC, is closely related to dissolved oxygen. Take note that curdlan production was apparently affected by agitation at a higher agitation speed (600 rpm), where the highest curdlan production (64.4 g/L) was obtained in 120 h of batch fermentation. However, curdlan production was not improved at the higher agitation speed (700 rpm; Lee, Kim et al. 1999) as agitation is associated with DO. Therefore, we explore in detail the effects of different DO on curdlan production. The optimum DO concentrations for curdlan production were 45–60% DO. The average curdlan yield, curdlan productivity, and glucose conversion efficiency into curdlan were enhanced by 80%, 66%, and 32%, respectively, compared to that at 15% DO (Zhang et al. 2012). Based on these results, an optimal multistage DO controlling strategy was developed with modeling. The optimal dissolved oxygen during curdlan production (20–120 h) was approximately 52% DO, and the maximum curdlan concentration of 48.5 g/L was obtained at this DO profile with 3.5 g/L cell concentration (Zhang et al. 2015).

Curdlan production can only happen when nitrogen is limited; nitrogen sources on curdlan production were studied; urea was found suitable for cell growth and curdlan accumulation (Jiang 2013). The initial ammonium chloride concentration can enhance cell concentration and curdlan production; however, cell growth and curdlan production cannot be enhanced by solely increasing the initial ammonium chloride level; the pH in the fermentation medium can also affect cell growth and curdlan production (Sun et al. 2005).

Detailed work on the effect of pH on curdlan production was also conducted. The ideal pH for cell growth is 7.0, but for curdlan production it is 5.6 (Lee, Lee et al. 1999; Jung et al. 2000). Therefore, an optimal procedure of pH that controls the increase cell concentration and curdlan production was developed. Ammonia water was fed during the first 24 h to keep pH at 7.0 and enhance cell concentration. The maximum value of cell concentration was achieved at 18.8 g/L. The pH was then adjusted to 5.6 for curdlan accumulation, and the maximum curdlan production of 72 g/L was acquired by feeding ammonia water for the first 14 h with the final cell density of 11.9 g/L (Wu et al. 2008). Another group also conducted some work on curdlan fermentation with high cell concentration. Jung found that the conversion rate and concentration of curdlan from 2% (w/v) glucose in the minimal salt medium with concentrated cells under nitrogen deletion was 67% and 13.4 g/L, respectively (Jung et al. 2001). A two-stage fermentation was developed to acquire high cell concentration. The cell in the first stage was cultured, and a high cell density of 29 g/L was attained. The cell in the second stage was fed with nutrient solution and high carbon/nitrogen ratio, and the pH was adjusted using NaOH solution. Productivity, conversion efficiency, and curdlan yield were at 0.98 g/(L h), 57% (w), and 67 g/L, respectively (Zheng et al. 2014).

3.3 Methods for Curdlan Oligosaccharide Production

Several methods have been used to produce CurdO: chemical degradation curdlan, enzymatic degradation curdlan, and fermentative degradation using curdlan as a substrate and microorganism producing glucanase.

3.3.1 Hydrolysis of Curdlan by Chemical Methods

3.3.1.1 Acid Hydrolysis Polysaccharides can undergo hydrolysis under acidic conditions. Formic acid (Larm and Seljelid 1985), sulfuric acid (Zong et al. 2014), trifluoroacetic acid (TFA; Ruiter et al. 1992; Fontaine et al. 2006), and HCl have been used to catalyze curdlan hydrolysis. The hydrolysis mechanism is shown in Figure 3.3.

Formic acid hydrolysis was one of the main hydrolysis conditions for curdlan. Lower molecular weight CurdO (DP 7, 16, 24, 39, 50, 82, and 125) was prepared from curdlan (DP, 540) and was produced by

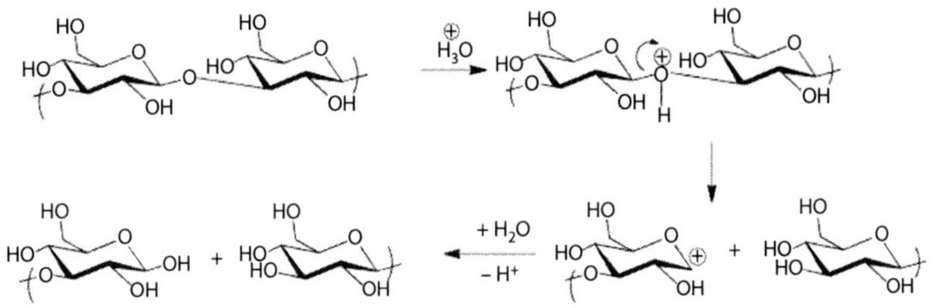

Figure 3.3 Curdlan oligosaccharide preparation using acid hydrolysis.

the cultivation of *Alcaligenes faecalis* IFO 13140. The optimal hydrolysis conditions are with 90% formic acid that is treated with curdlan at 95°C for 40 min (Sasaki et al. 1978).

Typical acid hydrolysis procedures are as follows (Figure 3.4). Curdlan is added to pure water at room temperature at a concentration of 10 g/L under strong agitation (500 rpm). The suspension is warmed up to 60°C. After 15 min, either sulfuric acid (H_2SO_4) or TFA, which are both final concentrations at 1 M, is added to the curdlan solution with a final volume of 50 mL at 60°C for 6, 24, 48, and 90 h. It is then cooled for 5 min on ice and neutralized with NaOH. The results revealed that the profile of oligosaccharides treated with sulfuric acid (H_2SO_4) or TFA is different: (1) minimal oligosaccharides are found but the DP of CurdO arrived at 5-mer; H_2SO_4 presents a quicker and stronger impact on the curdlan than that of TFA at identical concentration; (2) the DP of CurdO arrived at 17-mer using TFA, but only up to 5-mer can be observed in the batch that used sulfuric acid; (3) the final relative proportion of DP1–DP4 oligosaccharide fraction to the total obtained oligosaccharides increases gradually to 93% and 85% for TFA and H_2SO_4 at 90 h, respectively; however, only less than 25% yield of oligosaccharides can be reached with 1 M H_2SO_4 or TFA (Grandpierre et al. 2008). Most curdlan remains insoluble after being treated with 1 M H_2SO_4 or TFA for 90 h. Therefore, hydrolyzing

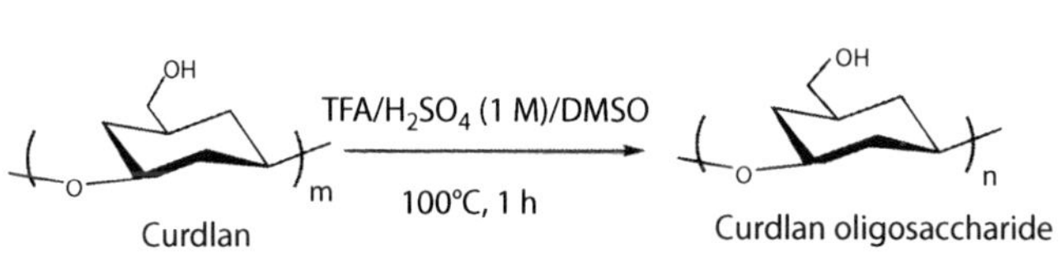

Figure 3.4 Curdlan hydrolysis using H_2SO_4 and TFA.

the part of insoluble curdlan is the key step to improve CurdO yield. Curdlan was easily degraded to oligosaccharides by degradation at 100°C by 90% formic acid for 20 min or 0.1 M TFA and 100°C for 60 min (Ohno et al. 1995). All of them continue to face this question.

The key step to hydrolyze the insoluble parts of curdlan is to understand the mechanism of curdlan resistant acid hydrolysis. The present study group conducted an experiment to explore the possible mechanism of curdlan resistance to acid. Curdlan has the ability to self-assemble. Thus, the ability of self-assembly may be responsible for curdlan resistance to acid hydrolysis. Curdlan can be dissolved in NaOH. Curdlan was dissolved into 1 M NaOH solution with a final concentration of 1% (w/v). About 1 M HCl was added to neutralize (pH 7) the curdlan–alkali solution. The mixture was centrifuged in 5,000×*g* for 20 min. The precipitate was washed for three times with deionized water and then used for scanning (Figure 3.5b). The precipitate was dissolved in 1 M H_2SO_4, treated for 18 h at 70°C, and then centrifuged in 5,000×*g* for 20 min. The precipitate was washed three times with deionized water and used for scanning (Figure 3.5c). Figure 3.5 shows that the size of alkali-neutralization-treated curdlan reduced to a small size compared with untreated curdlan (Figure 3.5a). However, the shape of curdlan residue (Figure 3.5c) altered distinctly in contrast to the alkali-neutralization-treated particles. The cube-shaped outer layer disappeared, whereas the inner ball-like core remained unchanged.

The above results show the mechanism of curdlan particle to acid. Curdlan formed a curdlan particle by self-assembly during curdlan extraction from fermentation. The curdlan changed from three-helix to single-chain conformation when the curdlan particle was redissolved in

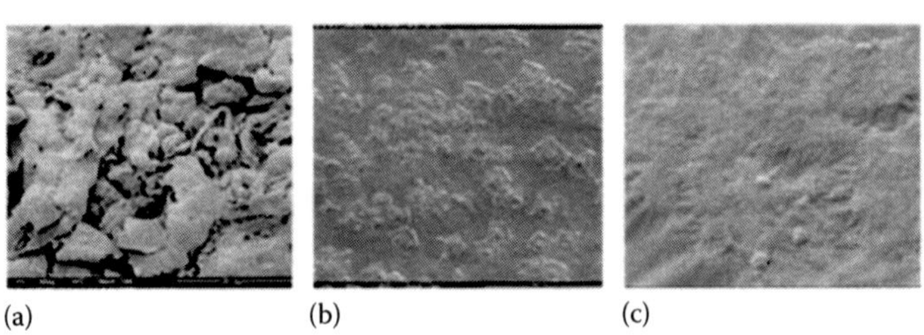

(a) (b) (c)

Figure 3.5 Scanning electron microscopy images of curdlan samples. (a) Untreated curdlan powders; (b) alkali-neutralization-treated (ANT) curdlan sample; (c) sample (b) further treated by 1 mol/L H_2SO_4 at 70°C for 18 h. All samples were magnified by 2400-fold.

NaOH solvent that contains water. The triple helix conformation of curdlan was reformed after it was neutralized using acid. It cross-linked with dissociated chains to form microparticles. The curdlan particle was finally formed. The outer layer of the curdlan particle can form a colloidal shell, which caused the core of the curdlan particle to be undissolved and water infiltration resistant. A new hydro layer can be formed on the outer layer of the curdlan undissolved compact core after the hydro layer is hydrolyzed, but the formation speed of the new hydro layer is very slow. Therefore, curdlan cannot be completely hydrolyzed by mild acid hydrolysis conditions at limited time (e.g., 6 and 10 h; Figure 3.6). However, curdlan can be completely hydrolyzed into glucose when using high concentration acid or high temperature (Prieto et al. 2011; Saito et al. 2014).

This study group conducted the following experiments: curdlan was dissolved in 1 M NaOH solution and then neutralized using 1 M HCl. The mixture was washed with water three times. The washed precipitate was dissolved in 1 M H_2SO_4 with the final curdlan

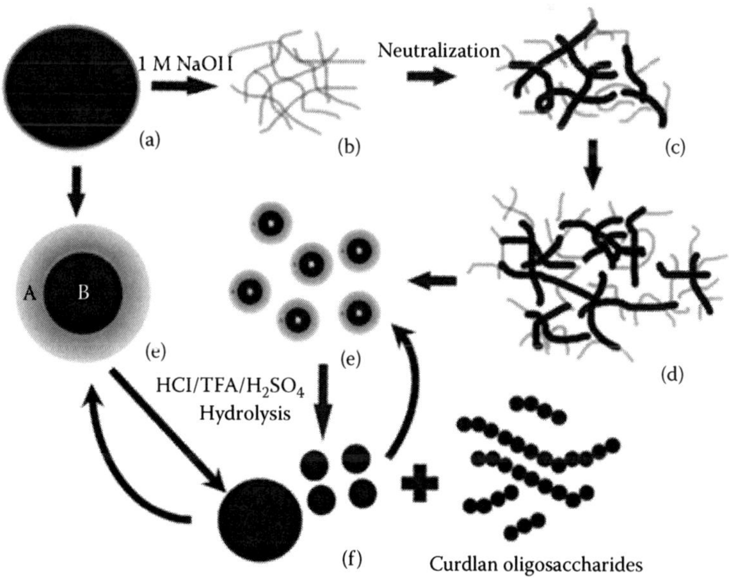

Figure 3.6 Schematic diagram of the structure of curdlan particle during hydrolysis. (a) Curdlan particle; (b) curdlan in single-chain conformation; (c) curdlan self-assembly to form microfibrils; (d) cross-link with dissociated chains to form microparticles; (e) hydro-layer formation on curdlan particle in water, and (f) curdlan particle after hydro layer was hydrolyzed. (A) Hydrated colloidal shell and (B) undissolved compact core.

of 1% concentration, incubated in water bath at 80°C, and shaken at 120 rpm. CurdO that is derived from the hydrolysate of alkali-neutralization-treated curdlan was increased by 25±5% compared with the hydrolysate products from untreated curdlan (Li et al. 2013). This result supported our inference that alkali-neutralization-treated curdlan can form a smaller size curdlan particle compared with untreated curdlan. About 40% of curdlan was hydrolyzed after incubating in 1 M H_2SO_4 for 30 min when the temperature was elevated up to 121°C. However, all the sugar is glucose (Li et al. 2013). High temperatures can accelerate the hydrolysis of curdlan. Therefore, a two-stage curdlan hydrolysis was developed using 1 M H_2SO_4. First, to acquire the DP of CurdO from 2–6-mer, the optimal condition for alkali-neutralization-treated curdlan in 1% was hydrolyzed using 1 M H_2SO_4 for 6 h at 70°C and then certificated. The precipitation was further hydrolyzed at 80°C for another 3 h. Second, to acquire the high DP of CurdO (7–10-mer), the time for the two-stage hydrolysis was 4 and 1 h, respectively (Zong et al. 2014). The hydrolysis ratio of curdlan by acid treatment was less than 12% because curdlan is not dissolved in water in neutralization or mild acid conditions. Therefore, overcoming the solubility of curdlan in acid hydrolysis solvent is the main barrier from further improving the curdlan hydrolysis ratio.

Curdlan can be completely dissolved in DMSO, 0.2 M NaOH, and formic acid. This study explored curdlan hydrolysis in DMSO solvent that contains 1 M HCl. Curdlan was dissolved in DMSO to prepare 83 g/L curdlan-DMSO solvent and then hydrolyzed in 60, 90, 120, 150, 180, and 240 min at 150°C (Figure 3.7). The optimal treatment conditions for curdlan hydrolysis in DMSO were 150°C for 150–180 min. The reaction mixture was subsided using acetone. The DP of CurdO was from 2-mer to 19-mer, but the main oligosaccharide is 5-mer (Fu, Liu et al. 2011).

Numerous studies on CurdOs derived from curdlan hydrolysis using mild acid have been performed. All products were mainly DP 2–10-mer. CurdOs larger than 10-mer are difficult to produce using curdlan hydrolysis methods. Improving the hydrolysis ratio should also be resolved because the hydrolysis ratio of curdlan is <15%. We performed a series of experiments that were mainly focused on optimizing hydrolysis conditions for curdlan in DMSO. First, the maximum concentration of curdlan (i.e., 150 g/L) in DMSO was achieved

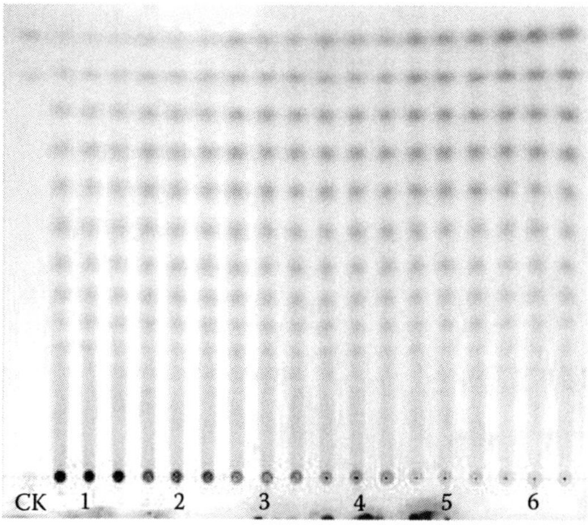

Figure 3.7 Best developing reagent on the development of linear β-1,3-glucooligosaccharides by unidirectional twice thin-layer chromatography using acetic acid/ethyl acetate/water = 2:2.4:1.4. CK, mixture of 0.5 g/L glucose and 0.25 g/L maltose; 1–6, curdlan hydrolysis product under 150°C for 60, 90, 120, 150, 180, and 240 min.

at 80°C. The optimal conditions for curdlan hydrolysis were 100 g/L curdlan dissolved in 1.0 mol/L HCl and treated at 92°C for 20 min. CurdO yield of 38.03 g/L was achieved (the hydrolysis ratio of curdlan was approximately 38%). The DP of curdlan oligosaccharides covered from 2-mer to 19-mer (Lu 2015).

In addition, acetolysis of curdlan was also performed to acquire CurdOs. Dried curdlan (1 g) was added to a mixture of glacial acetic acid (3.8 mL), acetic anhydride (3.8 mL), and sulfuric acid (0.8 mL), previously kept at four different temperatures (40°C, 50°C, 60°C, and 70°C). The reaction mixture was poured into ice-water after reaction under vigorous stirring for 15 min to 3 h. The acetolysate was extracted with chloroform after neutralization with sodium carbonate and saponified with 0.1 N potassium hydroxide in toluene–methanol (1:3, v/v) as described previously. The water-soluble curdlan oligosaccharide was obtained by repeated extraction with distilled water followed by deionization with Dowex 50×8 (H form) and Dowex 1×8 (acetate form) resin columns jointed in tandem. The optimum temperature to prepare a mixture of curdlan oligosaccharide was 60°C, and the DP of final oligosaccharides is from 2-mer to 7-mer (Azuma et al. 1989). In addition, dry formic acid containing 25% sulfuric acid was used to hydrolyze curdlan at 50–55°C for

5 days. The oligosaccharide mainly produced α-curdlan-biose octa-acetate in 27% yield (Wang et al. 1991).

Oligosaccharides have been prepared from curdlan by acid degradation. However, the hydrolysis rate was low (10–16%) or the main hydrolysis products exhibited DP < 5.

3.3.2 Enzymatic Hydrolysis of Curdlan

Enzymes hydrolyzing β-1,3-glucans have been observed in various organisms (Hasamudin and Kennedy 1994). Several glucanase enzymes produced by fungi (Liang et al. 1994), yeast (Fleet and Phaff 1975), or bacteria (Kanzawa et al. 1994) can degrade glucan (Table 3.1). These β-1,3-glucanases are divided into two groups according to their possible mechanisms of action (Hrmova and Fincher 2007): (1) exo-β-1,3-glucanases (or β1,3-glucan glucohydrolase, EC 3.2.1.58) hydrolyze β-glucans by sequentially cleaving glucose residues from the nonreducing end and only glucose as hydrolysis products; (2) endo-β-1,3-glucanases (EC 3.2.1.6, or β-1,3-glucan glucanohydrolase EC 3.2.1.39) cleave β-linkages at random sites along the polysaccharide chain releasing short-chain oligosaccharides (Cruz et al. 1996). Glucans are polysaccharides consisting of glucose with different linkages, and curdlan belongs to β-1,3-glucan. Therefore, the degradation of curdlan to produce CurdOs using enzyme has been widely explored.

3.3.2.1 Curdlan Oligosaccharide Production Using Exo-β-1,3-glucanases Exo-β-1,3-glucosidase (EC 3.2.1.58) can successfully hydrolyze β-D-glucose units from the nonreducing ends of β-1,3-glucans and produce glucose (Barras and Stone 1969). Thus, the number of glycoside hydrolases family reached 135. Exo-1,3-β-glucosidase activities were observed in GH families of 3, 5, 17, and 55 (Henrissat and Bairoch 1993). Glucose is the final product of sugar from curdlan hydrolysis treated with the majority of exo-β-1,3-glucosidase. Curdlan from *A. faecalis* is treated with exo-β-1,3-glucosidase from *Trichoderma viride* AZ36, and the final hydrolysis products of sugar is only glucose (Kulminskaya et al. 2002).

Exo-β-1,3-glucosidase can also produce curdlan-biose. The exo-β-1,3-glucanase purified from *Bacillus circulans* YK9 can hydrolyze curdlan. Then, 20 mg of curdlan in 20 mL of reaction mixture was incubated with 2 units of the purified exo-β-1,3-glucanase enzyme. The enzyme hydrolyzed approximately 75% of curdlan in 2 h. Further hydrolysis for 6 h resulted in over 95% yield of curdlan-biose (Kanzawa et al. 1994). In addition, curdlan was hydrolyzed with the β-1,3-glucanase system of *Streptomyces* sp. K27-4. The hydrolysate mainly consisted of glucose and curdlan-biose in an approximate ratio of 1:1 by weight (Kusama et al. 1984; Satoru et al. 2014). The exo-β-1,3-glucanase from *Rhizoctonia solani*, which is distinct from the former exo-1,3-β-glucosidase, mainly released glucose from curdlan. However, the ratios of glucose/curdlan-biose for the latter were 6:11, 7.8:0.5, and 18:0.9 at 360, 720, and 1080 min, respectively, when the purified enzyme directly acted on curdlan powder. This result clearly indicated that the enzyme could directly act on curdlan powder and hydrolyze curdlan into glucose and curdlan-biose (Vijayendra and Kashiwagi 2009).

3.3.2.2 Curdlan Oligosaccharide Production Using Endo-β-1,3-glucanases for Curdlan Ideal hydrolysis conditions for curdlan oligosaccharide production using curdlan should maximize polysaccharide breakdown and minimize monosaccharide production. We mainly focused on reviewing the oligomerization of curdlan using endo-β-1,3-glucanases because curdlan treated with exo-glucanase mainly produces glucose and laminari-biose.

Endo-β-1,3-glucanases are classified into the two categories of EC3.2.1.6 and EC3.2.1.39. Moreover, glycoside hydrolase family could be classified into 135 based on substrate specificities and catalytic patterns (Henrissat and Bairoch 1993). Endo-β-1,3-glucanase (EC3.2.1.6) activity was found in the GH families of 9 and 16. Endo-β-1,3-glucanase (EC3.2.1.39) was found in the GH families of 16, 17, 55, 64, 81, and 128. CurdOs in different DPs can be produced because of the randomness of the hydrolysis position in the chain of curdlan using endo-β-1,3-glucanase. These products can be categorized into three groups for endo-β-1,3-glucanase: (1) the products contained CurdOs, such as glucose, for example, DP 1-2 (Kikuchi et al. 2005), DP 1–3 (Hua et al. 2011), DP 1–4 (Fontaine et al. 1997), and DP 1–7 (Genta et al. 2003); (2) the hydrolysis products of curdlan oligosaccharides, for example, DP 2–4 (Zakharenko et al. 2011) and DP 2–6 (Akiyama et al. 2004), were larger than 2-mer; (3) the hydrolysis products mainly contained one oligosaccharide. For example, laminaribiose (DP2) was the main product of curdlan treated with endo-β-1,3-glucanase from *Streptomyces* sp. Mo (Kurakake et al. 2013), and DP 5 was the only product of curdlan treated with LPHase from *Streptomyces matensis* DIC-108 (Nishimura et al. 2001). In this section, endo-β-1,3-glucanase enzymes were reviewed and classified based on the final products of oligosaccharides using curdlan treated with endo-β-1,3-glucanase.

β-1,3-Glucanase from *Streptomyces* sp. K27-4 was used to produce curdlan-biose prepared from the hydrolysis products of curdlan. The β-1,3-glucanase activity in the culture filtrate peaked at 45–50 h after the beginning of cultivation and showed an activity of 1.56 units/mL. Curdlan (100 g, equivalent to 93 g based on glucose after acid hydrolysis) was hydrolyzed with 1400 mL (2180 units) of the enzyme solution produced by *Streptomyces* sp. K27-4 at pH 5.7 and 55°C for 27 h. Finally, 31 g of crystalline curdlan-biose was obtained (Kusama et al. 1984). β-1,3-Glucanases produced by *Trichoderma harzianum Rifai* PAMB-86 cultivated on botryosphaeran in a bench fermenter can also attack paramylon β-1,3-D-glucan to the extent of ~15–20% in 30 min releasing glucose and curdlan-biose as major end products (Giese et al. 2011). Endo-β-1,3-glucanase from *Streptomyces* sp Mo was found to hydrolyze curdlan powder. Then, 5% curdlan powder and 0.4 U/mL β-1,3-glucanase were reacted. Curdlan powder was liquefied, and DP 2–5 oligosaccharides were detected after 6 h of reaction. Trisaccharide was the major product of this reaction. By contrast, a 60-h reaction resulted in the release of sugar that was almost exclusively disaccharide, and a small amount of trisaccharide and glucose was detected. Sugar composition from the complete hydrolysis of curdlan was glucose 13%, disaccharide 81.7%, trisaccharide 4.8%, tetrasaccharide 0.4%, and pentasaccharide 0.1%. The disaccharide (major product) was identified as curdlan-biose. Therefore, enzymatic hydrolysis of curdlan powder resulted in the accumulation of curdlan-biose (yield 81.7%) (Kurakake et al. 2013).

Enzymes can hydrolyze curdlan, and the hydrolysis rate can reach 60–90%. We performed a series of experiments on the production, purification of endo-β-1,3-glucanase, and hydrolysis of curdlan using purified endo-β-1,3-glucanase. Endo-β-1,3-glucanase (Endo23) was purified from a *Trichoderma reesei* GIMCC 3.498 fermentation broth using anion exchange. The crude enzyme solution with a SuperQ-650M (55 mm × 80 cm; Tosoh Corp., Tokyo, Japan) anion column was used to separate endo-β-glucanase from the filtered solution using a linear gradient of NaCl. The optimal conditions for hydrolyzing curdlan were pH 5.0 and 50°C. The main hydrolytic products were curdlan-biose and curdlan-triose (Li et al. 2014).

Sufficient commercial enzymes are needed to prepare large-scale CurdOs. Commercial Zymolyase-20T was recently developed. This enzyme is a mixture of several enzymes, such as β-1,3-glucan laminaripentaose hydrolase and β-1,3-glucanase. Curdlan (10 g) was suspended in 900 mL of water and solubilized by adding 10 mL of 1 M sodium hydroxide to prepare oligosaccharides. The solution was stirred for 10 min and neutralized by adding 50 mL of 1 M Tris-buffer (pH 7.0) and 10 mL of 1 M hydrochloric acid. Then, 100 mg of Zymolyase-20T and 1 mL of 1 M dithiothreitol were added to the mixture. The mixture was incubated overnight with vigorous stirring at 37°C. Finally, 1.7 g of curdlan-biose and 1.2 g of curdlan-triose were obtained from 10 g of curdlan (Isono et al. 2013).

An endo-β-1,3-glucanase purified from a cell wall autolysate of *Aspergillus fumigates* was explored to hydrolyze curdlan to acquire large CurdOs. The influences of different parameters, such as pH, temperature, and substrate concentration, on oligosaccharide production using this enzyme were determined. The effect of temperature on activity was measured at pH 7.0 in 100 mM imidazole/acetic acid from 4°C to 80°C during 30 min of incubation. Substrate concentration ranged from 0.11 to 1.28 mg/mL to reduce curdlan. The optimum pH of 7.0 and optimum temperature of 60°C were determined. The hydrolysis conditions for curdlan hydrolysis were as follows: enzyme preparation (40 μL; 10 μg protein/mL) was added into curdlan (0.16 mg) in 100 mM imidazole acetic acid (pH 7.0) to obtain a final volume of 250 μL. The mixture was treated at 37°C. The products obtained after 15–60 min incubation contained a series of curdlan oligosaccharides (from 1-mer to 20-mer). The main products were only 2-mer to 4-mer of curdlan oligosaccharides (Fontaine, Hartland et al. 1997).

Thus, the hydrolysates using most β-1,3-glucan hydrolases were glucose and small-sized oligosaccharides (DP 2–4). Hydrolases of endo-β-1,3-glucanase using glucan as substrates larger than DP4 are required to produce long-chain oligosaccharides.

GH64, which is known as a family of yeast lytic enzyme, produced laminaripentaose (L5) and did not produce small-sized oligosaccharides with DP < 5. This characteristic is suitable to produce oligosaccharides with comparatively large chain lengths. However, hydrolysis activity toward insoluble curdlan was considerably low compared with soluble substrates such as laminarin. LPHase from *S. matensis* DIC-108 also showed that the enzyme belongs to the GH-64 family. Curdlan 5% (w/v) was incubated with LPHase for 6 h at 55°C in 50 mM acetate buffer (pH 5.5). The DP of CurdO is DP5 as shown by the results of the hydrolysis step (Wu et al. 2009).

The gene of laminaripentaose-producing enzyme (KfGH64) (Gen Bank accession 104 number: CP001736) from *Kribbella flavida* DSM 17836 was expressed in *Escherichia coli*. The effects of temperature (50°C, 70°C, 90°C, or 100°C), heating time (0.5, 2.0, or 4.0 h), and curdlan concentration (1.0%, 2.5%, or 5.0%) on the heat-treatment procedure were explored. The most efficient pretreatment was 1 g of curdlan dissolved in 20 mL of water and heated at 90°C for 0.5 h. Hydrolysis was performed in 1% pretreated curdlan and 20 μg/mL KfGH64 at 37°C, pH 5.5, 10 μg of KfGH64/1 g of curdlan, and 100 rpm mixing speed in 24 h. Approximately 60% of curdlan was converted into soluble saccharides under optimized enzyme reaction conditions. The resulting products were predominantly curdlan-pentaose and a small amount of β-(1,3)-glucans with an average DP of 13 and 130 (Kumagai et al. 2016).

Curdlan particles can also affect the hydrolysis of endo-1,3-β-glucanase. Therefore, CurdOs were prepared using pretreated curdlan and different methods. *Cellulosimicrobium cellulans* strain E4-5 producing endo-β-1,3-glucanase was cultured and harvested after 72 h to obtain endo-β-1,3-glucanase. The maximum enzymatic activity reached 33.52 U/L. The fermentation broth was centrifuged at 6000 rpm for 10 min. Then, 25 mL of enzyme was added to 75 mL of deionized water containing 0.5 g of heated curdlan gel and alkali-neutralized curdlan, respectively. Curdlan was gradually degraded, and the ratio was 90.2% at 96 h. The products of alkali-neutralization pretreatment ranged from DP 2 to DP 10. However, heated pretreatment produced oligosaccharides with lower values of DP (≤6) (Fu et al. 2015).

In addition, several other endo-β-1,3-glucanases can also produce CurdOs. However, these enzymes prefer to use soluble β-1,3-glucan, known as laminarin, not curdlan (Table 3.2).

CurdOs were successfully prepared by hydrolyzing curdlan with commercial α-amylase. Curdlan (5 g) was dispersed in distilled water to make a suspension with a concentration of 1% (w/v), and the pH was adjusted to 5.98 using 1 M HCl. α-Amylase was added and then maintained in a thermostatic water bath at 55.92°C for 30 min. The products were composed of CurdOs of DP 2–9. The hydrolysates were filtered, concentrated to 20% (w/v), precipitated at a ratio of 1:5 (concentrated hydrolysis/ethanol), and then freeze-dried. CurdO content in the product and yield were 97.7% and 97.6% (w/w), respectively (Qian et al. 2012).

3.3.3 Curdlan Oligosaccharide Production Using Fermentation Methods

Compared with acid hydrolysis, enzymatic hydrolysis methods can be applied for large-scale mild production of oligosaccharides. Most oligosaccharides produced in industrial scale use enzymatic catalytic methods. For example, enzymatic catalysis is ideal for curdlan-oligosaccharide production. A previous study showed that all curdlan-oligosaccharides isolated from curdlan using enzymatic hydrolysis methods always utilized purified endoglucanase. However, this enzyme was only used in the laboratory scale but not in industrial scale, because of the high cost of enzyme purification, and insufficient enzyme can be acquired in industrial scale. Thus, large-scale production of endoglucanase remains the key question for industrial-scale production of curdlan-oligosaccharide.

Curdlan was directly hydrolyzed using the fermentation broth of *Trichode Fma harzianum* CECT 2413 containing glucanase. Curdlan-oligosaccharide was acquired at low cost. Curdlan-oligosaccharide production resulted in high yield, but the DP of curdlan-oligosaccharide was lower than 6-mer. We recently developed methods for enzymatic hydrolysis of curdlan to acquire curdlan-oligosaccharide using fermentation methods. In our laboratory, *T. harzianum* was used for glucanase production. *T. harzianum* can simultaneously produce endoglucanase and exoglucanase. However, only endoglucanase can be used for curdlan-oligosaccharides production and mainly produced

Table 3.2 Enzymatic Properties of Endo-β-1,3-glucanases

SOURCE	MW (KDA)	PI	OPTIMUM SUBSTRATE	*T* (°C)	pH	HYDROLYSIS ABILITY	REFERENCE
Streptomyces sioyaensis	44.6		Curdlan	75	5.5	Strong	(Hong et al. 2002)
Paenibacillus sp.	44	4.3	Laminarin	55	5.0	Weak	(Hong and Meng 2003)
Lentinula edodes	26	3.87	Laminarin	50	4.0	Weak	(Sakamoto et al. 2011)
Oryza sativa L. cv Yukihikari	34	8.38	Laminarin			Weak	(Akiyama and Pillai 2001)
O. sativa	29	4.0	Laminaran		5.0	Weak	(Akiyama et al. 1997)
Trichoderma viride	70.1	7.5	Laminaran	45	5.0	No	(Nobe et al. 2003)
Saccharomyces cerevisiae	–	–	Laminarin	55	6.5	Weak	(Martín-Cuadrado et al. 2008)
Schizosaccharomyces pombe	–	–	Laminarin	50–55	6.0	Weak	(Martín-Cuadrado et al. 2008)
Rhizomucor miehei	39.7	4.45	Barley β-glucan	60	3.0	Strong	(Boyce and Walsh 2007)
Talaromyces emersonii	40.7	4.4	Barley β-glucan	80	4.8	No	(Murray et al. 2001)

glucose when curdlan was treated with exoglucanase. Therefore, improving the relative content of endoglucanase in total glucanase produced by *T. harzianum* is an efficient way.

Different morphologies of *T. harzianum* could be formed in the different stages and conditions during our fermentation process. The different conditions apparently affected the relative content of endoglucanase in the fermentation broth. Endoglucanase reached 48.5 ± 2.8 U/mL when the morphology of *T. harzianum* is as shown in Figure 3.8a. By contrast, endoglucanase reached 69.5 ± 3.7 U/mL when the morphology of *T. harzianum* is as shown in Figure 3.8b, in which the endoglucanase content was 143.3% than that in the morphology of *T. harzianum* after 144 h of fermentation is shown in Figure 3.8c, in which the endoglucanase was significantly decreased.

We developed efficient fermentation methods to acquire enough endoglucanase in 5 L fermenter for the large-scale production of CurdO. The fermentation profile of *T. harzianum* for endoglucanase production is shown in Figure 3.9. The biomass reached maximum concentration of 17.8 ± 0.6 g/L. Endoglucanase/total enzyme increased until 168 h of fermentation, although the biomass concentration was reduced. Endoglucanase/total enzyme reached the maximum value of 0.43. These results revealed that endoglucanase content in this method was lower than our results in the shake flask method.

We also observed the morphology of *T. harzianum* because this characteristic can affect the endoglucanase content (Figure 3.10). *T. harzianum* showed hair-like structure, compared with that in Figure 3.8. We infer that the impeller sheer force of the fermenter destroyed the structure of *T. harzianum*, reducing endoglucanase production.

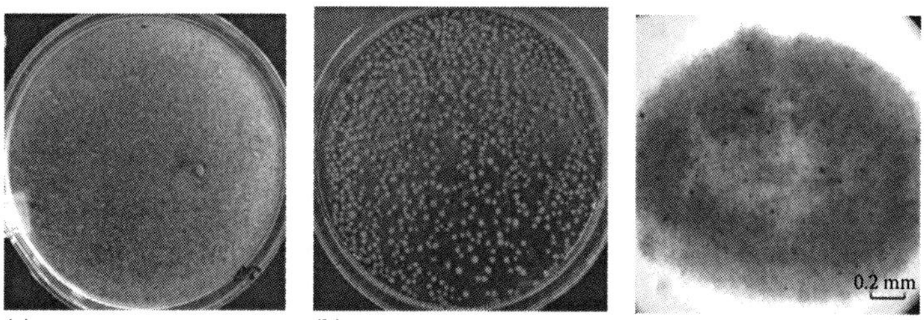

Figure 3.8 Different morphologies of *Trichoderma harzianum* at different stages. (a) Mycelium (72 h); (b) pellet (72 h); (c) aging pellet (144 h).

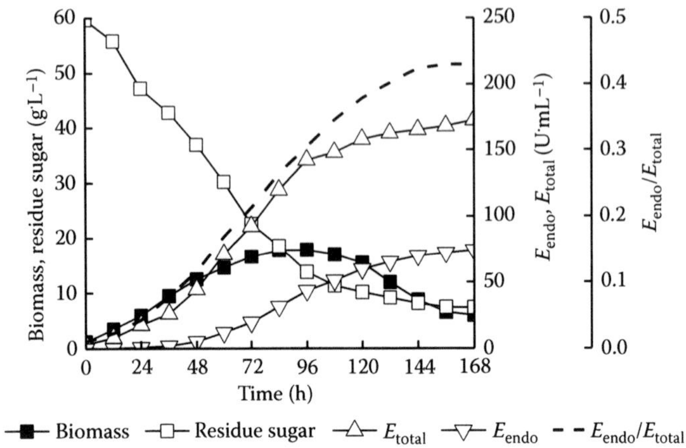

Figure 3.9 Growth curve of *T. harzianum* in mechanical agitation fermenter.

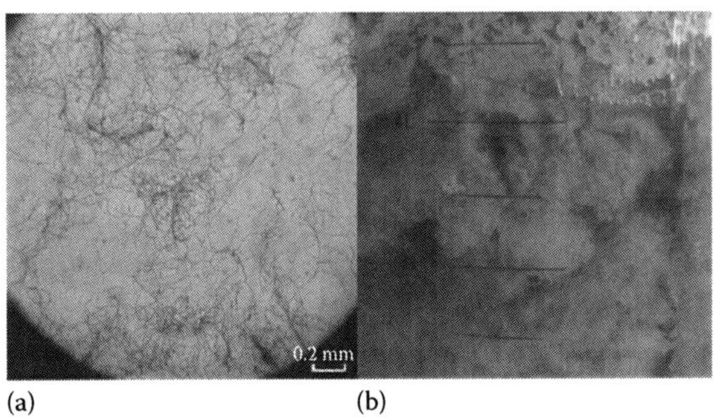

Figure 3.10 Fungal morphology of *T. harzianum* in the mechanical agitation fermenter. (a) Mycelia morphology and (b) clump of mycelia morphology in the mechanical agitation fermenter.

An airlift reactor was used for *T. harzianum* fermentation to further enhance the value of endoglucanase/total enzyme. The effect of ventilator capacity on morphology of *T. harzianum* was explored in a 65 L airlift reactor. The results are shown in Figure 3.11. The biomass reached maximum value of 23.8 g/L at 84 h, whereas E_{endo} reached 115.4 ± 6.7 U/mL at 144 h. E_{endo}/E_{total} reached 0.61, which increased by 55.6% comparing with that from the mixing-type fermenter. This result indicated that ventilatory capacity was proportional to the sheer force.

The effects of aeration ratio on the cell growth of *T. harzianum* and on the endo-β-1,3-glucanase activity of *T. harzianum* were explored. The results are shown in Figure 3.12. The results indicated that low ventilatory capacity was advantageous during the early stage of fermentation to

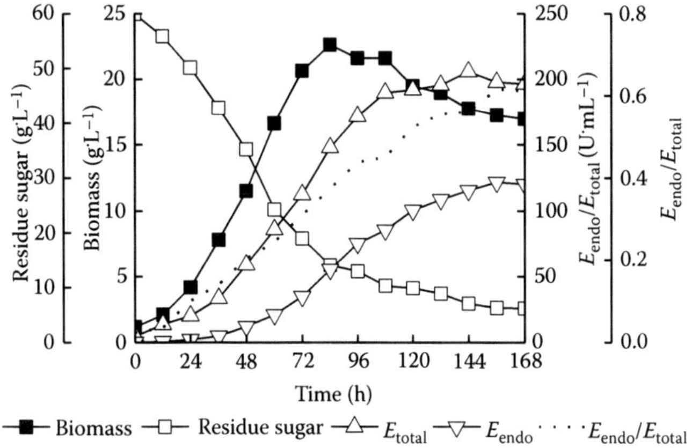

Figure 3.11 Growth curve of *T. harzianum* in airlift reactor.

maintain good morphology. Higher ventilation capacity was beneficial during the late stage to improve β-1,3-glucanase activity (Table 3.3).

Therefore, a two-stage control strategy was proposed to resolve the ventilation problem. Ventilatory capacity was maintained at 1.0 vvm during the early stage of fermentation (before 30 h), which was increased to 1.6 vvm during the middle periods of fermentation (30–168 h). Finally, the highest biomass reached 25.6 ± 0.8 g/L, and β-1,3-glucanase activity reached 165.1 ± 7.5 U·m/L and increased by 20% compared with the results under constant ventilation (Figure 3.13).

We previously showed that ANT curdlan can apparently improve the curdlan-oligosaccharides yield (Li et al. 2013). Moreover, another previous study found that the most efficient pretreatment for curdlan was heating curdlan at 90°C for 0.5 h. The ratio of curdlan converting into soluble saccharides can reach approximately 60% under optimized enzyme reaction conditions (pH 5.5, 37°C, 100 rpm mixing speed, 24 h, and 10 µg of KfGH64/1 g of curdlan; Kumagai et al. 2016). The curdlan-oligosaccharide products were predominantly curdlan-pentaose and only a small amount of β-(1,3)-glucans with an average DP of 13 and 130. The products did not contain small oligosaccharides (DP < 5). Therefore, two pretreatment methods were used to process curdlan before hydrolysis, that is, by direct dissolution in water and alkali neutralization of the two curdlans. Moreover, two sources of curdlan were compared. These sources were commercial curdlan, which was heat-treated to dry it to obtain powder form, and fresh curdlan,

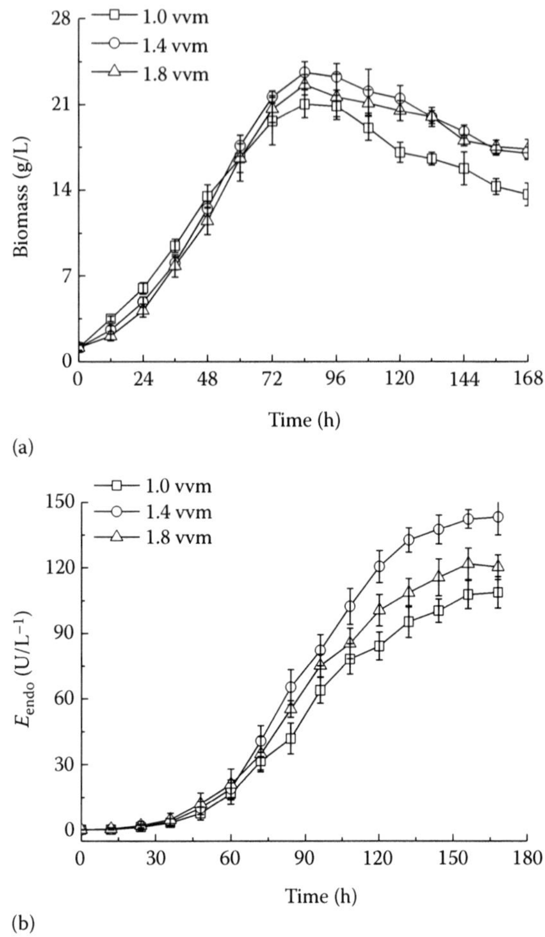

Figure 3.12 Effect of aeration ratio on (a) cell growth of *T. harzianum* and on the (b) endo-β-1,3-glucanase activity of *T. harzianum.*

which was prepared with *Agrobacterium* sp. in our laboratory. Scanning electron micrographs of the curdlan samples using curdlan processed with different pretreatment methods are shown in Figure 3.14. The particle sizes of the pretreated curdlans apparently varied. The particle sizes of the two curdlan sources after alkali-neutralization pretreatment were smaller than that under original curdlan. The particle of commercial heat-treated curdlan was looser than that of fresh curdlan. These results revealed that ANT curdlan exhibited good solubility, which could improve the enzyme–curdlan contact area. Moreover, the specific surface area of enzymatic hydrolysis could be increased by

Table 3.3 Effect of Aeration Ratio on the Fermentation Result of *T. harzianum*

VENTILATION (vvm)	RATIO IN MORPHOLOGY TO TOTAL BIOMASS (%)	AVERAGE DIAMETER (mm)	DIAMETER OF RELATIVE STANDARD DEVIATION (%)	BIOMASS (g/L)	ENDO-ENZYME ACTIVITY (U/mL)
1.0	88.4	0.81	35.3	21.0	100.3
1.4	61.1	0.68	28.0	23.6	137.5
1.8	37.6	0.32	21.6	22.6	115.4

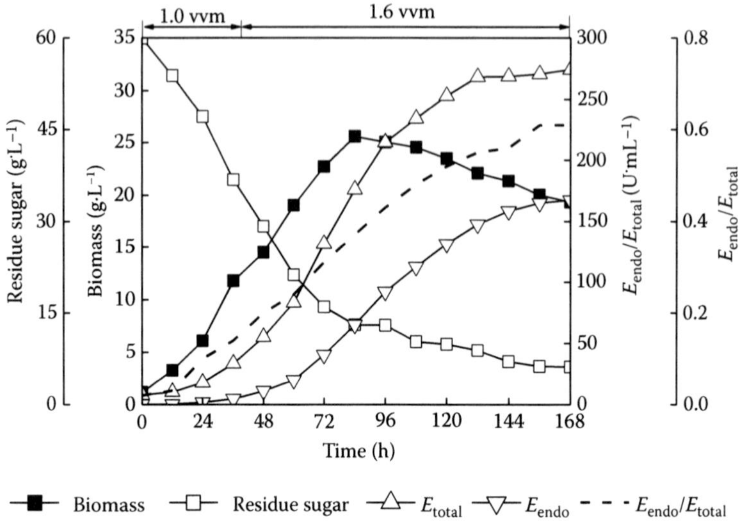

Figure 3.13 Time course of *T. harzianum* fermentation implemented by two-stage aeration ratio control.

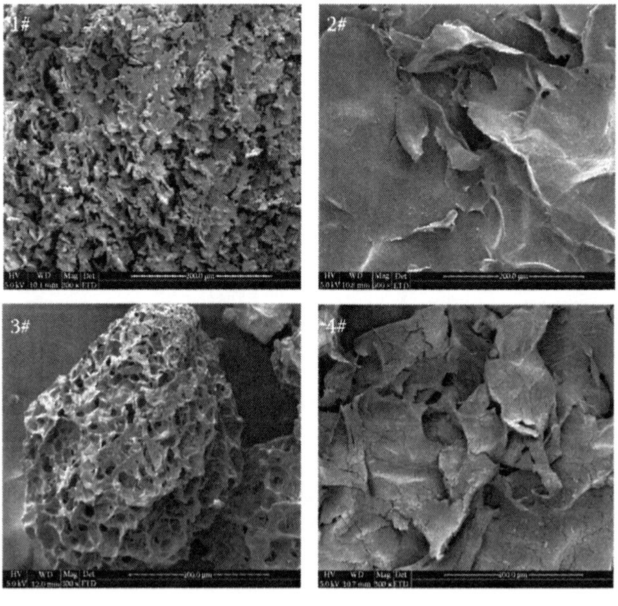

Figure 3.14 Scanning electron micrographs of curdlan samples. (1) Fresh curdlan dissolved in water; (2) alkali-neutralization-treated (ANT); (3) heat-treated; and (4) ANT and heat-treated curdlan.

decreasing the particle size of the disperse enzyme. The relationship between oligosaccharide content and hydrolysis time was investigated by hydrolyzing curdlan (Figure 3.15). The results showed that fresh curdlan, which was pretreated by alkali neutralization, was easier to hydrolyze than by direct dissolution in water, and the hydrolyzation rate reached 74.5%. In addition, high concentration of oligosaccharides was easier to achieve when hydrolyzation time was 2 h.

The hydrolysis products of curdlan contain glucose and curdlan-biose in a ratio of approximately 1: 1 by weight, and glucanase in *Streptomyces* sp. R27-4 was used to hydrolyze curdlan (Kanzawa et al. 1994). Then, a method for laminari-biose production was developed using this enzyme in cooperation with yeast to consume glucose. Our group also explored the methods for removing glucose using fermentation method. Five types of microbial strains (*Pichia membranaefaciens, Pichia membranaefaciens* M1, *Candida utilis, Saccharomyces cerevisiae*, and *Escherichia coli* JM109) were applied to remove glucose. Monosaccharide consumption rate, separability of microorganisms, and hydrolysate were used as indicators. The mutant *P. membranaefaciens* M1 was found to have the best separability (reaching 90% after precipitation for 5 min). The monosaccharide consumption rate was 4.3 g/L/h when the *P. membranaefaciens* M1 concentration was 8 g/L. The survival rate was over 80% after 10 repeated uses. The final monosaccharide removal rate was 88.4–91.0%, and the recovery rate was 83.7–88.5% (Figure 3.16).

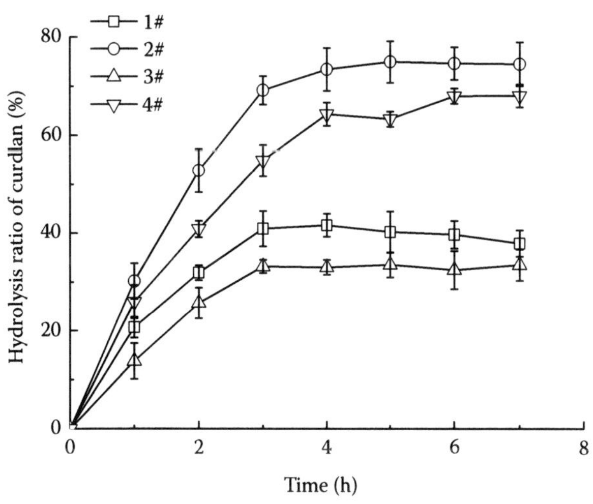

Figure 3.15 Time course of the hydrolysis ratio of curdlan by various substrate treatments.

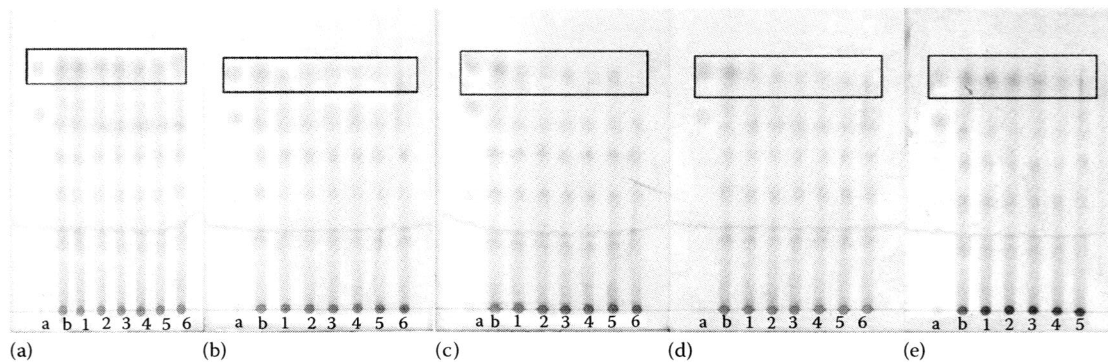

Figure 3.16 Comparison of removal ability of monosaccharide by various microorganisms. (a) Standard marker of glucose and sugar; (b) hydrolysate of curdlan; 1–6: biotransformation resultant after 10, 20, 30, 40, 50, and 60 min reaction. (a) *Pichia membranaefaciens*; (b) *Pichia membranaefaciens* M1; (c) *Candida utilis*; (d) *Saccharomyces cerevisiae*; and (e) *Escherichia coli* JM109.

3.3.4 Curdlan Degradation Using Hydrogen Peroxide

Curdlan was subjected to oxidative degradation using hydrogen peroxide (H_2O_2). The optimal hydrolysis conditions were determined as follows: reaction time, 40 min; temperature, 60°C; H_2O_2 concentration, 1.5% (v/v); and NaOH concentration, 2.5 M. Maximum dextrose equivalent value of 13.49% was obtained under the optimized conditions. The composition and structure of the hydrolysates were characterized by high-performance liquid chromatography and Fourier transform infrared spectroscopy, respectively. The hydrolysates were filtered, neutralized with HCl, concentrated to 12% (w/v), desalted, and freeze-dried to yield a water-soluble, white powder. The β-1,3-glucan oligosaccharide content of the product was 98.6%, and yield was 91.4% (w/w) (Wu et al. 2012).

3.3.5 Chemical Synthesis of Curdlan Oligosaccharides

β-1,3-glucan, including curdlan, were found to promote cancer cell death and active immune system by recognizing Dectin-1, CR3, and DC-SIGN. Therefore, increasing studies have focused on exploring the length and number of β-1,6-branch in relation to the structural function of glucan. Determining the molecular weight and the number of branching of polysaccharides remains a challenge. Thus, analyzing the interaction between protein-oligosaccharides isolated from glucan is feasible to assess glucan–protein interaction. An efficient HPLC method to separate mixture oligosaccharides larger than 7-mer to acquire purified oligosaccharides is not available, although degradation of curdlan can yield CurdOs in the mixture. Glucan oligosaccharides are synthesized via chemical methods to produce pure CurdOs.

Several methods have been developed to synthesize CurdOs in this section. We will review the chemical synthesis of CurdOs (also named oligo-β-1,3-glucans). The typical chemical synthesis of CurdOs contains a glycosyl donor and a glycosyl acceptor. C2, C4, and C6 were protected using a protecting group, but the C1 position was activated with an ester-type group (acetyl and benzoyl were the most widespread) or an ether-type group (benzyl, acetal, and silyl). The CurdOs consisted of glucopyranose presenting interglycosidic connections between the anomeric center of a glucosyl entity and the oxygen atom at C-3 position of the next glucopyranose residue. Therefore, the protection group in 1- and 3-position of glucosyl residue plays an important role to efficiently synthesize curdlan oligosaccharides, especially to chemically synthesize oligosaccharides. The most important objects are constructed of structurally well-defined oligosaccharides products in glycosylation reactions, especially for α and β anomers. This result not only deduced the yield of desired product but also increased the extra work in separation and purification process. The C-2 position is the key for deciding the α or β anomers. Hence, the protecting group of C-2 and C-3 position should be different. Moreover, all groups should be easily removed, and finally, the special protecting group in C-2 position should remain a donor in α or β anomers.

Iterative approach for CurdOs synthesis has been successfully developed based on the chemistry of the methylnaphthyl group (NAP). Diacetone glucose 1 was as an original material to acquire donor 2 via deprotection, protection, and activation steps. The resulting donor 2 was used to further glycosidic couplings to acquire CurdOs. Donor 2 reacted first with benzyl alcohol under activation by N-iodosuccinimide (NIS) and a catalytic amount of triethylsilyl trifluoromethanesulfonate (TESOTf), producing the required β-glucoside 3 in 89% yield. Then, β-glucoside 3 was removed from the Fmoc group and conjugated with donor 2. The deprotection and glycosylation steps were repeated to acquire linear CurdOs from 2-mer to 6-mer β-1,3-glucosyl residues (Figure 3.17).

The grafting method to chemically synthesize oligosaccharides and synthesize CurdOs (Mo et al. 2009; Elsaidi et al. 2015; Yashunsky et al. 2016) has been used to synthesize an even number of CurdO or an odd number of CurdO. For example, 4,6-o-benzylidene-1-thio-β-D-glucopyranoside was used as starting materials to chemically synthesize an even number of CurdO (Figure 3.18). First, selective silylation was performed at the C-3 hydroxyl, followed by benzoylation at C-2 hydroxyl group and then hydrolysis. The hydrolysis products were treated with NEt_3/CH_2Cl_2. Coupling of the monosaccharide acceptor 5 and thioglucoside 4 using NIS and AgOTf as the promoter afforded disaccharide 6 in 95% yield. Glucosyl imidate 7 coupled with acceptor 8 upon standard activation by catalytic TMSOTf resulted in disaccharide donor 9 in 92% yields. The achievement of disaccharide donor and acceptor facilitated

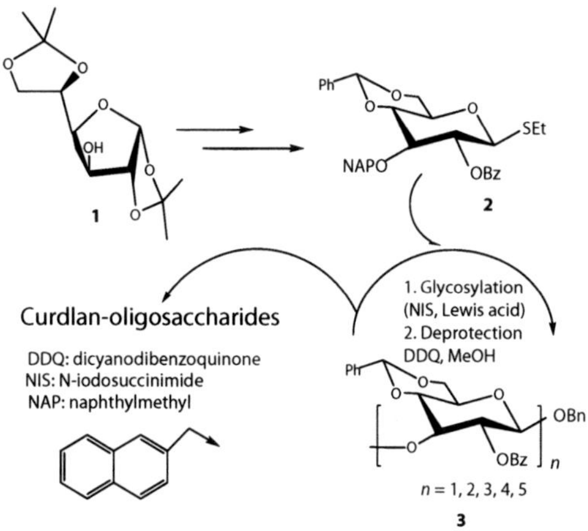

Figure 3.17 Chemical synthesis strategy of curdlan oligosaccharides using iterative approach. (Jamois, F. et al., *Glycobiology*, 2005. 15(4), 393–407, by permission of Oxford University Press.)

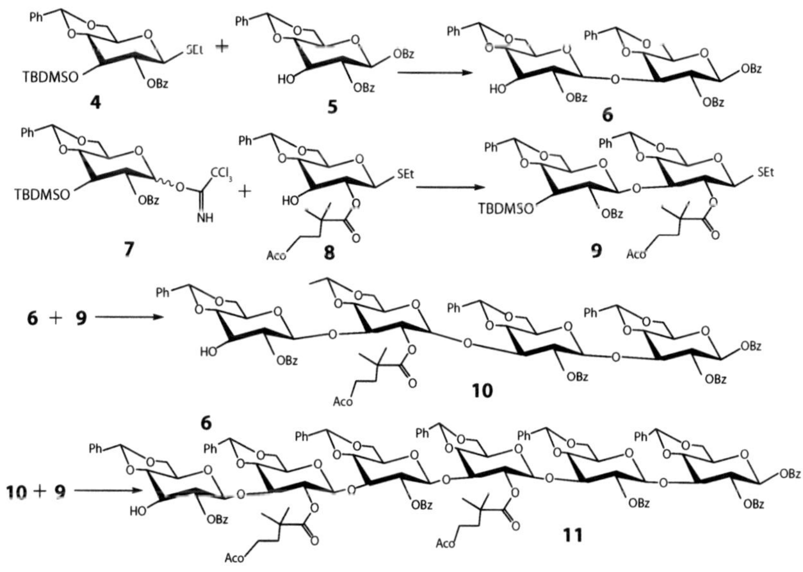

Figure 3.18 Chemical synthesis of an even number of curdlan oligosaccharides. (Reprinted from *Carbohydrate Research*, 344/(4), Mo, K.-F. et al., Synthesis of the beta-1,3-glucan, laminarahexaose: NMR and conformational studies, 439–447, Copyright 2009, with permission from Elsevier.)

the iterative synthesis of 6 and 9 with NIS and AgOTf as the promoter, providing tetrasaccharide 10. Coupling of donor 9 and acceptor 10 using NIS/AgOTf provided hexasaccharide 11. Complete deprotection of 11 was accomplished by a three-step process. The TBDMS group was removed by treatment with HF:Py:THF, and the benzylidene groups were removed by catalytic hydrogenolysis using 10% Degussa Pd/C in methanol/tetrahydrofuran/acetic acid (8:4:2.4) to provide hexasaccharide. The acquisition of oligosaccharides in purified α or β anomers is difficult to chemical synthesis of oligosaccharides. A series of experiments have been conducted to resolve this concern (Yu et al. 2005). The strategy to chemically synthesize odd-numbered CurdO is shown in Figure 3.19. This process will not be described in detail in this study.

Automated solid-phase synthesis of CurdOs was recently developed by Seeberger et al. (Figure 3.20). β-D-Glucose pentaacetate 12 was used as the original material to synthesize thioglycoside 15 via a seven-step process. Then, thioglycoside 15 was conjugated with functionalized Merrifield resin 16. The products removed Fmoc group at the C3 position with piperidine (20% v/v) in DMF at RT for 15 min.

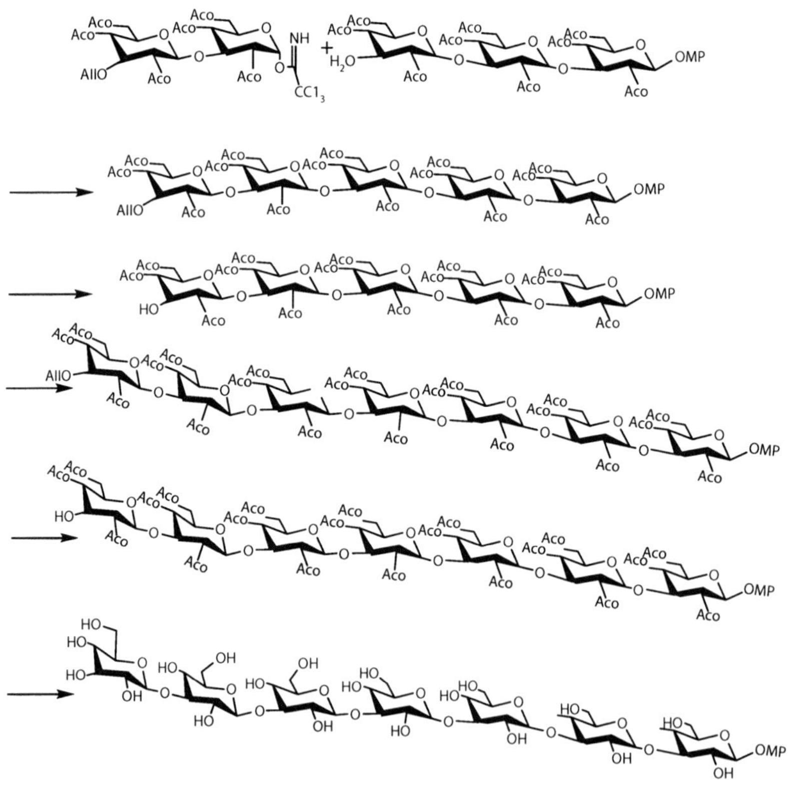

Figure 3.19 Chemical synthesis of an odd number of curdlan oligosaccharides. (Reprinted from *Carbohydrate Research*, 338/(20), Zeng, Y. and Kong, F., Highly efficient synthesis of alternate alpha- and beta-(1-->3)-linked glucose hepta- and octasaccharides, 2047–2056, Copyright 2003, with permission from Elsevier.)

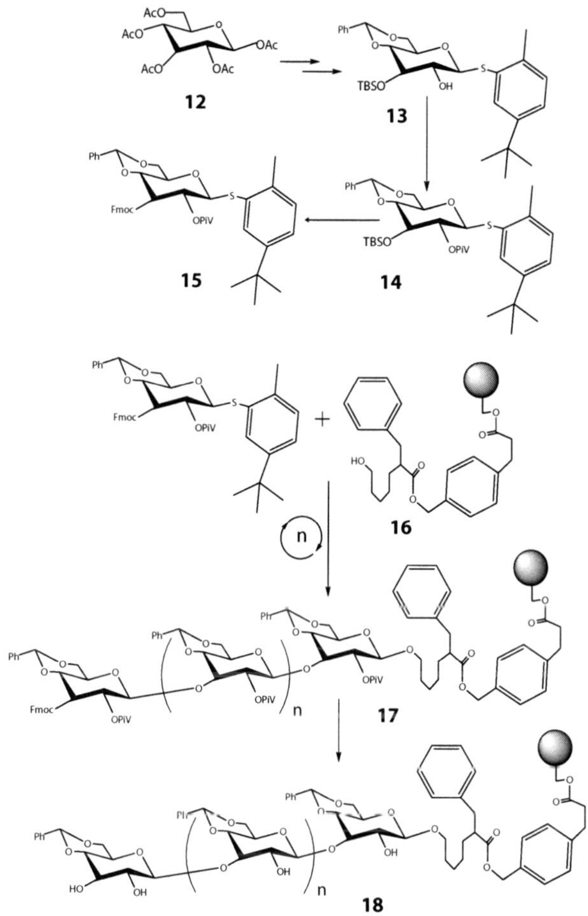

Figure 3.20 General scheme for the automated solid-phase synthesis of curdlan oligosaccharides. (From Weishaupt, M. W. and Matthies, S., Automated solid-phase synthesis of a ß-(1,3)-glucan dodecasaccharide. *Chemistry—A European Journal.* 2013. Vol. 19(37). Pp. 12497–12503. Copyright Wiley-VCH Verlag GmbH & Co. KGaA. Reproduced with permission.)

Then, iteration of glycosylation and deprotection was repeated to obtain different DP linear CurdOs 18.

3.4 Functional Applications of CurdOs

3.4.1 Agricultural Applications

Fast response defenses, such as extracellular pH shift, hydrogen peroxide (H_2O_2) production, and nitric oxide (NO) release, were the main parameters for evaluating the response defenses valence.

Similar to curdlan, which is β-1,3 glucan, laminarin also is a linear β-1,3 glucan (mean DP of 33) extracted from the brown alga *Laminaria digitata*. Laminarin has very good solubility in water. Thus, the effect of laminarin as elicitor activity on tobacco (*Nicotiana tabacum*) was evaluated. A dose of 200 mg/mL laminarin induced within a few minutes pH 1.9 unit alkalinization of the extracellular medium and a transient release of H_2O_2. Several hours of reaction resulted in a strong stimulation of Phe ammonia-lyase, caffeic acid O-methyltransferase, and lipoxygenase activities, as well as accumulation of salicylic acid. Laminarin did not induce tissue damage or cell death. No induction of the accumulation of the typical tobacco phytoalexin capsidiol was observed. Laminaripentaose was the smallest elicitor-active structure. Infiltration of 200 mg/mL laminarin in tobacco leaves triggered accumulation within 48 h of the four families of antimicrobial pathogenesis-related proteins investigated in accordance with its strong stimulating effect on defense responses in tobacco cells. Challenge of laminarin-infiltrated leaves 5 days after treatment with the soft rot pathogen *Erwinia carotovora* subsp. *carotovora* resulted in a strong reduction of the infection compared with water-treated leaves (Klarzynski et al. 2000).

N-p-Coumaroyloctopamine (PCO) has been identified as an induced phenolic substance in potato (*Solanum tuberosum*) tuber disks treated with a hyphal wall component from a blight fungus, *Phytophthora infestans*. The induction was also caused by treating the disks with laminarin, an algal β-1,3-glucooligosaccharide, in a dose-dependent manner. Maximum level of induction was approximately 150 μg (500 nmol)/g tissue. A purified β-1,3-glucosaccharide with a DP of 7 was significantly active at 1.2 mM, although the amount of induced amide was smaller than that induced by laminarin itself. The fatty acid elicitor, arachidonic acid, was not as effective for inducing PCO. This response of potato tuber disks may elucidate the action mechanism of oligosaccharides that can modify plant secondary metabolism (Miyagawa et al. 1998; Kobayashi et al. 2014).

Compared with laminarin, curdlan is cheap and can also be produced in large scale. Moreover, both compounds are β1,3-linked glucose. Laminaripentaose was the smallest elicitor-active structure. Therefore, the activity of CurdOs as elicitor was explored. CurdOs triggered faster extracellular alkalinization than laminarin and also induced cells to generate NO. Moreover, CurdOs exhibited more intensive stomatal movement than laminarin. However, protection against TMV implied that laminarin played a longer-term role in defense responses than CurdOs. These results suggested that β-1,3-glucan with low DP played a vital role in the rapid responses, whereas high DP β-1,3-glucan was responsible for the long-term effects (Fu et al. 2011). The effect of nine purified consecutive CurdO monomers upon H_2O_2-release in plants was evaluated. CurdO was found to have stronger DP-induced defense responses in plants, thereby elucidating the relationship between CurdO and biological activities (Fu et al. 2015).

The defense responses of CurdOs on *S. tuberosum L. cv. McCain* G1 were investigated at biochemical and proteomic levels. The results indicated that the CurdOs exhibited activation effect on the early- and late-defense responses in potato leaves. In addition, glucopentaose was proved to be the shortest active curdlan molecule based on the accumulation of H_2O_2 and salicylic acid and the activities of phenylalanine amino-lyase, β-1,3-glucanase, and chitinase using 2D-PAGE analysis. Furthermore, the potato yield was also enhanced compared with the control (Li et al. 2014).

3.4.2 Applications in Biomedicine

Recent studies revealed that commensal *Bifidobacterium* (also named gut microbiota) plays an important role in human health. This species can enhance anticancer immunotherapy by CTLA-4 blockade, which relies on the gut microbiota (Sivan et al. 2015; Vetizou et al. 2015). Bifidobacteria can also affect the brain (Smith 2015), participate in the synthesis of nutrient substance in host (Yano et al. 2015), and can especially realize host–gut microbiota metabolic interactions (Nicholson et al. 2012). The effect of curdlan feed on the growth and metabolites of gut microbiota was investigated. The results implied that significant increases in the amounts of short-chain fatty acids and lactate and in the numbers of bifidobacteria were observed in the cecum of curdlan-fed rats. The in vitro proliferation of five species of bifidobacteria was markedly increased in the cultures containing the supernatant obtained from the cecal contents of curdlan-fed rats compared with the control group. Curdlan can be degraded into CurdOs under acid condition; the pH is low in the stomach of humans. Therefore, bifidus factors produced in the curdlan-fed rats should be caused by CurdOs derived from curdlan (Shimizu et al. 2001).

In addition, glucan can enhance host immunity (Cheung and Engstad 2009). Therefore, the effect of CurdOs on the immune response was explored. IFN-γ and TNF-α, CurdOs induced production of GM-CSF and IFN-g, but not TNF-α, compared with scleroglucan, which stimulated splenocytes in DBA/2 mice to produce GM-CSF. However, the amounts of GM-CSF and IFN-γ were lower than those produced in response to scleroglucan. The effect of scleroglucan on TNF-α production was partially inhibited by CurdOs. CurdOs induced the production of TNF-α and IL-6 in bone marrow-derived dendritic cells (Hida et al. 2009). CurdOs also induced monocytes to produce TNF-α (Miyanishi et al. 2003) and stimulated the secretion of interleukin 1β (Jamois et al. 2005). Oligosaccharides with DP > 4 possess biological activity and function as bioactive materials. Therefore, these results suggest that CurdOs can induce the production of cytokines and then enhance the immune response (Wang et al. 2010).

PRR is involved in the activation of the immune system (Brubaker et al. 2015; Iwasaki and Medzhitov 2015). CurdOs can bind to Dectin-1 and then activate the immune system. The minimum DP of CurdOs to bind to Dectin-1 is 10-mer (Palma et al. 2006). We found that CurdOs can bind to DC-SIGN and DC-SIGNR (Palma et al. 2015). DC-SIGN and DC-SIGNR can remarkably enhance transduction and infection of pathogenic bacterium or virus via the interaction between DC-SIGN/DC-SIGNR and carbohydrates from pathogenic bacterium or virus (Shimojima et al. 2014; Zhang et al. 2014). CurdOs can bind to DC-SIGN and DC-SIGNR, suggesting that CurdOs can act as inhibitors to inhibit the interaction between DC-SIGN/DC-SIGNR and pathogens. Synthesis of a β-1,3-glucan hexasaccharide, similar to CurdOs, can act as a Dectin-1 ligand and functionalize to permit covalent attachment to glycoconjugate vaccines, thereby facilitating improved antigen processing by dendritic cells (Tanaka et al. 2012; Elsaidi et al. 2015). Laminarin-derived penta-, hexa-, and hepta-glucooligosaccharides were used to determine the minimum unit ligand bind to glucan receptors on human U937 cells. The results indicated that a heptasaccharide is the smallest unit ligand recognized by macrophage glucan receptors (Lowe et al. 2001). The affinity interaction between Dectin-1 and CurdOs is dependent on primary structure and is influenced by the DP of CurdOs (Adams et al. 2008).

CurdOs have also been used clinically. A therapeutic method comprising administration of a composition, comprising an amount of oligo-β-1,3-glucan and a pharmaceutically acceptable carrier, was used to treat a disease selected from the group consisting of a tumor, cancer, viral disease, bacterial disease, fungal disease, disease of the immune system, autoimmune disease, or a disease related to a deficiency in immunostimulation, in which the amount of oligo-β-1,3-glucan is effective to treat the disease (Yvin et al. 2005). Combining CurdO with monoclonal antibody to treat against cancer has been reported, and a patent for this compound has been applied for (Yvin et al. 2006).

The synthesis of potent anti-HIV-active sulfated alkyl curdlan-oligosaccharides composed of glucosidic residues of 5–9 was investigated. The anti-HIV activity and anticoagulant activity of these sulfated alkyl curdlan oligosaccharides were assessed. The anticoagulant activity increased as the number of sugar units increased from 5 to 9 (Katsuraya et al. 1994).

A previous study revealed that β,1,3-glucan-CRM197 can confer protection against *Candida albicans* in mice (Bromuro et al. 2010). Further studies have suggested that only linear 15-mer CurdO conferred protection against *C. albicans.* Recent bioactivity analysis against *C. albicans* using curdlan oligosaccharide conjugated with CRM 197 was performed. The results implied that CRM 197 conjugated with small CurdO work better at high loading than that conjugated with large CurdO. That is, CurdO 7-mer-CRM 197 have a loading degree of approximately 17 compared with a value of 7.5 for the Lam-CRM17 conjugate (Adamo et al. 2011) (Figure 3.21).

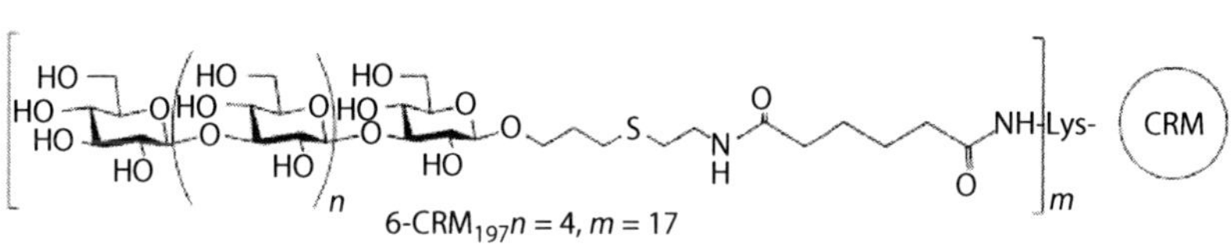

Figure 3.21 Structure of curdlan oligosaccharide-CRM 197.

3.5 Explorations of the New Functionality of CurdOs and Other Glucan Oligosaccharides Using Carbohydrate Microarray

Carbohydrate-mediated molecular recognition is involved in many biological aspects, such as cellular adhesion, immune response, blood coagulation, inflammation, and infection. Synthetic saccharide-based systems have emerged as powerful tools to understand the protein–carbohydrate interactions by considering the crucial importance of such biological events, which normally involved proteins (Lopez-Jaramillo et al. 2012). The identification of these interactions is a priority therefore in biomedical sciences. Carbohydrate microarray technology is a new development at the frontier of glycomics that has revolutionized the study of carbohydrate–protein interactions and the elucidation of their specificities in endogenous biological processes, immune defense mechanisms, and microbe–host interactions (Liu and Feizi 2008). Thus, carbohydrate microarray has been used to understand immunity (Arthur et al. 2014), explore HIV glycobiology (Ratner and Seeberger 2007), and so on.

We explored the interaction between curdlan oligosaccharide–protein from the immune system and found that CurdO larger than 10-mer can bind to Dectin-1 (Palma et al. 2015), which can direct T helper cell differentiation by controlling noncanonical NF-kB activation (Gringhuis et al. 2009), and also participate in antifungal infection (Marakalala et al. 2011). We also recently found that DC-SIGN and DC-SIGNR can recognize CurdOs. We will explore the novel functionalities as food or medicine of curdlan and other glucan oligosaccharides by carbohydrate microarrays in our future studies.

3.6 Future Trends

Several aspects on curdlan oligosaccharide production and application need further study. The hydrolysis rate during the preparation of oligosaccharides from curdlan by acid or enzyme degradation was low (10–16%) or the main hydrolysis products possessed DP less than 5. Although degradation of alkali-formed curdlan gel by a fungus enzyme cocktail was successful in a high hydrolysis rate, the hydrolytic products were mainly small-sized oligosaccharides with DP $<$ 5, and the quantity of oligosaccharides was mainly in a small scale. We recently developed methods for curdlan-oligosaccharide production from curdlan using fermentation methods that can resolve the production scale. However, the DP of main products was smaller than 10-mer. Therefore, the preparation methods for large quantities of CurdOs with a DP > 4 would be in demand.

In addition, the products of curdlan oligosaccharides mainly exist as mixtures. Studies on CurdOs mainly focus on hydrolysis. Only limited related works have separated these compounds using HPLC in the laboratory. The resulting quantities are not sufficient for application in medicine or food. The large-scale separation of these compounds to acquire large quantities of CurdOs still needs be resolved.

The functions of CurdOs have been detected, but the molecular mechanisms remain unclear. Understanding the molecular mechanisms of CurdO is beneficial in developing more efficient functional units from curdlan and other glucan. Oligosaccharides interact with target protein, and therefore, searching for proteins targeted by curdlan oligosaccharides is the first step to understanding the mechanism of the function of CurdOs. Carbohydrate microarray is a powerful tool that can be employed to explore proteins for CurdO reorganization.

References

Adamo, R., Tontini, M., Brogioni, G., Romano, M. R., Costantini, G., Danieli, E., Proietti, D., Berti, F. and Costantino, P. (2011). Synthesis of laminarin fragments and evaluation of a β-(1,3) glucan hexasaccharide-CRM 197 conjugate as vaccine candidate against *Candida albicans*. *Journal of Carbohydrate Chemistry*. 30(4–6), 249–280.

Adams, E. L., Rice, P. J., Graves, B., Ensley, H. E., Yu, H., Brown, G. D., Gordon, S., Monteiro, M. A., Papp-Szabo, E., Lowman, D. W., Power, T. D., Wempe, M. F. and Williams, D. L. (2008). Differential high-affinity interaction of dectin-1 with natural or synthetic glucans is dependent upon primary structure and is influenced by polymer chain length and side-chain branching. *Journal of Pharmacology and Experimental Therapeutics*. 325(1), 115–123.

Akiyama, T. and Pillai, M. A. (2001). Molecular cloning, characterization and in vitro expression of a novel endo-1,3-β-glucanase up-regulated by ABA and drought stress in rice (*Oryza sativa* L.). *Plant Science*. 161(6), 1089–1098.

Akiyama, T., Pillai, M. A. and Sentoku, N. (2004). Cloning, characterization and expression of OsGLN2, a rice endo-1,3-β-glucanase gene regulated developmentally in flowers and hormonally in germinating seeds. *Planta*. 220(1), 129–139.

Akiyama, T., Shibuya, N., Hrmova, M. and Fincher, G. B. (1997). Purification and characterization of a (1→3)-β-D-glucan endohydrolase from rice (*Oryza sativa*) bran. *Carbohydrate Research*. 297(4), 365–374.

Arthur, C. M., Cummings, R. D. and Stowell, S. R. (2014). Using glycan microarrays to understand immunity. *Current Opinion in Chemical Biology*. 18(5), 55–61.

Azuma, J. I., Imamura, T. and Okamura, K. (1989). Preparation of laminarin-oligosaccharides: By acetolysis of curdlan. *Bulletin of the Kyoto University Forests*. 61, 338–349.

Barras, D. R. and Stone, B. A. (1969). β1,3-Glucan hydrolases from *Euglena gracilis*: I. The nature of the hydrolases. *Biochimica et Biophysica Acta*. 191(2), 329–341.

Boyce, A. and Walsh, G. (2007). Production, purification and application-relevant characterisation of an endo-1,3(4)-β-glucanase from *Rhizomucor miehei*. *Applied Microbiology and Biotechnology*. 76(4), 835–841.

Bromuro, C., Romano, M., Chiani, P., Berti, F., Tontini, M., Proietti, D., Mori, E., Torosantucci, A., Costantino, P. and Rappuoli, R. (2010). Beta-glucan-CRM197 conjugates as candidates antifungal vaccines. *Vaccine*. 28(14), 2615–2623.

Brown, G. D. and Gordon, S. (2005). Immune recognition of fungal beta-glucans. *Cellular Microbiology*. 7(4), 471–479.

Brubaker, S. W., Bonham, K. S., Zanoni, I. and Kagan, J. C. (2015). Innate immune pattern recognition: A cell biological perspective. *Annual Review of Immunology*. 33, 257–290.

Cheung, N. K. V. and Engstad, R. E. (2009). Immune response enhancing glucan, US. US20090053221.

Cruz, J., De La, Pintor-Toro, J. A., Benítez, T., Llobell, A. and Romero, L. C. (1996). A novel endo-beta-1,3-glucanase, BGN13.1, involved in the mycoparasitism of *Trichoderma harzianum*. *Journal of Bacteriology*. 177(23), 6937–6945.

Elsaidi, H. R., Paszkiewicz, E. and Bundle, D. R. (2015). Synthesis of a 1,3 beta-glucan hexasaccharide designed to target vaccines to the dendritic cell receptor, Dectin-1. *Carbohydrate Research*. 408, 96–106.

Fang, Y., Shuang-zhi, Z., Li, S., Yan-hao, Z. and Hong-ming, G. (2013). Optimization of fermentation medium for curdlan by response surface methodology. *Science and Technology of Food Industry*. 34(24), 150–154.

Fleet, G. H. and Phaff, H. J. (1975). Glucanases in Schizosaccharomyces. Isolation and properties of an exo-β-glucanase from the cell extracts and culture fluid of *Schizosaccharomyces japonicus* var. *versatilis*. *Biochimica et Biophysica Acta (BBA)—Enzymology*. 410(2), 318–332.

Fontaine, T., Hartland, R., Beauvais, A., Diaquin, M. and Latge, J.-P. (1997). Purification and characterization of an endo-1,3-β-glucanase from *Aspergillus fumigatus*. *European Journal of Biochemistry*. 243(1–2), 315–321.

Fontaine, T., Hartland, R., Mouyna, I. and Latge, J. P. (2006). Method for sorting antifungal molecules acting on the glucanosyltransferase activity, US, US6551811B1.

Footrakul, P., Suyanandana, P., Amemura, A. and Harada, T. (1981). Extracellular polysaccharides of *Rhizobium* from the Bangkok MIRCEN collection. *Journal of Fermentation Technology*. 59(1), 9–14.

Fu, Y., Cheng, L., Meng, Y., Li, S., Zhao, X., Du, Y. and Yin, H. (2015). Cellulosimicrobium cellulans strain E4-5 enzymatic hydrolysis of curdlan for production of (1→3)-linked β-d-glucan oligosaccharides. *Clinical Science*. 134(5), 740–744.

Fu, Y., Liu, Q., Li, S., Zhang, J., Bai, X., Zhao, X. and Du, Y. (2011). Preparation and component analysis of curdlan oligomers. *Food Science*. 32(3), 6–9.

Fu, Y., Wang, M., Wang, W., Tuo, Y., Guo, Z., Du, Y. and Yin, H. (2015). (1→3)-β-D-glucan oligosaccharides monomers purification and its H2O2 induction effect study. *International Journal of Biological Macromolecules*. 81, 1069–1073.

Fu, Y., Yin, H., Wang, W., Wang, M., Zhang, H., Zhao, X. and Du, Y. (2011). β-1,3-Glucan with different degree of polymerization induced different defense responses in tobacco. *Carbohydrate Polymers*. 86(2), 774–782.

Gao, J., Xu, H., Li, Q. J., Feng, X. H. and Li, S. (2010). Optimization of medium for one-step fermentation of inulin extract from Jerusalem artichoke tubers using *Paenibacillus polymyxa* ZJ-9 to produce R, R-2,3-butanediol. *Bioresource Technology*. 101(18), 7087–7093.

Genta, F. A., Terra, W. R. and Ferreira, C. (2003). Action pattern, specificity, lytic activities, and physiological role of five digestive β-glucanases isolated from *Periplaneta americana*. *Insect Biochemistry and Molecular Biology*. 33(11), 1085–1097.

Ghai, S. K., Hisamatsu, M., Amemura, A. and Harada, T. (1981). Production and chemical composition of extracellular polysaccharides of *Rhizobium*. *Microbiology*. 122(1), 33–40.

Giese, E. C., Dekker, R. F. H., Barbosa, A. M., da Silva, M. L. C. and da Silva, R. (2011). Production of β-(1,3)-glucanases by *Trichoderma harzianum* Rifai: Optimization and application to produce gluco-oligosaccharides from paramylon and pustulan. *Fermentation Technology*. 1(1), 1–5.

Grandpierre, C., Janssen, H. G., Laroche, C., Michaud, P. and Warrand, J. (2008). Enzymatic and chemical degradation of curdlan targeting the production of β-(1→3) oligoglucans. *Carbohydrate Polymers*. 71(2), 277–286.

Gringhuis, S. I., den Dunnen, J., Litjens, M., van der Vlist, M., Wevers, B., Bruijns, S. C. and Geijtenbeek, T. B. (2009). Dectin-1 directs T helper cell differentiation by controlling noncanonical NF-kappaB activation through Raf-1 and Syk. *Nature Immunology*. 10(2), 265–270.

Gummadi, S. N. and Kumar, K. (2005). Production of extracellular water insoluble β-1,3-glucan (curdlan) from *Bacillus* sp. SNC07. *Biotechnology & Bioprocess Engineering*. 10(6), 546–551.

Harada, T., Masada, M., Hidaka, H. and Takada, M. (1966). Production of firm, resilient gel-forming polysaccharide in natural medium by a mutant of *Alcaligenes faecalis* var. *myxogenes* 10C3. *Journal of Fermentation Technology*. 44, 20–24.

Hasamudin, W. H. W. and Kennedy, J. F. (1994). *Review of Chemistry and Biology of (1→3)-β-glucans*, edited by B. A. Stone and A. E. Clarke. *Carbohydrate Polymers*. 25(1), 61–62.

Henrissat, B. and Bairoch, A. (1993). New families in the classification of glycosyl hydrolases based on amino-acid sequence similarities. *Biochemical Journal*. 293, 781–788.

Hida, T. H., Ishibashi, K., Miura, N. N., Adachi, Y., Shirasu, Y. and Ohno, N. (2009). Cytokine induction by a linear 1,3-glucan, curdlan-oligo, in mouse leukocytes in vitro. *Inflammation Research*. 58(1), 9–14.

Hong, T.-Y., Cheng, C.-W., Huang, J.-W. and Meng, M. (2002). Isolation and biochemical characterization of an endo-1,3-β-glucanase from *Streptomyces sioyaensis* containing a C-terminal family 6 carbohydrate-binding module that binds to 1,3-β-glucan. *Microbiology*. 148(4), 1151–1159.

Hong, T.-Y. and Meng, M. (2003). Biochemical characterization and antifungal activity of an endo-1,3-β-glucanase of *Paenibacillus* sp. isolated from garden soil. *Applied Microbiology and Biotechnology*. 61(5–6), 472–478.

Hrmova, M. and Fincher, G. B. (2007). Dissecting the catalytic mechanism of a plant β-D-glucan glucohydrolase through structural biology using inhibitors and substrate analogues. *Carbohydrate Research*. 342(12), 1613–1623.

Hua, C., Yi, H. and Jiao, L. (2011). Cloning and expression of the endo-1,3 (4)-β-glucanase gene from *Paecilomyces* sp. FLH30 and characterization of the recombinant enzyme. *Bioscience Biotechnology and Biochemistry*. 75, 1807–1812.

Isono, N., Kimoto, T. and Hisamatsu, M. (2013). A simple method for preparing laminaribiose and laminaritriose from the enzymatic hydrolysates of curdlan. *The Bulletin of the Graduate School of Bioresources, Mie University*, 55–60.

Iwasaki, A. and Medzhitov, R. (2015). Control of adaptive immunity by the innate immune system. *Nature Immunology*. 16(4), 343–353.

Jamois, F., Ferrieres, V., Guegan, J. P., Yvin, J. C., Plusquellec, D. and Vetvicka, V. (2005). Glucan-like synthetic oligosaccharides: Iterative synthesis of linear oligo-beta-(1,3)-glucans and immunostimulatory effects. *Glycobiology*. 15(4), 393–407.

Jiang, L. (2013). Effect of nitrogen source on curdlan production by *Alcaligenes faecalis* ATCC 31749. *International Journal of Biological Macromolecules*. 52(1), 218–220.

Jung, D. Y., Cho, Y. S., Chung, C. H., Jung, D. I., Kim, K. and Lee, J. W. (2001). Improved production of curdlan with concentrated cells of *Agrobacterium sp. Biotechnology & Bioprocess Engineering*. 6(2), 107–111.

Jung, D. Y., Seo, H. P., Lee, N. K., Ji, M. K., Hong, K. P., Cho, Y. S. and Jin, W. L. (2000). Influence of inoculum size and initial pH on production of curdlan by *Agrobacterium sp.* ATCC 31749. *European Actuarial Journal*. 247(247), 1–5.

Kalyanasundaram, G. T., Doble, M. and Gummadi, S. N. (2012). Production and downstream processing of (1→3)-beta-D-glucan from mutant strain of *Agrobacterium sp.* ATCC 31750. *AMB Express*. 2(1), 31.

Kanegae, Y., Yutani, A. and Nakatsui, I. (1996). Mutant strains of agrobacterium for producing β-1,3-glucan, US, US5508191.

Kanzawa, Y., Kurasawa, T., Kanegae, Y., Harada, A. and Harada, T. (1994). Purification and properties of a new exo-(1→3)-beta-D-glucanase from *Bacillus circulans* YK9 capable of hydrolysing resistant curdlan with formation of only laminari-biose. *Microbiology*. 140 (Pt 3), 637–642.

Katsuraya, K., Shoji, T., Inazawa, K., Nakashima, H., Yamamoto, N. and Uryu, T. (1994). Synthesis of sulfated alkyl laminara-oligosaccharides having potent anti-HIV activity and the relationship between structure and biological activities. *Macromolecules*. 27(23), 6695–6699.

Kikuchi, T., Shibuya, H. and Jones, J. T. (2005). Molecular and biochemical characterization of an endo-β-1,3-glucanase from the pinewood nematode *Bursaphelenchus xylophilus* acquired by horizontal gene transfer from bacteria. *Biochemical Journal*. 389, 117–125.

Kim, M. K., Lee, I. Y., Ko, J. H., Rhee, Y. H. and Park, Y. H. (1999). Higher intracellular levels of uridine monophosphate under nitrogen-limited conditions enhance metabolic flux of curdlan synthesis in *Agrobacterium* species. *Biotechnology and Bioengineering*. 62(3), 317–323.

Klarzynski, O., Plesse, B., Joubert, J. M., Yvin, J. C., Kopp, M., Kloareg, B. and Fritig, B. (2000). Linear beta-1,3 glucans are elicitors of defense responses in tobacco. *Plant Physiology*. 124(3), 1027–1037.

Kobayashi, A., Tai, A., Kanzaki, H. and Kawazu, K. (2014). Elicitor-active oligosaccharides from algal laminaran stimulate the production of antifungal compounds in *Alfalfa*. *Zeitschrift für Naturforschung C*. 48(7–8), 575–579.

Konno, A. and Harada, T. (1992). Thermal properties of curdlan in aqueous suspension and curdlan gel. *Food Hydrocolloids*. 5(5), 427–434.

Kulminskaya, A. A., Thomsen, K. K., Shabalin, K. A., Sidorenko, I. A., Eneyskaya, E. V., Savel'Ev, A. N. and Neustroev, K. N. (2002). Isolation, enzymatic properties, and mode of action of an exo-1,3-beta-glucanase from *Trichoderma viride*. *European Journal of Biochemistry*. 268(23), 6123–6131.

Kumagai, Y., Okuyama, M. and Kimura, A. (2016). Heat treatment of curdlan enhances the enzymatic production of biologically active β-(1,3)-glucan oligosaccharides. *Carbohydrate Polymers*. 146,396–401.

Kurakake, M., Yamanouchi, Y., Kinohara, K. and Moriyama, S. (2013). Enzymatic properties of β-1,3-glucanase from *Streptomyces* sp Mo. *Journal of Food Science*. 78(4), C502–C506.

Kusama, S., Kusakabe, I., Zama, M., Muralami, K. and Yasui, T. (1984). Studies on the BETA-1,3-glucanase system of Streptomyces. Part I. Enzymatic preparation of crystalline laminaribiose from curdlan. *Agricultural and Biological Chemistry*. 48(6), 1433–1440.

Kwak, S. and Lafleur, M. (2003). Self-diffusion of macromolecules and macroassemblies in curdlan gels as examined by PFG-SE NMR technique. *Colloids & Surfaces A Physicochemical & Engineering Aspects*. 221(1–3), 231–242.

Larm, O. and Seljelid, R. (1985). Water-soluble aminated beta-1,3-bound D-glucan and composition containing same, EP, EP014737.

Lee, I. Y., Kim, M. K., Lee, J. H., Seo, W. T., Jung, J. K., Lee, H. W. and Park, Y. H. (1999). Influence of agitation speed on production of curdlan by *Agrobacterium* species. *Bioprocess Engineering*. 20(4), 283–287.

Lee, J. H. and Lee, I. Y. (2000). Optimization of uracil addition for curdlan (β-1→3-glucan) production by *Agrobacterium* sp. *Biotechnology Letters*. 23(14), 1131–1134.

Lee, J. H., Lee, I. Y., Kim, M. K. and Park, Y. H. (1999). Optimal pH control of batch processes for production of curdlan by *Agrobacterium* species. *Journal of Industrial Microbiology & Biotechnology*. 23(2), 143–148.

Li, J., Zhu, L., Lu, G., Zhan, X. B., Lin, C. C. and Zheng, Z. Y. (2014a). Curdlan β-1,3-glucooligosaccharides induce the defense responses against *Phytophthora infestans* infection of potato (*Solanum tuberosum* L. cv. McCain G1) leaf cells. *PLoS One*. 9(5), e97197.

Li, J., Zhu, L., Zhan, X. B., Xu, M., Lin, C. C., Zheng, Z. Y. and Li, W. J. (2014b). Purification and characterization of a new endo-β-1,3-glucanase exhibiting high specificity for curdlan to produce β-1,3-glucan oligosaccharides. *Food Science & Biotechnology*. 23(3), 799–806.

Li, J., Zhu, L., Zheng, Z. Y., Zhan, X. B., Lin, C. C., Zong, Y. and Li, W. J. (2013). A new effective process for production of curdlan oligosaccharides based on alkali-neutralization treatment and acid hydrolysis of curdlan particles in water suspension. *Applied Microbiology and Biotechnology*. 97(19), 8495–8503.

Liang, Z. C., Hseu, R. S. and Wang, H. H. (1994). Partial purification and characterization of a 1,3-β-d-glucanase from *Ganoderma tsugae*. *Journal of Industrial Microbiology & Biotechnology*. 14(1), 5–9.

Liu, Y. and Feizi, T. (2008). Microarrays—A key technology for glycobiology. In B. O. Fraser-Reid, K. Tatsuta and J. Thiem (eds.), *Glycoscience*. Berlin: Springer Verlag, pp. 2121–2132.

Lopez-Jaramillo, F. J., Ortega-Muñoz, M., Megia-Fernandez, A., Hernandez-Mateo, F. and Santoyo-Gonzalez, F. (2012). Vinyl sulfone functionalization: A feasible approach for the study of the lectin-carbohydrate interactions. *Bioconjugate Chemistry*. 23(4), 846–855.

Lowe, E., Rice, P., Ha, T., Li, C., Kelley, J., Ensley, H., Lopezperez, J., Kalbfleisch, J., Lowman, D. and Margl, P. (2001). A (1→3)-beta-D-linked heptasaccharide is the unit ligand for glucan pattern recognition receptors on human monocytes. *Microbes & Infection*. 3(10), 789–797.

Lu, G. (2015). Efficient preparation of linear β-1,3-glucooligosaccharides by directed hydrolysis of curdlan and characterization of its biological activity. Masters thesis, Jiangnan University.

Marakalala, M. J., Kerrigan, A. M. and Brown, G. D. (2011). Dectin-1: A role in antifungal defense and consequences of genetic polymorphisms in humans. *Mammalian Genome*. 22(1–2), 55–65.

Martín-Cuadrado, A.-B., Fontaine, T., Esteban, P.-F., del Dedo, J. E., de Medina-Redondo, M., del Rey, F., Latgé, J. P. and Aldana, C. R. (2008). Characterization of the endo-β-1,3-glucanase activity of *S. cerevisiae* Eng2 and other members of the GH81 family. *Fungal Genetics and Biology*. 45(4), 542–553.

Martinez, C. O., Ruiz, S. P., Nogueira, M. T., Bona, E., Portilho, M. and Matioli, G. (2015). Effective immobilization of *Agrobacterium* sp. IFO 13140 cells in loofa sponge for curdlan biosynthesis. *Molecules*. 20(5), 7957–7973.

McIntosh, M., Stone, B. A. and Stanisich, V. A. (2005). Curdlan and other bacterial (1→3)-β-D-glucans. *Applied Microbiology and Biotechnology*. 68(2), 163–173.

Miyagawa, H., Ishihara, A., Lim, C. H., Ueno, T. and Furuichi, N. (1998). Induction of N-p-coumaroyloctopamine in potato tuber disks by b-1,3-glucooligosaccharide. *Journal of Pesticide Science*. 23, 49–53.

Miyanishi, N., Iwamoto, Y., Watanabe, E. and Odaz, T. (2003). Induction of TNF-alpha production from human peripheral blood monocytes with beta-1,3-glucan oligomer prepared from laminarin with beta-1,3-glucanase from *Bacillus clausii* NM-1. *Journal of Bioscience and Bioengineering*. 95(2), 192–195.

Mo, K.-F., Li, H., Mague, J. T. and Ensley, H. E. (2009). Synthesis of the β-1,3-glucan, laminarahexaose: NMR and conformational studies. *Carbohydrate Research*. 344 (4):439–447.

Murray, P. G., Grassick, A., Laffey, C. D., Cuffe, M. M., Higgins, T., Savage, A. V., Planas, A. and Tuohy, M. G. (2001). Isolation and characterization of a thermostable endo-β-glucanase active on 1,3-1,4-β-D-glucans from the aerobic fungus *Talaromyces emersonii* CBS 814.70. *Enzyme and Microbial Technology*. 29(1), 90–98.

Na, K., Park, K. H., Kim, S. W. and Bae, Y. H. (2000). Self-assembled hydrogel nanoparticles from curdlan derivatives: Characterization, anti-cancer drug release and interaction with a hepatoma cell line (HepG2). *Journal of Controlled Release*. 69(2), 225–236.

Nakanishi, I., Kimura, K., Suzuki, T., Ishikawa, M., Banno, I., Sakane, T. and Harada, T. (1976). Demonstration of curdlan-type polysaccharide and some other P-1,3-glucan in microorganisms with aniline blue. *Journal of General & Applied Microbiology*. 22(1), 1–11.

Nicholson, J. K., Holmes, E., Kinross, J., Burcelin, R., Gibson, G., Jia, W. and Pettersson, S. (2012). Host-gut microbiota metabolic interactions. *Science*. 336(6086), 1262–1267.

Nishimura, T., Bignon, C., Allouch, J., Czjzek, M., Darbon, H., Watanabe, T. and Henrissat, B. (2001). *Streptomyces matensis* laminaripentaose hydrolase is an 'inverting' β-1,3-glucanase. *FEBS Letters*. 499(1), 187–190.

Nobe, R., Sakakibara, Y., Fukuda, N., Yoshida, N., Ogawa, K. and Suiko, M. (2003). Purification and characterization of laminaran hydrolases from *Trichoderma viride*. *Bioscience Biotechnology and Biochemistry*. 67(6), 1349–1357.

Novak, M. and Vetvicka, V. (2008). Beta-glucans, history, and the present: Immunomodulatory aspects and mechanisms of action. *Journal of Immunotoxicology*. 5(1), 47–57.

Ohno, N., Terui, T., Chiba, N., Kurachi, K., Adachi, Y. and Yadomae, T. (1995). Resistance of highly branched (1→3)-beta-D-glucans to formolysis. *Chemical & Pharmaceutical Bulletin*. 43(6), 1057–1060.

Palma, A. S., Feizi, T., Zhang, Y., Stoll, M. S., Lawson, A. M., Diaz-Rodriguez, E., Campanero-Rhodes, M. A., Costa, J., Gordon, S., Brown, G. D. and Chai, W. (2006). Ligands for the beta-glucan receptor, Dectin-1, assigned using "designer" microarrays of oligosaccharide probes (neoglycolipids) generated from glucan polysaccharides. *Journal of Biological Chemistry*. 281(9), 5771–5779.

Palma, A. S., Liu, Y., Zhang, H., Zhang, Y., McCleary, B. V., Yu, G., Huang, Q., Guidolin, L. S., Ciocchini, A. E. and Torosantucci, A. (2015). Unravelling glucan recognition systems by glycome microarrays using the designer approach and mass spectrometry. *Molecular & Cellular Proteomics*. 14(4), 974–988.

Phillips, K. R., Pik, J., Lawford, H. G., Lavers, B., Kligerman, A. and Lawford, G. R. (1983). Production of curdlan-type polysaccharide by *Alcaligenes faecalis* in batch and continuous culture. *Canadian Journal of Microbiology*. 29(10), 1331–1338.

Prieto, M. A., Vázquez, J. A. and Murado, M. A. (2011). Hydrolysis optimization of mannan, curdlan and cell walls from *Endomyces fibuliger* grown in mussel processing wastewaters. *Process Biochemistry*. 46(8), 1579–1588.

Puertas, A., Dueñas, M. and Marieta, C. (2014). Supramolecular structure and renaturation of a (1→3)-β-d-glucan compared with curdlan and scleroglucan. *Fibers*. 2(3), 255–263.

Qian, Z., Wu, S., Pan, S. and Xia, Z. (2012). Preparation of (1→3)-β-d-glucan oligosaccharides by hydrolysis of curdlan with commercial α-amylase. *Carbohydrate Polymers*. 87(3), 2362–2364.

Rafigh, S. M., Yazdi, A. V., Vossoughi, M., Safekordi, A. A. and Ardjmand, M. (2014). Optimization of culture medium and modeling of curdlan production from *Paenibacillus polymyxa* by RSM and ANN. *International Journal of Biological Macromolecules*. 70, 463–473.

Ratner, D. M. and Seeberger, P. H. (2007). Carbohydrate microarrays as tools in HIV glycobiology. *Current Pharmaceutical Design*. 13(2), 173–183.

Ruiter, G. A. D., Schols, H. A., Voragen, A. G. J. and Rombouts, F. M. (1992). Carbohydrate analysis of water-soluble uronic acid-containing polysaccharides with high-performance anion-exchange chromatography using methanolysis combined with TFA hydrolysis is superior to four other methods. *Analytical Biochemistry*. 207(1), 176–185.

Saito, H., Misaki, A. and Harada, T. (2014). A comparison of the structure of curdlan and pachyman. *Agricultural and Biological Chemistry*. 32(10), 1261–1269.

Sakamoto, Y., Nakade, K. and Konno, N. (2011). Endo-β-1,3-glucanase GLU1, from the fruiting body of *Lentinula edodes*, belongs to a new glycoside hydrolase family. *Applied and Environmental Microbiology*. 77(23), 8350–8354.

Salah, R. B., Jaouadi, B., Bouaziz, A., Chaari, K., Blecker, C., Derrouane, C., Attia, H. and Besbes, S. (2011). Fermentation of date palm juice by curdlan gum production from Rhizobium radiobacter ATCC 6466™: Purification, rheological and physico-chemical characterization. *LWT—Food Science and Technology*. 44(4), 1026–1034.

Saraswat-Ohri, S., Vashishta, A., Vetvicka, V., Descroix, K., Jamois, F., Yvin, J. C. and Ferrieres, V. (2011). Biological properties of (1→3)-beta-D-glucan-based synthetic oligosaccharides. *Journal of Medicinal Food*. 14(4), 369–376.

Sasaki, T., Abiko, N., Sugino, Y. and Nitta, K. (1978). Dependence on chain length of antitumor activity of (1→3)-beta-D-glucan from Alcaligenes faecalis var. myxogenes, IFO 13140, and its acid-degraded products. *Cancer Research*. 38(2), 379–383.

Shimizu, J., Tsuchihashi, N., Kudoh, K., Wada, M., Takita, T. and Innami, S. (2001). Dietary curdlan increases proliferation of bifidobacteria in the cecum of rats. *Bioscience, Biotechnology, and Biochemistry*. 65(2), 466–469.

Shimojima, M., Takenouchi, A., Shimoda, H., Kimura, N. and Maeda, K. (2014). Distinct usage of three C-type lectins by Japanese encephalitis virus: DC-SIGN, DC-SIGNR, and LSECtin. *Archives of Virology*. 159(8), 2023–2031.

Sivan, A., Corrales, L., Hubert, N., Williams, J. B., Aquino-Michaels, K., Earley, Z. M., Benyamin, F. W., Lei, Y. M., Jabri, B., Alegre, M. L., Chang, E. B. and Gajewski, T. F. (2015). Commensal *Bifidobacterium* promotes antitumor immunity and facilitates anti-PD-L1 efficacy. *Science*. 350(6264), 1084–1089.

Smith, P. A. (2015). The tantalizing links between gut microbes and the brain. *Nature*. 526(7573), 312–314.

Sun, Y. S., Lei, W. and Zhan, X. B. (2005). Influence of nitrogen source NH_4Cl Concentration on curdlan production in *Alcaligenes faecalis*. *Chinese Journal of Biotechnology*. 21(2), 328–331.

Tada, T., Tamai, N., Matsumoto, T. and Masuda, T. (2001). Network structure of curdlan in DMSO and mixture of DMSO and water. *Biopolymers*. 58(2), 129–137.

Tanaka, H., Kawai, T., Adachi, Y., Hanashima, S., Yamaguchi, Y., Ohno, N. and Takahashi, T. (2012). Synthesis of β(1,3) oligoglucans exhibiting a Dectin-1 binding affinity and their biological evaluation. *Bioorganic & Medicinal Chemistry*. 20(12), 3898–3914.

Thornton, B. P., Vetvicka, V., Pitman, M., Goldman, R. C. and Ross, G. D. (1996). Analysis of the sugar specificity and molecular location of the beta-glucan-binding lectin site of complement receptor type 3 (CD11b/CD18). *Journal of Immunology*. 156(3), 1235–1246.

Toshifumi, F. H. and Kozo, O. (1999). Viscometry of curdlan, a linear (1→3)-β-D-glucan, in DMSO or alkaline solutions. *Bioscience, Biotechnology & Biochemistry*. 63(8), 1481–1483.

Unno, T., Nakakuki, T., Fujimoto, Y., Okada, G., Kainuma, S. and Goda, T. (2005). Industrial production and higher application of functional BETA-glucooligosaccharides having a bitter taste. *Journal of Applied Glycoscience*. 52, 59–64.

Verma, M. S. and Gu, F. X. (2012). 1,3-β-glucans: Drug delivery and pharmacology. In D. N. Karunaratne (ed.), *Complex World of Polysaccharides*. Rijeka, Croatia: Intech, pp. 555–572.

Vetizou, M., Pitt, J. M., Daillere, R., Lepage, P. et al. (2015). Anticancer immunotherapy by CTLA-4 blockade relies on the gut microbiota. *Science*. 350(6264), 1079–1084.

Vijayendra, S. V. N. and Kashiwagi, Y. (2009). Characterization of a new acid stable exo-β-1,3-glucanase of *Rhizoctonia solani* and its action on microbial polysaccharides. *International Journal of Biological Macromolecules*. 44(1), 92–97.

Wang, J., Dong, S., Liu, C., Wang, W., Sun, S., Gu, J., Wang, Y., Boraschi, D. and Qu, D. (2010). beta-Glucan oligosaccharide enhances CD8(+) T cells immune response induced by a DNA vaccine encoding hepatitis B virus core antigen. *Journal of Biomedicine and Biotechnology*. 2010, 645213.

Wang, L. X., Sakairi, N. and Kuzuhara, H. (1991). Peracetylated laminaribiose: Preparation by specific degradation of curdlan and its chemical conversion into N-acetylhyalobiuronic acid. *Carbohydrate Research*. 219(1), 133–148.

Weishaupt, M. W. and Matthies, S. (2013). Automated solid-phase synthesis of a β-(1,3)-glucan dodecasaccharide. *Chemistry—A European Journal*. 19(37), 12497–12503.

West, T. P. (2006). Pyrimidine base supplementation effects curdlan production in *Agrobacterium* sp. ATCC 31749. *Journal of Basic Microbiology*. 46(2), 153–157.

Wu, H. M., Liu, S. W., Hsu, M. T., Hung, C. L., Lai, C. C., Cheng, W. C., Wang, H. J., Li, Y. K. and Wang, W. C. (2009). Structure, mechanistic action, and essential residues of a GH-64 enzyme, laminaripentaose-producing β-1,3-glucanase. *Journal of Biological Chemistry*. 284(39), 26708–26715.

Wu, J., Zhan, X. and Hui, L. (2008). Enhanced production of curdlan by *Alcaligenes faecalis* by selective feeding with ammonia water during the cell growth phase of fermentation. *Chinese Journal of Biotechnology*. 24(6), 1035–1039.

Wu, S., Cai, R. and Sun, Y. (2012). Degradation of curdlan using hydrogen peroxide. *Food Chemistry*. 135(135), 2436–2438.

Yano, J. M., Yu, K., Donaldson, G. P., Shastri, G. G., Ann, P., Ma, L., Nagler, C. R., Ismagilov, R. F., Mazmanian, S. K. and Hsiao, E. Y. (2015). Indigenous bacteria from the gut microbiota regulate host serotonin biosynthesis. *Cell*. 161(2), 264–276.

Yashunsky, D. V., Tsvetkov, Y. E., Grachev, A. A., Chizhov, A. O. and Nifantiev, N. E. (2016). Synthesis of 3-aminopropyl glycosides of linear β-(1→3)-d-glucooligosaccharides. *Carbohydrate Research*. 419, 8–17.

Yu, H., Williams, D. L. and Ensley, H. E. (2005). 4-Acetoxy-2,2-dimethylbutanoate: A useful carbohydrate protecting group for the selective formation of β-(1→3)-d-glucans. *Tetrahedron Letters*. 46(19), 3417–3421.

Yu, L., Wu, J., Liu, J., Zhan, X., Zheng, Z. and Lin, C. C. (2011). Enhanced curdlan production in *Agrobacterium sp.* ATCC 31749 by addition of low-polyphosphates. *Biotechnology & Bioprocess Engineering*. 16(1), 34–41.

Yvin, J.-C., Jamois, F. and Vetvicka, V. (2005). Therapeutical treatment with oligo-beta-(1,3)-glucans, drugs used in said treatment, US, US20050065114.

Yvin, J.-C., Panak, E. and Vetvicka, V. (2006). Treating breast cancer by administering a monoclonal antibody with an oligo-beta-(1,3)-glucan and a carrier. US, US7,070,778. 4.

Zakharenko, A. M., Kusaykin, M. I., Kovalchuk, S. N., Anastyuk, S. D., Bui, M. L., Sova, V. V., Rasskazov, V. A. and Zvyagintseva, T. N. (2011). Enzymatic and molecular characterization of an endo-1,3-β-D-glucanase from the crystalline styles of the mussel *Perna viridis*. *Carbohydrate Research*. 346(2), 243–252.

Zeng, Y. and Kong, F. (2003). Highly efficient synthesis of alternate α- and β-(1→3)-linked glucose hepta- and octasaccharides. *Carbohydrate Research*. 338(20), 2047–2056.

Zhan, X. B., Lin, C. C. and Zhang, H. T. (2012). Recent advances in curdlan biosynthesis, biotechnological production, and applications. *Applied Microbiology and Biotechnology*. 93(2), 525–531.

Zhang, F., Ren, S. and Zuo, Y. (2014). DC-SIGN, DC-SIGNR and LSECtin: C-type lectins for infection. *International Reviews of Immunology*. 33(1), 54–66.

Zhang, H. (2011). Enhancement of β-1,3-glucan production by Agrobacterium and exploration of its mini-functional oligosaccharides. Ph.D. thesis, Jiangnan University.

Zhang, H. T., Zhan, X. B., Zheng, Z. Y., Wu, J. R., English, N., Yu, X. B. and Lin, C. C. (2012). Improved curdlan fermentation process based on optimization of dissolved oxygen combined with pH control and metabolic characterization of *Agrobacterium sp.* ATCC 31749. *Applied Microbiology and Biotechnology*. 93(1), 367–379.

Zhang, H. T., Zhu, L., Liu, D., Zhan, X. B., Ding, J. and Lin, C. C. (2015). Model-based estimation of optimal dissolved oxygen profile in *Agrobacterium sp.* fed-batch fermentation for improvement of curdlan production under nitrogen-limited condition. *Biochemical Engineering Journal*. 103, 12–21.

Zheng, Z. Y., Jiang, Y., Zhan, X. B., Ma, L. W., Wu, J. R., Zhang, L. M. and Lin, C. C. (2014). An increase of curdlan productivity by integration of carbon/nitrogen sources control and sequencing dual fed-batch fermentors operation. *Applied Biochemistry & Microbiology*. 50(1), 44–51.

Zong, Y., Li, J., Zhu, L., Zhan, X.-B., Zheng, Z.-Y. and Zhang, H.-T. (2014). Preparation of β-1,3-glucan oligosaccharides by two-step sulfuric acid hydrolysis of curdlan. *Industrial Microbiology*. 44(1), 22–27.

4

α-CYCLODEXTRIN

Enzymatic Production and Applications

JING WU AND SHENG CHEN

Contents

4.1 Introduction

Cyclodextrins are a family of cyclic oligosaccharides typically containing six (α-cyclodextrin), seven (β-cyclodextrin), or eight (γ-cyclodextrin) 1,4-linked d-glucose units (Ermolinsky et al. 2013). Because the glucose units adopt the chair conformation, the cyclodextrins are shaped like a hollow truncated cone with a hydrophilic outer surface (Pavlov et al. 2009), which makes them water-soluble. The central cavity of the cone is lined by the skeletal carbon atoms and ethereal oxygen atoms of the glucose residues, which gives it a lipophilic character (Jiang et al. 2011). This combination of a hydrophilic exterior with a hydrophobic interior enables cyclodextrins to form inclusion complexes with hydrophobic guest molecules (Dijkhuizen 2000a; Jiang et al. 2011). As a result, the physical and chemical properties of the guest molecule can be greatly modified, mostly in terms of water solubility (Messner et al. 2011). This is the primary reason why cyclodextrins have attracted great interest in a variety of industries, including those related to food, pharmaceuticals, cosmetics, chemicals, and agriculture (Davis and

Brewster 2004; Szente and Szejtli 2004; Li et al. 2007; Tahir and Lee 2013; Kim et al. 2014).

Cyclodextrins are produced from starch or starch derivatives by means of an enzymatic conversion catalyzed by cyclodextrin glycosyltransterase (CGTase, EC 2.4.1.19) (Armbruster et al. 1972; Flaschel et al. 1984; Szejtli 1988; van der Veen et al. 2000). The product obtained from the enzymatic conversion is usually a mixture of α-, β-, and γ-cyclodextrins, containing trace amounts of cyclodextrins with more than nine d-glucose units (Terada et al. 1997). A relatively small internal cavity and high resistance to enzymatic hydrolysis make α-cyclodextrin ideally suited to special applications in many fields, especially in the food industry. However, the market share of α-cyclodextrin is currently much smaller than that of β-cyclodextrin due to its low production yield and high price. Substantial efforts have been undertaken to improve the α-cyclodextrin production processes by modifying the properties of CGTases. With the production costs coming down, the availability of more affordable α-cyclodextrin will increase significantly in the next decade.

4.2 Structure and Properties of α-Cyclodextrin

All three cyclodextrins have similar structures, apart from the structural necessities of accommodating a different number of glucose units. The truncated cone-shaped cyclodextrin molecules are stiffened by hydrogen bonding between the 3-OH and 2-OH groups around the wider rim. The flexible 6-OH hydroxyl groups around the narrower rim are also capable of forming hydrogen bonds, but they are easily dissociated in aqueous solution and are not commonly found in cyclodextrin crystals. α-Cyclodextrin has the lowest hydrogen bond strengths.

The cavities of the three cyclodextrins have different diameters, depending on the number of glucose units. The side rim depth is the same for all three (at about 7.9 Å). The diameter of the internal cavity of α-cyclodextrin, 4.7–5.3 Å, is much smaller than that of β- or γ-cyclodextrins. As a result, the cavity volume of α-cyclodextrin is about 66% or 41% of that of β- or γ-cyclodextrins, respectively (Li et al. 2007). The decisive factor for the formation of inclusion complexes is that the guest molecule must be able to fit into the internal cavity of the cyclodextrin (Szejtli 1982). Based on the cavity dimensions, β-cyclodextrin should be able to form complexes with aromatics or heterocycles, γ-cyclodextrin should be able to accommodate macrocycles or steroids, while α-cyclodextrin typically forms inclusion complexes with benzene derivatives (Del Valle 2004). Nevertheless, geometry is not the sole factor for the formation of stable inclusion complex, since previous studies have shown that some guest molecules that were well compatible with α-cyclodextrin could not fit satisfactorily into the larger internal cavies of β- or γ-cyclodextrins (Szejtli 1982; Ahuja 1991). It has generally been thought that van der Waals interactions are the main driving force for inclusion complex formation between cyclodextrins and guest molecules. This has been fully supported by molecular mechanics calculations performed on complexes of α- or β-cyclodextrin with guest molecules (Linert et al. 1992; Alvira, Cativiela et al. 1995; Alvira, Mayoral et al. 1995).

Although α-, β-, and γ-cyclodextrins are water soluble, the solubility of cyclodextrins in water follows an irregular trend (Sabadini et al. 2006). α-Cyclodextrin has modest solubility in water that is almost eight times greater than that of β-cyclodextrin, but approximately 1.6 times lower than that of γ-cyclodextrin at 25°C (Li et al. 2007).

The three cyclodextrins are thermally stable (up to at least 200°C) and also stable in alkaline solutions (pH < 14) or moderately acidic solutions (pH > 3). Compared with β- and γ-cyclodextrins, α-cyclodextrin is considerably more resistant to hydrolysis in acid solutions. Even after 3 h at 100°C in extremely acidic conditions (pH $= 2.4$), no breakdown of α-cyclodextrin is discernible. The three cyclodextrins are stable in the presence of glucoamylase or β-amylase, but they can be hydrolyzed by some α-amylases. Although α-amylases from fungal or bacterial sources can hydrolyze relatively rigid α-cyclodextrin (Jodái et al. 1984; Saha and Zeikus 1992), α-cyclodextrin cannot be hydrolyzed by human salivary and pancreatic amylases to an obvious extent (Kondo et al. 1990). β-Cyclodextrin has similar resistance to enzymatic hydrolysis by human salivary and pancreatic amylases. By comparison, γ-cyclodextrin is quite flexible and can be rapidly and essentially completely digested by human salivary or pancreatic amylases (Kondo et al. 1990). As a result, α-cyclodextrin can be absorbed intact at a level of approximately 2% from the small intestine, but most absorption takes place after metabolism by the microflora in the caecum. Intact α-cyclodextrin that is absorbed is rapidly excreted in the urine.

An embryotoxicity/teratogenicity study found that no adverse effects were observed at α-cyclodextrin intakes of up to 20% of the diet, which was the highest dose level tested. At this dose, the rats consumed about 13 $g \cdot kg^{-1}$ bw/day (Waalkens-Berendsen and Bar 2004). Dietary α-cyclodextrin is generally well tolerated by pregnant rabbits. It has no adverse effect on maternal reproductive performance and is not embryotoxic, fetotoxic, or teratogenic at dietary concentrations of up to 20% (Waalkens-Berendsen and Bar 2004). On the basis of the available safety studies on α-cyclodextrin, and studies on the related β- and γ-cyclodextrin, the Joint FAO/WHO Expert Committee on Food Additives (JECFA) allocated an Acceptable Daily Intake of "not specified" to α-cyclodextrin for technological uses in food or dietary fiber (WHO 2002). Based on an evaluation of the available safety data and other pertinent information, the Food Standards Australia New Zealand (FSANZ) recommends approval of the use of α-cyclodextrin as a novel food with no specified limits of use (FSANZ 2004). Furthermore, α-cyclodextrin is generally recognized as safe by the U.S. Food and Drug Administration (FDA) (FDA 2004). The hydroxyl groups of α-cyclodextrin can be derivatized to modify the specificity, physical properties, and chemical properties (Kulkarni et al. 2013; Zhu et al. 2013). The 6-OH groups are most easily derivatized.

4.3 Extracellular Expression and Biochemical Characterization of α-CGTase from *Paenibacillus macerans*

CGTase is a multifunctional enzyme that can catalyze three transglycosylation reactions: disproportionation, cyclization, and coupling, and a hydrolysis reaction (van der Veen, Uitdehaag et al. 2000a; Dijkhuizen 2000b; van der Veen, Uitdehaag et al. 2000; Dijkstra, and Dijkhuizen 2000). A variety of bacteria and archaea produce CGTase as an extracellular enzyme. The most extensively studied CGTases are from *Bacillus* species, but examples are also produced by *Paenibacillus*, *Klebsiella*, *Thermoanaerobacterium*, *Thermoanaerobacter* species, and *Actinomycetes* (Tonkova 1998; Qi and Zimmermann 2005). CGTases have been further classified, according to their major cyclodextrin products, into α-, β-, and γ-CGTases (Penninga et al. 1995; Li et al. 2007). Table 4.1 summarizes some sources of CGTase producing α-cyclodextrin as a main product. The α-CGTases from *B. macerans* or *P. macerans* have been most commonly used in the commercial production of α-cyclodextrin.

The production of α-CGTase by wild-type strains is very similar to that of the other CGTases and has been described in detail elsewhere (Tonkova 1998; Gawande and Patkar 1999; Rosso et al. 2002; Gawande et al. 2003; Pinto et al. 2007; Kuo et al. 2009; Ahmed and El-Refai 2010). Since the production of α-CGTases from wild strains is relatively low, overexpression of α-CGTase genes in genetically engineered bacteria has been attempted, especially in *Escherichia coli*. However, previous reports have shown that α-CGTases expressed in *E. coli* often accumulate in either the cytosol, as inactive inclusion bodies, and/or the periplasm as soluble forms (Kwon et al. 2002; Jeang et al. 2005; Kim et al. 2005). Kwon et al. found that the coexpression of GroEL/ES (GroESL) with the α-CGTase from *B. macerans* in *E. coli* BL21 (DE3) increased the production of soluble CGTase by

Table 4.1 Sources of α-CGTases

BACTERIA SPECIES	ACCESSION NUMBER	MAIN PRODUCT	REFERENCE
Bacillus macerans	P31835	α-CD	(Takano et al. 1986)
B. macerans NRRL B388	P04380	α-CD	(Fujiwara et al. 1992)
Paenibacillus macerans JFB05-01	–	α- CD	(Li, Li et al. 2010)
P. macerans IFO3490	AAC04359	α-CD	(Kitahata et al. 1974)
P. graminis	–	α-CD	(Vollu et al. 2008)
Klebsiella pneumoniae M5a1	P08704	α-CD	(Binder et al. 1986) (Hashimoto et al. 2001)
Thermococcus sp. B1001	AB025721	α-CD	
Haloferax mediterranei	–	α-CD	(Bautista et al. 2012)
Thermococcus kodakaraensis KOD1	BAB78538	α/β-CD	(Rashid et al. 2002)
T. thermosulfurigenes EM1	P26827	α/β-CD	(Wind et al. 1995)
Thermoanaerobacter sp. ATCC 53627	Z35484	α/β-CD	(Jørgensen et al. 1997)
B. licheniformis	P14014	α/β-CD	(Hill et al. 1990)
B. stearothermophylus NO2	P31797	α/β-CD	(Fujiwara et al. 1992)

Source: Reprinted from *Trends in Food Science & Technology*, 35/2, Zhaofeng Li, Sheng Chen, Zhengbiao Gu, Jian Chen, and Jing Wu, Alphacyclodextrin: Enzymatic production and food applications, Copyright 2014, with permission from Elsevier.

Note: The abbreviation "CD" refers to cyclodextrin. "α/β" indicates that the CGTase produces an approximately equal mixture of α- and β-CD.

preventing aggregation. Various attempts have been made to facilitate extracellular secretion of recombinant α-CGTase from *E. coli*.

Secretion of enzymes has generally been shown to be beneficial to protein folding, product stability, and solubility, as well as downstream processing.

In the present study, the α-CGTase from *P. macerans* strain JFB05-01 was expressed extracellularly in *E. coli*. The recombinant enzyme was then purified and characterized with respect to the cyclization activity, including optimum reaction conditions, the effects of metal ions, and enzyme kinetics. Some properties were compared between the recombinant and native enzyme (Liu et al. 2013).

4.3.1 Molecular Cloning of cgt Gene

The 2061 bp open reading frame of *cgt* gene encoding a polypeptide of 687 amino acid residues was amplified from the genomic DNA of *P. macerans* strain JFB05-01 by PCR and subcloned into the T7-driven expression vector pET-20b(+), resulting in the plasmid *cgt*/pET-20b(+). Next, the gene without its stop codon was placed downstream of a DNA sequence encoding *pelB* signal peptide and fused in frame to six histidine codons and a stop codon already present in the vector. The plasmid was used for the production of the recombinant α-CGTase in *E. coli* BL21 (DE3) under the control of T7 promoter.

4.3.2 Extracellular Expression of α-CGTase in E. coli

Culture conditions for extracellular expression of the recombinant α-CGTase in *E. coli* BL21 (DE3) were optimized. The optimum culture medium and IPTG concentration were TB (Figure 4.1a) and 0.01 mM (Figure 4.1b), respectively. To investigate the effect of induction temperature on extracellular expression, the time courses of α-CGTase activity at different temperatures were measured (Figure 4.2). The results showed that, at 25°C, although very few recombinant α-CGTase was secreted into the culture medium within the first 40 h of induction, subsequently the amount of the recombinant enzyme in the culture medium increased rapidly and the cyclization activity of α-CGTase reached 22.5 U·mL^{-1} after 90 h of induction, which was approximately 42-fold higher than that from the parent strain,

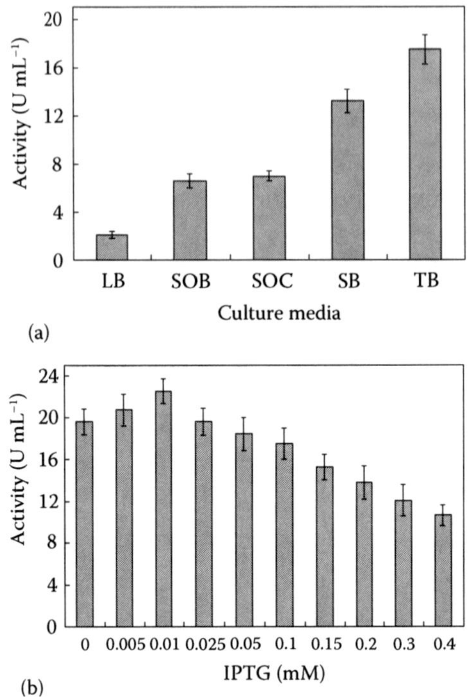

Figure 4.1 Effects of different culture media (a) and IPTG concentration (b) on extracellular expression of α-CGTase. (a) *E. coli* cells were cultured in each medium with 0.1 mM IPTG at 25°C for 90 h. (b) *E. coli* cells were cultured in TB medium with different concentrations of IPTG at 25°C for 90 h. Error bars correspond to the standard deviation of three measurements. (From Li Z et al., *Carbohydrate Research*, 354:886–892, 2010.)

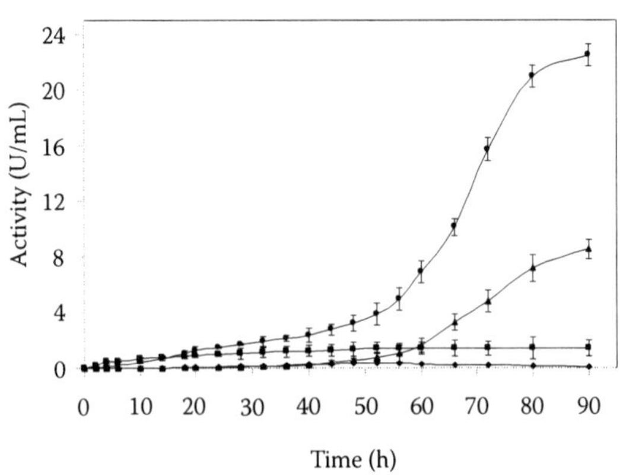

Figure 4.2 Time courses of α-CGTase activity at different induction temperatures. *E. coli* cells were cultured in TB medium with 0.01 mM IPTG at 20°C (▲), 25°C (●), 30°C (■), and 37°C (◆) for 90 h. Error bars correspond to the standard deviation of three measurements. (From Li Z et al., *Carbohydrate Research*, 354:886–892, 2010.)

P. macerans strain JFB05-01, after 70 h of culture. However, the extracellular secretion of the recombinant enzyme was depressed at 30°C or higher, while the secretion rate was too slow at 20°C or lower. Thus, the optimum induction temperature was 25°C.

4.3.3 Purification of α-CGTases

The recombinant α-CGTase carried a C-terminal fusion peptide containing a His_6 sequence that provides metal-binding affinity. It was therefore purified by one-step affinity chromatography on Ni-NTA resin. A single protein band observed on the SDS-PAGE gel demonstrated homogeneity (Figure 4.3, Lane 1), but the typical yield of purified enzyme was low and >95% of the recombinant enzyme was present in the flow through fraction, probably suggesting that the C-terminal His_6-tag was partially inaccessible as a result of being partially buried or bound to the surface of the protein. To overcome this problem, the recombinant enzyme was purified by a combination of chromatographic separations, including Q-Sepharose and phenyl-Superose chromatography. The purified protein was determined to be homogeneous by SDS-PAGE (Figure 4.3, Lane 4) and the relative high yield was obtained. The native α-CGTase was also purified to apparent homogeneity by the two-step purification system (data not shown).

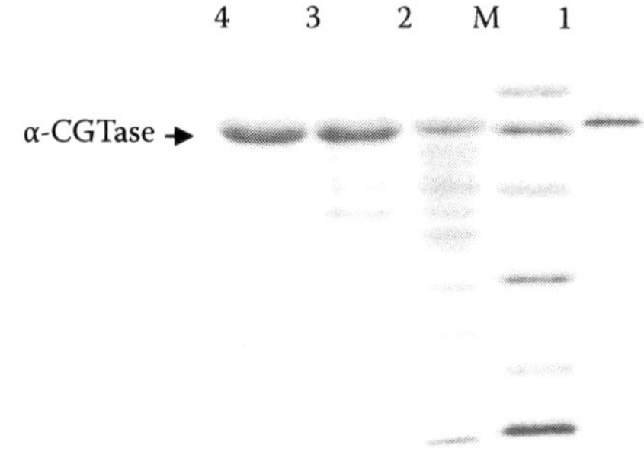

Figure 4.3 SDS-PAGE analysis of the purified recombinant α-CGTase. M, molecular weight markers. Lane 1, purified fraction of the recombinant enzyme after nickel affinity chromatography (10 μL). Lane 2, culture supernatant (10 μL). Lane 3, partially purified fraction of the recombinant enzyme after ion-exchange chromatography (10 μL). Lane 4, the purified recombinant enzyme after hydrophobic interaction chromatography (10 μL). (From Li Z et al., *Carbohydrate Research*, 354:886–892, 2010.)

4.3.4 Physical Properties of α-CGTase

The recombinant α-CGTase migrated as a single band on SDS-PAGE with a molecular mass of approximately 72 kDa (Figure 4.3). The molecular mass of the recombinant enzyme in its native form was 75 kDa as determined by analytical gel filtration chromatography. Since the molecular mass under native conditions was almost equal to that under denatured conditions, the α-CGTase was predicted to have a monomeric structure in solution.

4.3.5 Optimum Temperature and Thermostability

The optimum temperature of the recombinant α-CGTase was 45°C for the cyclization activity, which was below that of the native enzyme (50°C; Figure 4.4a). In the temperature range between 40°C and 50°C, the recombinant enzyme exhibited more than 95% of its maximal cyclization activity. The stabilities of the recombinant enzyme at 40°C, 45°C, 50°C, and 60°C (pH 6.0) are shown in Figure 4.4b. It retained 50% of its initial cyclization activity after incubation for approximately 8 h at 40°C, 1.25 h at 45°C, and 0.5 h at 50°C and was inactivated after 30 min of incubation at 60°C. The half-life time of the native enzyme at 50°C was 0.8 h (Figure 4.4b).

4.3.6 Optimum pH and pH Stability

The recombinant and native α-CGTases both showed the highest cyclization activity at pH 5.5 (Figure 4.5a). At pH values below or above the optimum, the activity of α-CGTase decreased rapidly (Figure 4.5a). The enzymes lost their cyclization activity at extreme pH values (below 4 and above 9.5). It should be noted that the enzymes were more active in glycine/NaOH buffer compared with that in Na_2HPO_4/NaH_2PO_4 buffer at the same pH. The stabilities of both α-CGTases for the cyclization activity were also examined in the pH ranging from 3 to 11 (Figure 4.5b). The recombinant and native α-CGTases were quite stable in the pH ranging from 6 to 9.5 and 6 to 10, respectively. Their stabilities decreased significantly at pH below 5.5 or above 10.

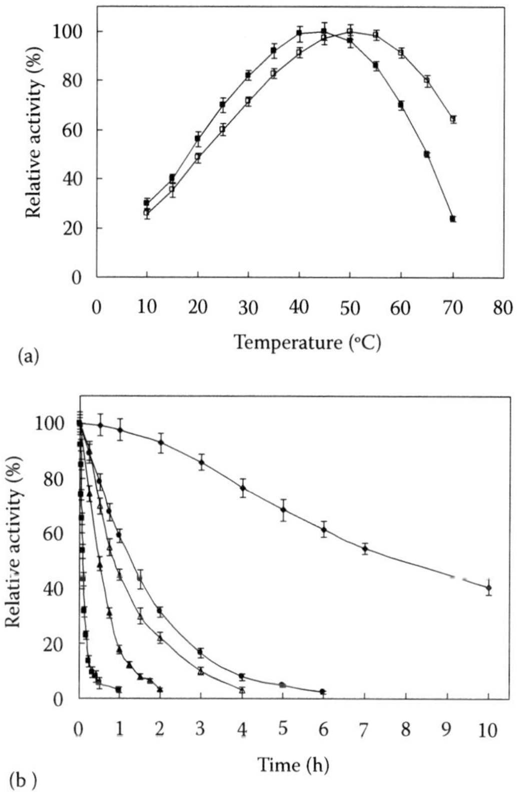

Figure 4.4 Optimum temperature (a) and thermostability (b) of α-CGTase. (a) Cyclization activities of the recombinant (■) and native (□) α-CGTases were determined at 10–70°C. The activity at optimum temperature was taken as 100%. (b) Recombinant enzyme was incubated at 40°C (◆), 45°C (●), 50°C (▲), and 60°C (■), respectively. The native enzyme was incubated at 50°C (△). The activity of the enzyme without preincubation was taken as 100%. Error bars correspond to the standard deviation of three measurements. (From Li Z et al., *Carbohydrate Research*, 354:886–892, 2010.)

4.3.7 Metal Requirement

To determine whether metal(s) was required for the cyclization activity of α-CGTase, the recombinant enzyme was incubated with metal chelator (EDTA) and then the residual activity was assayed. The result showed that the enzyme was not inhibited by the addition of EDTA. However, the cyclization activity of α-CGTase was inhibited almost completely by 1 mM Hg^{2+}, moderately by 1 mM Ni^{2+}, and slightly by 1 mM Fe^{2+} or Co^{2+} (Table 4.2). On the other hand, the enzyme could be activated by some bivalent metal ions, such as Ca^{2+}, Ba^{2+}, and Zn^{2+},

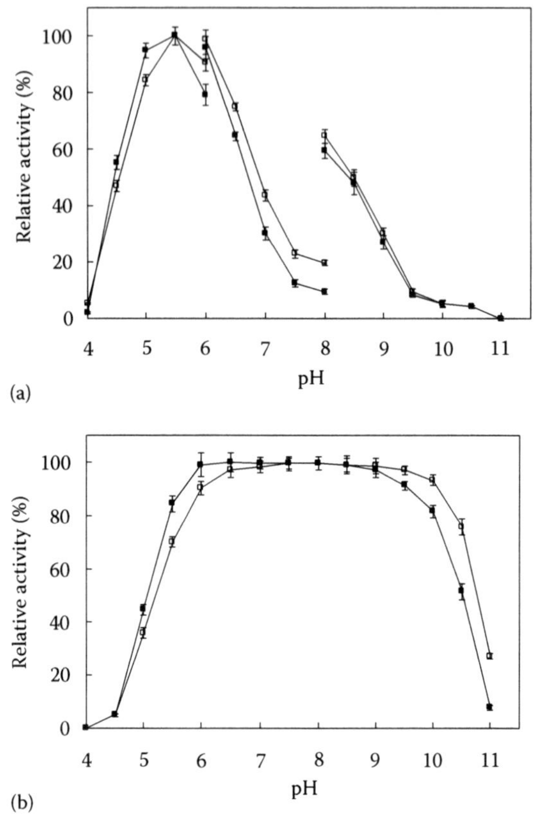

Figure 4.5 Optimum pH (a) and pH stability (b) of α-CGTase. (a) Cyclization activities of the recombinant (■) and native (□) α-CGTases were measured at 40°C in the buffers with different pH values. The activity at optimum pH was taken as 100%. (b) Recombinant (■) and native (□) α-CGTases were incubated in the buffers with different pH values for 1 h at 4°C. The activity of the enzyme without preincubation was taken as 100%. Error bars correspond to the standard deviation of three measurements. (From Li Z et al., *Carbohydrate Research*, 354:886–892, 2010.)

in a concentration-dependent manner (Table 4.2); 2 mM Ca^{2+}, 2 mM Ba^{2+}, or 0.1 mM Zn^{2+} activated the cyclization activity of α-CGTase up to 123.0%, 128.1%, or 118.9%, respectively. The effects of metal ions and metal chelator on the recombinant and native α-CGTases were similar (data not shown).

4.3.8 Cyclodextrin Production and Product Specificity

At the initial stage of the reaction, α-cyclodextrin was the main product, while small amounts of β- and γ-cyclodextrin were produced.

Table 4.2 Effects of Metal Chelator and Bivalent Metal Ions on a-CGTase Activity

METAL CHELATOR OR METAL IONS	CONCENTRATION (Qi and Zimmermann 2005)	RELATIVE ACTIVITY (%)[a]
None	0	100 ± 2.2
EDTA	10	105.2 ± 4.8
Hg^{2+}	1	5.2 ± 0.3
Ni^{2+}	1	57.7 ± 4.6
Fe^{2+}	1	89.7 ± 4.2
Co^{2+}	1	90.7 ± 2.1
Mg^{2+}	1	97.9 ± 1.8
Pb^{2+}	1	97.9 ± 2.5
Mn^{2+}	1	102.1 ± 3.0
Cu^{2+}	1	104.1 ± 5.1
Cr^{2+}	1	109.3 ± 6.2
Ca^{2+}	0.2	111.5 ± 2.2
	0.5	116.7 ± 2.0
	1	120.8 ± 2.4
	1.5	122.9 ± 2.8
	2	123.0 ± 3.0
Ba^{2+}	0.2	121.9 ± 2.1
	0.5	125.0 ± 2.0
	1	127.1 ± 2.4
	1.5	128.0 ± 2.7
	2	128.1 ± 2.5
Zn^{2+}	0.02	112.5 ± 3.4
	0.05	117.7 ± 2.2
	0.1	118.9 ± 2.5
	0.2	119.0 ± 2.7
	1	119.1 ± 2.9

Source: Li Z et al., *Carbohydrate Research*, 354:886–892, 2010.

[a] The activity of the enzyme in the absence of the added ion or chelator was taken as 100%.

Subsequently, the proportion of α-cyclodextrin in the total cyclodextrins decreased while the proportion of β-cyclodextrin increased. After prolonged incubation, β-cyclodextrin was the main product (Figure 4.6). At each time point, the proportion of α-cyclodextrin in total cyclodextrins produced by the recombinant α-CGTase was slightly higher than that of the native enzyme (Figure 4.6). After 10 or 40 h of incubation, the conversion rates of starch into cyclodextrins by the recombinant enzyme were 36.8% or 42.3%, respectively, which were almost the same as those of the native enzyme (37.2% or 42.4%, respectively).

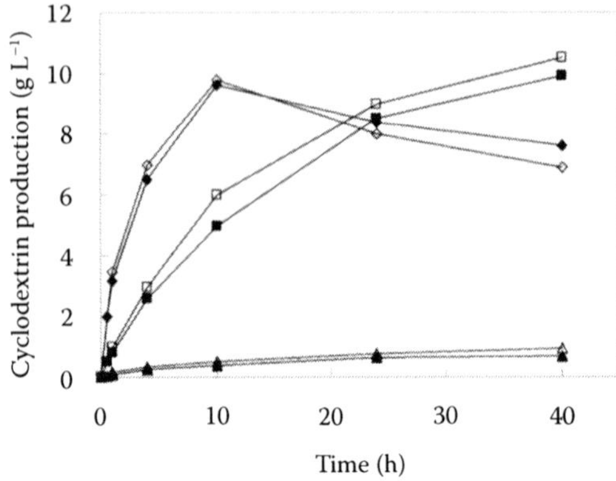

Figure 4.6 Production of cyclodextrins during incubation of α-CGTase with 5% (w/v) soluble starch at pH 6.0 and 40°C. The recombinant enzyme: α-cyclodextrin (◆), β-cyclodextrin (■), γ-cyclodextrin (▲); the native enzyme: α-cyclodextrin (◇), β-cyclodextrin (□), γ-cyclodextrin (△). (From Li Z et al., *Carbohydrate Research*, 354:886–892, 2010.)

4.3.9 Kinetic Properties

To determine the kinetic properties of α-CGTase, the cyclization activities of the recombinant and native α-CGTases were assayed by monitoring the formation of α-cyclodextrin over a range of concentrations of soluble starch. The curve fitting results revealed that the kinetics of cyclization reaction could be better described by the Hill equation than the Michaelis–Menten model (data not shown). Analysis of the data according to the Hill equations of the recombinant and native α-CGTases gave the Hill coefficients of 1.485 and 1.257, respectively.

4.4 Mutations in *Paenibacillus macerans* CGTase to Enhance α-Cyclodextrin Specificity

It is well known that wild-type CGTases produce a mixture of α-, β-, and γ-cyclodextrins when incubated with starch (Li et al. 2007; Schmid 2009). To isolate pure cyclodextrins from the reaction mixture, a series of additional steps are required, including selective crystallization of β-cyclodextrin (which is relatively poorly water-soluble) and selective complexation of α- or γ-cyclodextrins with organic solvents (Wind et al. 1998; van der Veen, Uitdehaag et al. 2000). However, the additional procedures will increase significantly the costs of cyclodextrin production, especially for α- or γ-cyclodextrins. Moreover, the use of organic solvents also limits cyclodextrin applications involving human consumption (Penninga et al. 1995; van der Veen, Uitdehaag et al. 2000). Therefore, the availability of CGTase capable of producing an increased ratio of one particular type of cyclodextrin is highly desired (van der Veen, Uitdehaag et al. 2000). This situation has strongly stimulated the constructions of mutant CGTases with improved cyclodextrin product specificity.

So far, large numbers of site-directed mutations in CGTases from different sources have been made (Fujiwara et al. 1992; Shieh and Hedges 1994; Penninga et al. 1995; van der Veen, Uitdehaag et al. 2000). Most of these mutations were based upon amino acid residues located in the active center cleft, which contains at least nine sugar-binding subsites with the catalytic site between subsites +1 and −1 (Strokopytov et al. 1996). Recently, some evidence supported that the amino acid residues contributing to the interactions with the glucoses bound at subsites −3 may be important for cyclodextrin product specificity of CGTase (Wind et al. 1998; van der Veen, Uitdehaag et al. 2000b; van der Veen, Uitdehaag et al. 2000; Kim et al. 2005).

The enzyme used in our studies, CGTase from *Paenibacillus macerans*, was most commonly used in the commercial production of α-cyclodextrin (Wind et al. 1998). Although it produces mainly α-cyclodextrin during the initial stage of starch conversion and is one of few α-CGTases identified, the proportion of α-cyclodextrin in the total cyclodextrin products was even lower than that of β-cyclodextrin after prolonged incubation under conditions resembling industrial production nonsolvent process. Thus, mutants of this CGTase capable of producing an increased ratio of α-cyclodextrin are of high industrial interest for the production of α-cyclodextrin.

The amino acid residues at position 372 and 89 (*P. macerans* CGTase numbering) are part of subsite −3 (Strokopytov et al. 1996). Moreover, it has been proved by x-ray studies of some CGTases that both the residues at the corresponding positions had the interactions with the glucose residue bound at subsite −3 (Wind et al. 1998; van der Veen, Uitdehaag et al. 2000b). Here we thus chose Asp372 and Tyr89 in the CGTase from *P. macerans* strain JFB05-01 as targets for site-directed mutagenesis (Li et al. 2009).

4.4.1 Cyclization Activities of the Wild-Type and Mutant CGTases

Different cyclodextrin forming activities of the (mutant) CGTases were shown in Table 4.3. The D372K mutation resulted in a slight increase in α-cyclodextrin forming activity but a two-fold decrease in β-cyclodextrin forming activity when compared to the wild-type CGTase. γ-Cyclodextrin forming activity was not affected significantly by the mutation. In addition, the D372K mutant showed a slight decrease in the total cyclization activity.

The Y89L mutation had almost no effect on different cyclodextrin forming activities, while the mutations Y89D and Y89N resulted in 15% and 11% increases in α-cyclodextrin forming activity, respectively, and slight increases in β-cyclodextrin forming activity when compared to the wild-type enzyme. The mutants Y89K and Y89R showed 16% and 18% increases in α-cyclodextrin forming activity but 1.3- and 2.3-fold decreases in β-cyclodextrin forming activity, respectively. The contribution of γ-cyclodextrin forming activity in total cyclization activity was not affected significantly by these mutations at position 89. It was noted that the mutants at position 89, except for the Y89L mutant, showed 7–14% increases in the total cyclization activity.

Compared to the wild-type enzyme, the double mutant D372K/Y89R showed a 24% increase in α-cyclodextrin forming activity but a threefold decrease in β-cyclodextrin forming activity. In addition, the double mutant showed a 10% increase in the total cyclization activity.

4.4.2 Cyclodextrin Product Ratio of the Wild-Type and Mutant CGTases under Conditions Resembling Industrial Production Processes

The production of cyclodextrins in the reaction mixture containing a (mutant) CGTase protein (0.2 units/ml cyclization activity) and 5% (wet basis, w/v) soluble starch incubated at 40°C was analyzed in time. For the wild-type CGTase, at the initial stage of the reaction, α-cyclodextrin was the main product, while small amounts of β- and γ-cyclodextrin were produced. At the later stages, the proportion of α-cyclodextrin in the total cyclodextrin products decreased while the proportion of β-cyclodextrin increased (Figure 4.7a). After 40 h incubation, the wild-type enzyme produced a mixture of α-, β-, and γ-cyclodextrin at a ratio of 41.8:54.4:3.8 (Table 4.4).

Table 4.3 Cyclodextrin Forming Activities and Stabilities (at 40°C) of the Wild-Type and Mutant CGTases from *P. macerans* Strain JFB05-01

CGTASE	α-CYCLODEXTRIN FORMING ACTIVITY (UNITS/mg)	β-CYCLODEXTRIN FORMING ACTIVITY (UNITS/mg)	γ-CYCLODEXTRIN FORMING ACTIVITY (UNITS/mg)	TOTAL (UNITS/mg)	STABILITY $T_{1/2}$ (h)
Wild type	190.4 (85.1)	32.2 (14.4)	1.1 (0.5)	223.7	8.0
D372K	202.5 (92.4)	15.6 (7.1)	1.0 (0.5)	219.1	7.5
Y89L	180.0 (85.0)	30.8 (14.5)	1.0 (0.5)	211.8	6.0
Y89D	210.7 (85.7)	33.6 (13.7)	1.6 (0.6)	245.9	7.6
Y89N	219.4 (86.1)	34.0 (13.3)	1.5 (0.6)	254.9	7.9
Y89K	220.1 (89.5)	24.6 (10.0)	1.1 (0.5)	245.8	7.8
Y89R	225.4 (93.9)	14.0 (5.8)	0.7 (0.3)	240.1	7.0
D372K/Y89R	235.6 (95.5)	10.5 (4.3)	0.5 (0.2)	246.7	6.8

Source: Li Z et al., *Applied Microbiology & Biotechnology*, 83(3):483–490.

Note: Each value represents the mean of three independent measurements. Numbers between brackets indicate the ratio in specific activities for formation of the different yclodextrins.

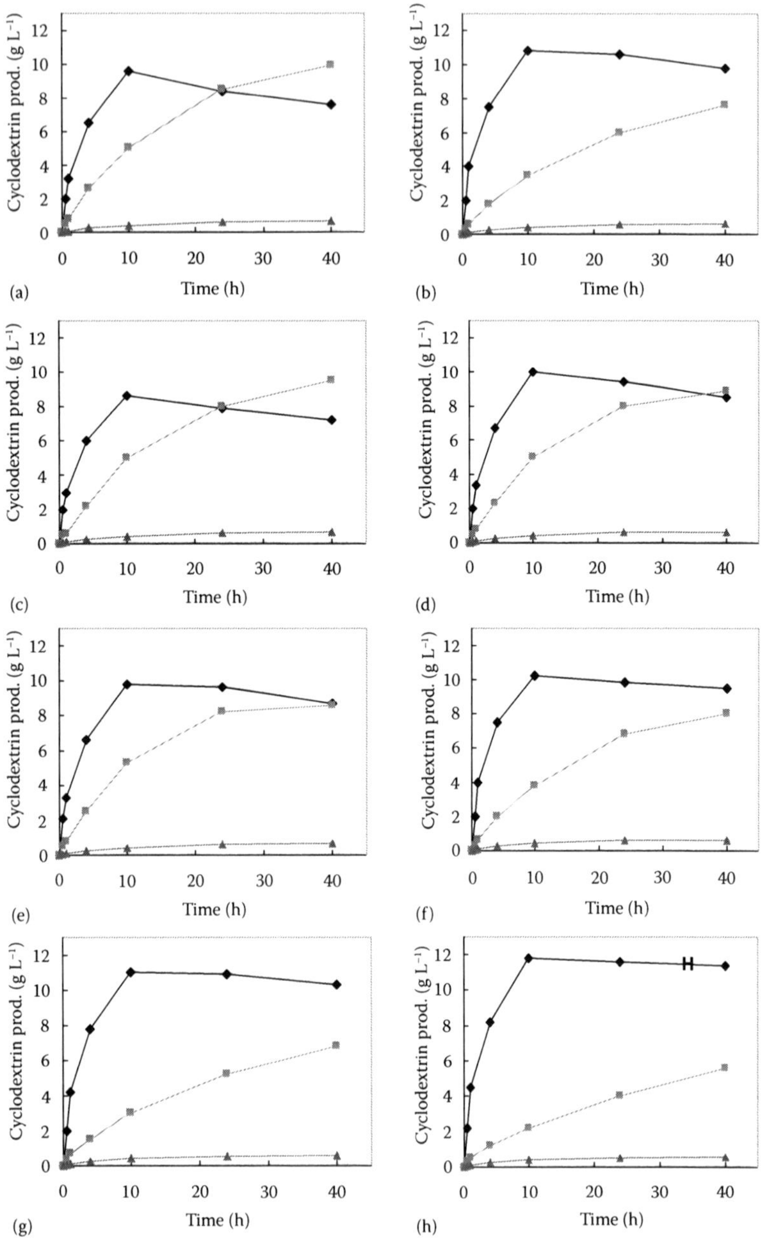

Figure 4.7 Cyclodextrins formed during the incubation of the (mutant) CGTases from *P. macerans* strain JFB05-01 (0.2 units/mL cyclization activity) with 5% (wet basis, w/v) soluble starch for 40 h at pH 6.0 and 40°C. Each value represents the mean of three independent measurements. ◆, α-cyclodextrin; ■, β-cyclodextrin; ▲, γ-cyclodextrin. (a) Wild-type CGTase; (b) mutant D372K; (c) mutant Y89L; (d) mutant Y89D; (e) mutant Y89N; (f) mutant Y89K; (g) mutant Y89R; (h) mutant D372K/Y89R. (From Li Z et al., *Applied Microbiology & Biotechnology*, 83(3):483–490.)

Table 4.4 Starch Conversions of the Wild-Type and Mutant CGTases from *P. macerans* Strain JFB05-01

CGTASE	CONVERSION OF STARCH INTO CYCLODEXTRINS (%)	PRODUCT ($g{\cdot}L^{-1}$)		
		α	β	γ
Wild type	42.3	7.6 (41.8)	9.9 (54.4)	0.69 (3.8)
D372K	42.0	9.8 (54.3)	7.6 (42.1)	0.65 (3.6)
Y89L	40.4	7.2 (41.4)	9.5 (54.7)	0.67 (3.9)
Y89D	42.0	8.5 (47.1)	8.9 (49.3)	0.64 (3.6)
Y89N	42.3	8.7 (48.4)	8.6 (47.8)	0.68 (3.8)
Y89K	42.2	9.5 (52.3)	8.0 (44.1)	0.65 (3.6)
Y89R	41.1	10.3 (58.2)	6.8 (38.4)	0.60 (3.4)
D372K/Y89R	40.8	11.4 (65.0)	5.6 (31.9)	0.55 (3.1)

Source: Li Z et al., *Applied Microbiology & Biotechnology*, 83(3):483–490.

Note: CGTase proteins (0.2 units/ml cyclization activity) were incubated with 5% (wet basis, w/v) soluble starch for 40 h at pH 6.0 and 40°C. Numbers between brackets indicate the ratio of the different cyclodextrins.

Compared to the wild-type enzyme, the D372K mutant produced more α-cyclodextrin at the expense of β-cyclodextrin, while the production of γ-cyclodextrin was not affected significantly (Figure 4.7b). After 40 h incubation, the amount of α-cyclodextrin in the reaction mixture was about 1.3-fold of that of the wild-type enzyme, while the amount of β-cyclodextrin decreased to 77% of that of the wild-type enzyme (Table 4.4).

The mutation Y89L had almost no effect on the production of cyclodextrins, so the Y89L mutant had similar cyclodextrin product ratio to the wild-type enzyme (Figure 4.7c and Table 4.4). For the mutants Y89D and Y89N, the production of α-cyclodextrin slightly increased while the production of β-cyclodextrin slightly decreased when compared to that of the wild-type enzyme (Figure 4.7d and e). After 40 h incubation, almost equal amounts of α- and β-cyclodextrin were produced by the two mutants (Table 4.4). The mutants Y89K and Y89R had significant increases in the production of α-cyclodextrin, with concomitant significant decreases in the production of β-cyclodextrin (Figure 4.7f and g). Eventually, the amounts of α-cyclodextrin in the reaction mixture were higher than those of β-cyclodextrin (Table 4.4). Especially for the Y89R mutant, the amount of α-cyclodextrin was about 1.4-fold of that of the wild-type enzyme, while the amount of β-cyclodextrin decreased to 70% of that of the wild-type enzyme

(Table 4.4). In addition, the production of γ-cyclodextrin was not affected significantly by all the mutations at position 89.

The changes in cyclodextrin product ratio for the single mutants D372K and Y89R were combined in the double mutant D372K/Y89R (Figure 4.7h). Eventually, the amount of α-cyclodextrin was about 1.5-fold of that of the wild-type enzyme, while the amount of β-cyclodextrin decreased to 57% of that of the wild-type enzyme, and no significant effect on the production of γ-cyclodextrin was observed. The double mutant had a final product ratio of 65:31.9:3.1 (Table 4.4).

Compared to the wild-type enzyme, the mutants Y89L, Y89R and D372K/Y89R had 3–5% decreases in the conversions of starch into cyclodextrins, whereas for other mutants, the conversions remained unaffected (Table 4.4).

4.5 Calcium and Glycine Lead to Increase in Extracellular Secretion of Recombinant α-CGTase in *Escherichia coli*

The secretion of recombinant CGTase into the culture medium of *E. coli* remained a challenging task since this enzyme was usually accumulated in the cytoplasm as biologically inactive inclusion bodies and/or partially in the periplasm as soluble forms. This greatly limited the industrial applications of recombinant CGTase. In the present study, we demonstrated significant enhancement of the extracellular secretion of recombinant α-CGTase by the synergistic effect of glycine and Ca^{2+} (Huang et al. 2003; Ilk et al. 2011; Li et al. 2012; Li et al. 2010).

4.5.1 Effect of Glycine and Ca^{2+} on Extracellular Secretion of α-CGTase

When *E. coli* cells were cultured in regular TB medium, very few recombinant α-CGTase was secreted into the culture medium within the first 40 h (Figures 4.8 and 4.9). SDS-PAGE analysis of α-CGTase distribution showed that many recombinant enzymes were accumulated in the periplasmic space, and a few insoluble inclusion bodies were formed at 40 h of culture (Figure 4.9). Furthermore, the N-terminal amino acid sequence of inclusion body proteins was determined to mainly be Met-Lys-Tyr-Leu-Leu, the same as that of *pelB* signal peptide, suggesting that most inclusion bodies were formed in the cytoplasm. In addition, no α-CGTase activity was detected in

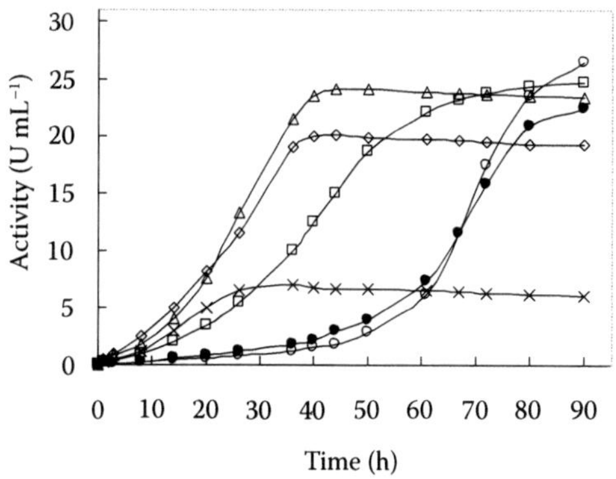

Figure 4.8 Effects of glycine or Ca^{2+} on extracellular secretion of recombinant α-CGTase. *E. coli* cells were grown in regular TB medium (●), or TB medium supplemented with glycine at 75 mM (□), 150 mM (△), 200 mM (◇), 300 mM (×), or Ca^{2+} at 20 mM (○), respectively. Each value represents the mean of three independent measurements, and the deviation from the mean is below 5%. Reprinted with permission from Li, Z. et al., *Journal of Agricultural and Food Chemistry.* 57, 6231–6237. Copyright 2009 American Chemical Society.

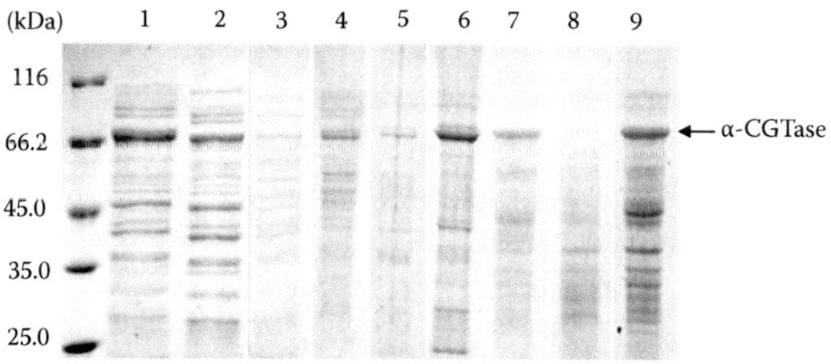

Figure 4.9 SDS-PAGE analysis of α-CGTase distribution. The distribution was analyzed after 40 h of culture. *E. coli* cells were grown in regular TB medium (lanes 3, 6, and 9), TB medium supplemented with 150 mM glycine (lanes 2, 5, and 8) or 150 mM glycine-20 mM Ca^{2+} (lanes 1, 4, and 7), respectively. Lane M, protein size marker; lanes 1, 2, and 3, extracellular fraction (10 μL); lanes 4, 5, and 6, periplasmic fraction (10 μL); lanes 7, 8, and 9, intracellular insoluble fraction (5 μL). Reprinted with permission from Li, Z. et al., *Journal of Agricultural and Food Chemistry.* 57, 6231–6237. Copyright 2009 American Chemical Society.

the soluble cytoplasmic fraction. After the lag phase of extracellular secretion, the activity of α-CGTase in a culture medium increased gradually and reached 22.5 U·mL^{-1} at 90 h of culture (Figure 4.8), suggesting that extracellular secretion of recombinant enzyme was limited to stationary or decline phase of cell growth. Due to the long lag time, the overall extracellular productivity of recombinant α-CGTase was relatively low. The maximum value of extracellular productivity was only 0.26 $U \cdot mL^{-1} \cdot h^{-1}$ at 80 h of culture.

When *E. coli* cells were cultured in a TB medium supplemented with 75–200 mM glycine, the enhancement in the extracellular secretion of recombinant α-CGTase was obviously observed (Figures 4.8 and 4.9). At the optimal glycine concentration of 150 mM, the α-CGTase activity in the culture medium reached 23.5 $U \cdot mL^{-1}$ at 40 h of culture, which was 11-fold higher than that of the culture in a regular TB medium at the same culture time. At 36 h of culture, the extracellular productivity of recombinant α-CGTase reached the maximum value of approximately 0.60 $U \cdot mL^{-1}\ h^{-1}$, which was 2.3-fold higher than that of the culture in a regular TB medium at 80 h of culture. Furthermore, SDS-PAGE analysis showed that, at 40 h of culture, no appreciable recombinant enzyme was accumulated in the periplasmic space and almost no insoluble inclusion body was formed (Figure 4.9), suggesting successful translocation of this protein across the cell membranes. It is worth noting that, at the glycine concentrations of above 200 mM, the enhancing effects of glycine on the extracellular activity and productivity of recombinant enzyme were significantly reduced (Figure 4.8), although neither recombinant enzyme was accumulated in the periplasmic space nor inclusion body was formed at 40 h of culture (data not shown).

When *E. coli* cells were cultured in a TB medium supplemented with 20 mM Ca^{2+}, no significant change in the extracellular secretion of recombinant α-CGTase was observed (Figure 4.8). This phenomenon was also apparent at other Ca^{2+} concentrations ranging from 5 to 30 mM (data not shown).

Very interestingly, when *E. coli* cells were cultured in a TB medium, supplemented with both glycine and Ca^{2+}, further increase in the glycine-enhanced extracellular secretion of recombinant α-CGTase was also obviously observed (Figures 4.8 and 4.9). As shown in Figure 4.3, the optimal Ca^{2+} concentration was 20 mM. In the culture with 150 mM glycine and 20 mM Ca^{2+}, the α-CGTase activity in the culture medium reached 35.5 $U \cdot mL^{-1}$ at 40 h of culture, which was approximately 1.5-fold higher than that in the culture with 150 mM glycine alone at the same culture time. At 36 h of culture, the extracellular productivity of recombinant α-CGTase reached the maximum value of approximately 0.90 $U \cdot mL^{-1}\ h^{-1}$, which was also 1.5-fold higher than that in the culture

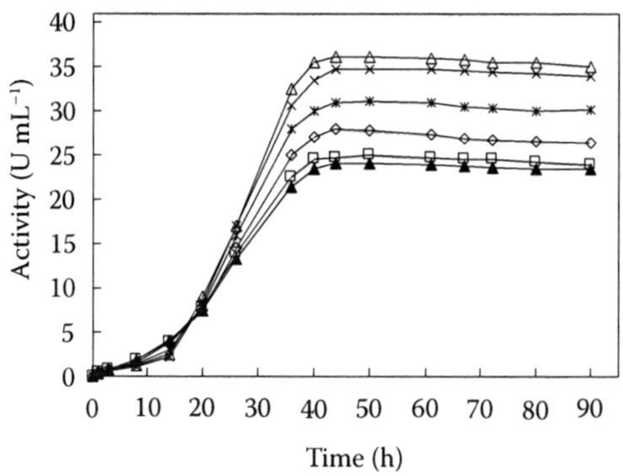

Figure 4.10 Synergistic effect of glycine and Ca^{2+} on extracellular secretion of recombinant α-CGTase. *E. coli* cells were grown in TB medium supplemented with 150 mM glycine (▲), 150 mM glycine-5 mM Ca^{2+} (◇), 150 mM glycine-10 mM Ca^{2+} (*), 150 mM glycine-20 mM Ca^{2+} (△), 150 mM glycine-30 mM Ca^{2+} (×), and 150 mM glycine–20 mM Ca^{2+}–20 mM EDTA (□), respectively. Each value represents the mean of three independent measurements, and the deviation from the mean is below 5%. Reprinted with permission from Li, Z. et al., *Journal of Agricultural and Food Chemistry.* 57, 6231–6237. Copyright 2009 American Chemical Society.

with 150 mM glycine alone. Furthermore, SDS-PAGE analysis showed very few periplasmic accumulation and inclusion body formation of recombinant enzyme at 40 h of culture (Figure 4.9), suggesting almost adequate secretion of this protein across the cell membranes.

To confirm that the above further increase in the glycine-enhanced extracellular secretion was exerted by Ca^{2+}, 20 mM EDTA was added into a TB medium supplemented with 150 mM glycine and 20 mM Ca^{2+}. It was found that the extracellular secretion of recombinant α-CGTase indeed returned to almost the same level as that in the culture with glycine alone (Figure 4.10).

4.5.2 Effect of Glycine and Ca^{2+} on Cell Growth

Although glycine stimulated the extracellular secretion of recombinant α-CGTase, it was found to inhibit the growth of *E. coli* cells in a dose-dependent manner. As shown in Figure 4.11, in the culture with glycine, the OD_{600} value at each culture time point was significantly lower than that of the culture in a regular TB medium, indicating a reduction in the total cell number. For example, in the culture with 75, 150, or 300 mM glycine, the OD_{600} value at 36 h of culture decreased by approximately 20%, 42%, or 72%, respectively.

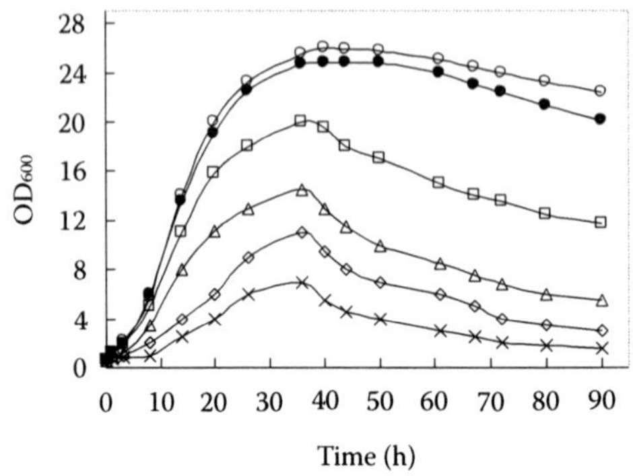

Figure 4.11 Effects of glycine or Ca^{2+} on cell density. *E. coli* cells were grown in regular TB medium (●), or TB medium supplemented with glycine at 75 mM (□), 150 mM (△), 200 mM (◇), 300 mM (×), or Ca^{2+} at 20 mM (○), respectively. Each value represents the mean of three independent measurements, and the deviation from the mean is below 5%. Reprinted with permission from Li, Z. et al., *Journal of Agricultural and Food Chemistry*. 57, 6231–6237. Copyright 2009 American Chemical Society.

In contrast, in the culture with 20 mM Ca^{2+}, the OD_{600} value at each time point was slightly higher than that of the culture in a regular TB medium (Figure 4.11). Furthermore, Ca^{2+} at other concentrations ranging from 5 to 30 mM also resulted in slightly increased OD_{600} values (data not shown).

Subsequently, the effects of glycine or Ca^{2+} on cell viability and possible cell lysis were analyzed. Colony forming unit (CFU) was used as an indicator of viable cell number. In the culture with 150 mM glycine, CFU per milliliter of culture per OD_{600} was relatively low (Figure 4.12). At 36 h of culture, it was 45% lower than that of the culture in a regular TB medium, indicating that glycine had a significantly negative effect on cell viability. G6PD is a cytoplasmic protein and the level of this enzyme in the culture medium is expected to be very low unless substantial cell lysis has occurred. Thus, G6PD activity in the culture medium can be used as an indicator of cell lysis. It was found that, in the culture without glycine and Ca^{2+}, G6PD activity in the culture medium was very low (Figure 4.12), while G6PD activity in the cytoplasmic fraction reached 2.23 $U{\cdot}mL^{-1}$ at 36 h of culture, indicating almost no cell lysis. In the culture with 150 mM glycine, G6PD activity in the culture medium increased gradually and reached 0.28 $U{\cdot}mL^{-1}$ at 36 h of culture (Figure 4.12), which was approximately 23% of G6PD activity in the cytoplasmic fraction (1.23 $U{\cdot}mL^{-1}$),

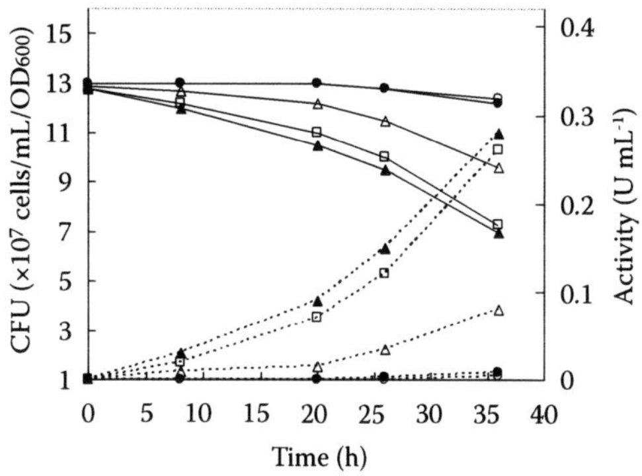

Figure 4.12 Effects of glycine and/or Ca^{2+} on cell viability and cell lysis. *E. coli* cells were grown in regular TB medium (●), or TB medium supplemented with 150 mM glycine (▲), 20 mM Ca^{2+} (○), 150 mM glycine–20 mM Ca^{2+} (△), or 150 mM glycine–20 mM Ca^{2+}–20 mM EDTA (□), respectively. Time courses of CFU per milliliter of culture per OD_{600} (solid line) and G6PD activity in the culture medium (dashed line) were analyzed. Each value represents the mean of three independent measurements, and the deviation from the mean is below 5%. Reprinted with permission from Li, Z. et al., *Journal of Agricultural and Food Chemistry*. 57, 6231–6237. Copyright 2009 American Chemical Society.

indicating that 150 mM glycine supplementation caused obvious cell lysis. On the other hand, in the culture with 20 mM Ca^{2+}, almost no change in cell viability and cell lysis was observed as compared with the culture in a regular TB medium (Figure 4.12). Furthermore, 5–30 mM Ca^{2+} had similar effects on cell viability and cell lysis (data not shown).

Although Ca^{2+} did not appear to obviously promote cell growth in the culture without glycine, it was found to significantly remedy cell growth inhibition induced by glycine (Figures 4.12 and 4.13). In the culture with 150 mM glycine and 20 mM Ca^{2+}, both OD_{600} value and CFU value at each culture time point were higher than those in the culture with 150 mM glycine alone, and close to those in the culture without glycine (Figures 4.12 and 4.13), indicating the increases of cell number and viability. At 36 h of culture, OD_{600} and CFU values were approximately 1.6- and 1.4-fold higher than those in the culture with 150 mM glycine alone, respectively. Similarly, in the culture with 150 mM glycine and 20 mM Ca^{2+}, G6PD activity in the culture medium at each culture time point was lower than that in the culture with 150 mM glycine alone (Figure 4.12). At 36 h of culture, G6PD activity in the culture medium (0.08 $U{\cdot}mL^{-1}$) had an approximately 71% reduction as compared with the culture with 150 mM

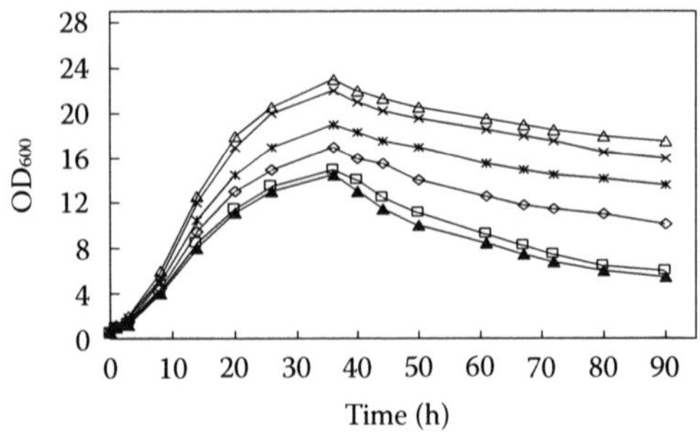

Figure 4.13 Effects of glycine and Ca^{2+} on cell density. *E. coli* cells were grown in TB medium supplemented with 150 mM glycine (▲), 150 mM glycine–5 mM Ca^{2+} (◇), 150 mM glycine–10 mM Ca^{2+} (*), 150 mM glycine–20 mM Ca^{2+} (△), 150 mM glycine–30 mM Ca^{2+} (×), or 150 mM glycine–20 mM Ca^{2+}–20 mM EDTA (□), respectively. Each value represents the mean of three independent measurements, and the deviation from the mean is below 5%. Reprinted with permission from Li, Z. et al., *Journal of Agricultural and Food Chemistry.* 57, 6231–6237. Copyright 2009 American Chemical Society.

glycine alone. Furthermore, it was only 4% of G6PD activity in the cytoplasmic fraction (2.01 $U{\cdot}mL^{-1}$), indicating no obvious cell lysis.

To further confirm that Ca^{2+} remedied cell growth inhibition induced by glycine, 20 mM EDTA was added to TB medium supplemented with 150 mM glycine and 20 mM Ca^{2+}. It was observed that the OD_{600} value, CFU value, and G6PD activity in the culture medium indeed returned to almost the same levels as those in the culture with 150 mM glycine alone (Figures 4.12 and 4.13).

In the meantime, the effects of glycine or Ca^{2+} on the morphology of *E. coli* cells were evaluated using scanning electron microscopy (SEM). *E. coli* cells in the culture without glycine exhibited regular shape and smooth surface (Figure 4.14a). However, glycine supplementation caused collapse and irregular shape of the cell granules giving the rough surface (Figure 4.14b), whereas Ca^{2+} partially counteracted the destructive effect of glycine on cell morphology (Figure 4.14c).

4.5.3 Effect of Glycine and Ca^{2+} on Cell Membrane Permeability

The hydrophobic fluorescent probe NPN was used as an indicator of outer membrane integrity. NPN has a low fluorescence quantum yield in aqueous solution but fluoresces strongly in the hydrophobic environment of a biological membrane. Normally, NPN is excluded from *E. coli* by a lipopolysaccharide layer of the outer membrane but can enter at points where

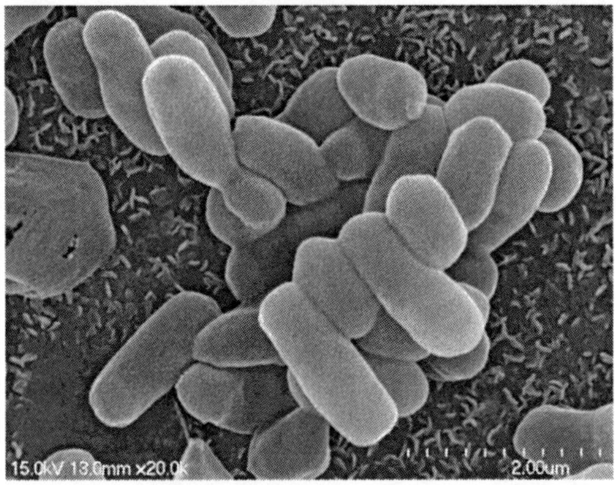

(a)

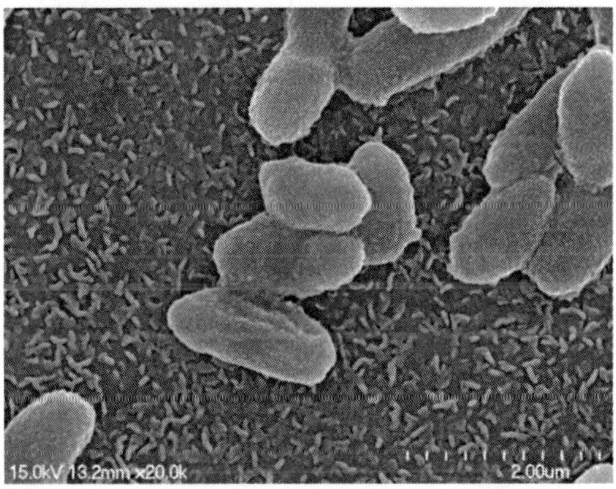

(b)

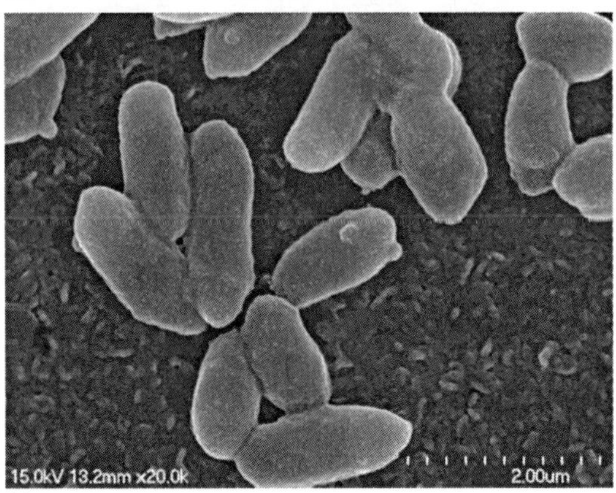

(c)

Figure 4.14 Effects of glycine and/or Ca^{2+} on cell morphology. *E. coli* cells were imaged by scanning electron microscopy when grown for 36 h in regular TB medium (a), or TB medium supplemented with 150 mM glycine (b) or 150 mM glycine–20 mM Ca^{2+} (c), respectively. Reprinted with permission from Li, Z. et al., *Journal of Agricultural and Food Chemistry.* 57, 6231–6237. Copyright 2009 American Chemical Society.

membrane integrity is compromised. Thus, fluorescence value and its increased rate indicated the outer membrane permeability. Permeability of the inner membrane was evaluated using the β-galactosidase substrate ONPG as a probe. When ONPG passes the inner membrane, it can be cleaved by β-galactosidase localized within the cytoplasm, resulting in the appearance of a yellow color. Thus, absorbency at 420 nm and its increased rate indicated the inner membrane permeability.

As shown in Figure 4.15, in the culture without glycine, both outer and inner membranes of *E. coli* cells had very low permeabilities

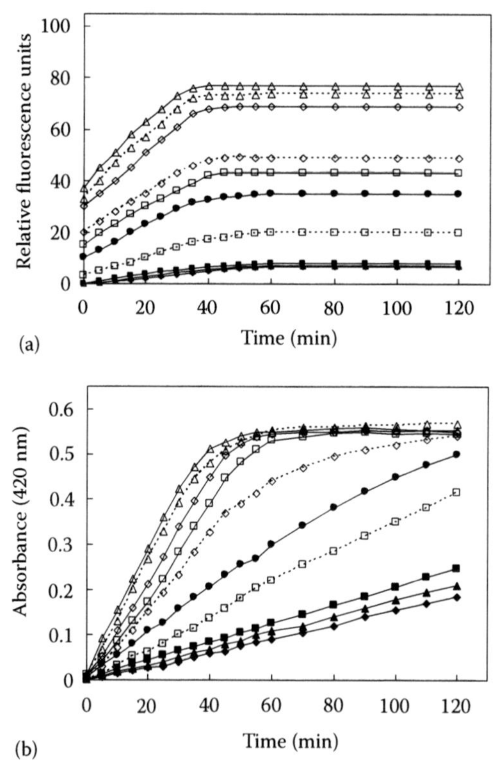

Figure 4.15 Effects of glycine and/or Ca^{2+} on the permeabilities of outer (a) and inner (b) membranes. After *E. coli* cells had been grown in regular TB medium for 12 h (■), 24 h (◆), 36 h (▲), and 60 h (●), respectively, and TB medium supplemented with 150 mM glycine (solid line) or 150 mM glycine–20 mM Ca^{2+} (dashed line) for 12 h (□), 24 h (◇), and 36 h (△), respectively, they were harvested, rinsed, and resuspended in 10 mM phosphate buffer (pH 7.4) to an OD_{600} of 0.5. NPN or ONPG was added into cell suspension. NPN fluorescence and the absorbance at 420 nm were monitored for 120 min, respectively. Each value represents the mean of three independent measurements, and the deviation from the mean is below 5%. Reprinted with permission from Li, Z. et al., *Journal of Agricultural and Food Chemistry.* 57, 6231–6237. Copyright 2009 American Chemical Society.

during the first 36 h of culture. Upon further culture, the membrane permeabilities gradually increased, which might be the main reason for the increase in extracellular secretion of recombinant α-CGTase at the later stage of culture. In the culture with 150 mM glycine, the outer and inner membranes of *E. coli* cells had high permeabilities soon after the culture started. Furthermore, the glycine-induced increases in the membrane permeabilities were dose-dependent (data not shown). Although, in the culture with 150 mM glycine and 20 mM Ca^{2+}, the outer and inner membranes of *E. coli* cells had relatively low permeabilities during the first 24 h of culture, their permeabilities kept on increasing and reached almost the same levels as those in the culture with 150 mM glycine alone at 36 h of culture.

4.6 High-Level Extracellular Production of α-Cyclodextrin Glycosyltransferase in 3 L Fermentor

The strategies in shaking flasks level on increasing secretion of α-CGTase to the liquid medium by *E. coli* was studied. In addition, in preliminary investigation in a 3 L fermentor, it was found that a synthetic medium was optimal to achieve high-cell-density culture for *E. coli*. When cultured in a synthetic medium and induced by 0.8 $g{\cdot}L^{-1}\ h^{-1}$ lactose, the α-CGTase activity in the medium could reach 113.0 $U{\cdot}mL^{-1}$ at 30 h.

In the present study, different cultivation conditions, such as feeding strategy, feeding solution, and induction methods, in a 3 L fermentor by high-cell-density culture were investigated to further promote α-CGTase secreting into the medium. The results showed that a two-stage feeding strategy was suitable for high-cell-density culture and protein production. When fed with complex nitrogen sources (peptone and yeast extract) in the feeding solution and induction at 25°C and intermediate cell density (DCW of 30 $g{\cdot}L^{-1}$) by lactose, the yield of α-CGTase in culture media could reach 275.3 $U{\cdot}mL^{-1}$, which represents the highest extracellular yield and productivity of α-CGTase reported so far (Bar et al. 2004; Cheng et al. 2011; Gawande et al. 1999; Lin and Jeang 1998; Tesfai et al. 2012).

4.6.1 Feeding Strategy for High-Cell-Density Culture

The yield of recombinant protein production depends on both the biomass concentration as well as the specific cellular productivity. To achieve high biomass concentration, the high-cell-density cultivation (HCDC) is preferred by appropriate feeding strategy in fed-batch cultivation. The feeding strategies include simple indirect feedback methods (such as pH-stat and DO-stat), predetermined feeding methods (such as exponential feeding), and feeding according to substrate uptake or demand. Exponential feeding is a simple method that allows cells to grow at a constant growth rate, which is below the critical value for acetate formation. Acetate is an extracellular by-product of aerobic fermentation, which usually inhibits growth and recombinant protein formation at a concentration of higher than 2.0 $g \cdot L^{-1}$. Usually, accumulation of acetate can be prevented by a limited supply of carbon source, and exponential feeding is a suitable strategy to achieve this. However, this strategy, without feedback, may lead to the accumulation of substrate in the medium in the process of fermentation of engineered cells, since the cellular physiology and metabolism may be changed after the recombinant protein is induced. Previously, it has been reported that the changes in cellular physiology and metabolism were dependent on temperature, methods of induction, and nature of recombinant protein, which all led to the difficulty in evaluating nutrition needs by cells. Thus, feeding strategy based on the substrate residues after induction should be applied.

Based on the above analysis and the specific expression system of the recombinant cell utilized, in the present study, a two-stage feeding strategy was applied. During the pre-induction phase, the glycerol feeding rate increased exponentially according to the exponential feeding method and the cell growth was controlled at a specific growth rate (μ_{set}) of 0.25 h^{-1}. When the DCW of the cell reached a certain value, the post-induction phase began and the feeding rate was shifted to a gradient decreasing method, in which an initial rate of 90% feeding rate was utilized at the induction point, and then reduced gradually based on glycerol residues (Figure 4.16a). Under this feeding strategy, the cells of *E. coli* were cultivated for 26 h from 0.14 to 94.4 $g \cdot L^{-1}$, and maintained the biomass without reduction for another 4 h (Figure 4.16a). Moreover, considerable accumulation

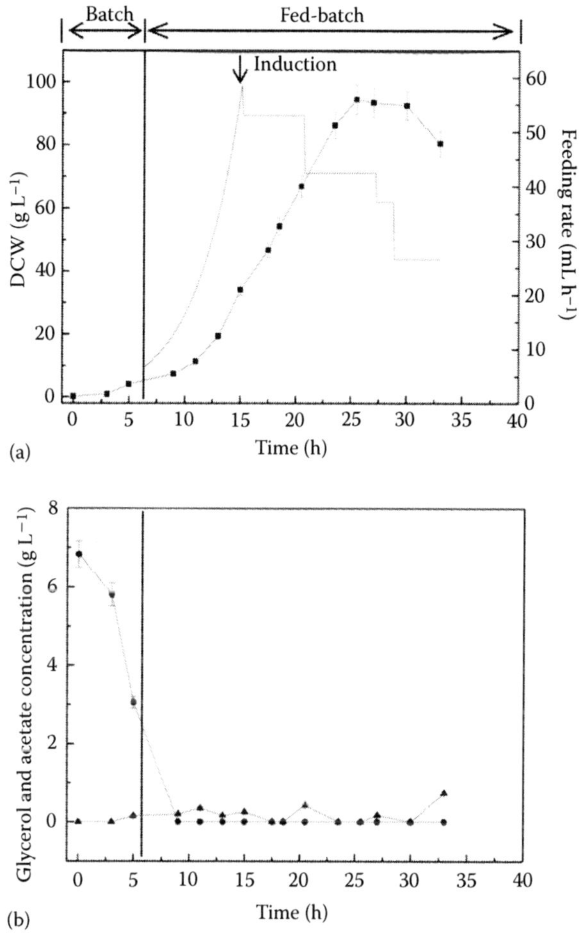

Figure 4.16 Fed-batch culture of recombinant *E. coli* BL21 (DE3). (a) Cell concentration (■) and glycerol feeding rate (solid line). (b) Glycerol (●) and acetate (▲). Arrows indicate the starting point of induction. (From Cheng J et al., *Journal of Agricultural and Food Chemistry*, 59(8):3797–3802.)

of glycerol or acetate (below 1.0 g·L^{-1}) in the whole process was not observed (Figure 4.16b).

4.6.2 Effect of Nitrogen Supplementation in Feeding Solution on Cell Growth and α-CGTase Production

Nitrogen is the critical medium component for bacteria growth and production performance. In the above HCDC, ammonia served as the sole nitrogen source as well as the base to control pH. With the aim to improve α-CGTase production by recombinant *E. coli*, the effect of nitrogen supplementation in the feeding solution on cell growth and enzyme production was investigated. Two kinds of nitrogen sources were chosen: inorganic (FS1) and complex nitrogen (FS2), and no extra nitrogen addition was used as control.

The results showed that the cell concentration increased obviously when supplemented with nitrogen sources in feeding solution. The highest DCW obtained in FS1 and FS2 were 82.1 and 94.4 $g \cdot L^{-1}$, respectively, which were 1.2- and 1.3-fold of that of control (Figure 4.17a). The extracellular activity obtained in FS2 was 105.1 $U \cdot mL^{-1}$, which was 2.3-fold of that of control (Figure 4.17b). However, the extracellular activity obtained in FS1 was 8.8 $U \cdot mL^{-1}$, which was only 19.0% of that of control (Figure 4.17b).

In our engineered *E. coli* cell, α-CGTase was first synthesized as precursors with OmpA signal peptide, which has a feature essential for secretion of the proteins into the periplasm of *E. coli*. During transport of pre-enzyme across the inner membrane, the signal sequence of OmpA was cleaved by signal peptidase to yield the mature α-CGTase. The mature enzyme was then transported into the medium by nonspecific periplasmic leakage. Thus, the extracellular activity of α-CGTase was dependent on both periplasmic transport and nonspecific periplasmic leakage.

In order to analyze the effect of nitrogen supplementation on extracellular enzyme production, the activity of α-CGTase in periplasm of above *E. coli* cells was measured, and the total activity of the expression was calculated as the sum of activities between culture medium and periplasm space. As seen from Figure 4.17c, the total activity obtained in FS2 reached 177.2 $U \cdot mL^{-1}$, which was 2.1-fold that of control. It is known that complex nitrogen, such as peptone and yeast extract, except for providing a nitrogen source, contains special factors, such as amino acids, trace metals, cofactors, and vitamins. In the present study, these factors may promote recombinant protein synthesis and translocation, which can lead to the enhanced total activity of α-CGTase in FS2. However, the total activity in FS1 was quite low, which accounted for 16.5% (14.1 $U \cdot mL^{-1}$) of that in FS1. The phenomenon of ammonium chloride suppressing the α-CGTase production has not been reported previously, and thus further experiments are needed to explore the underlying mechanism.

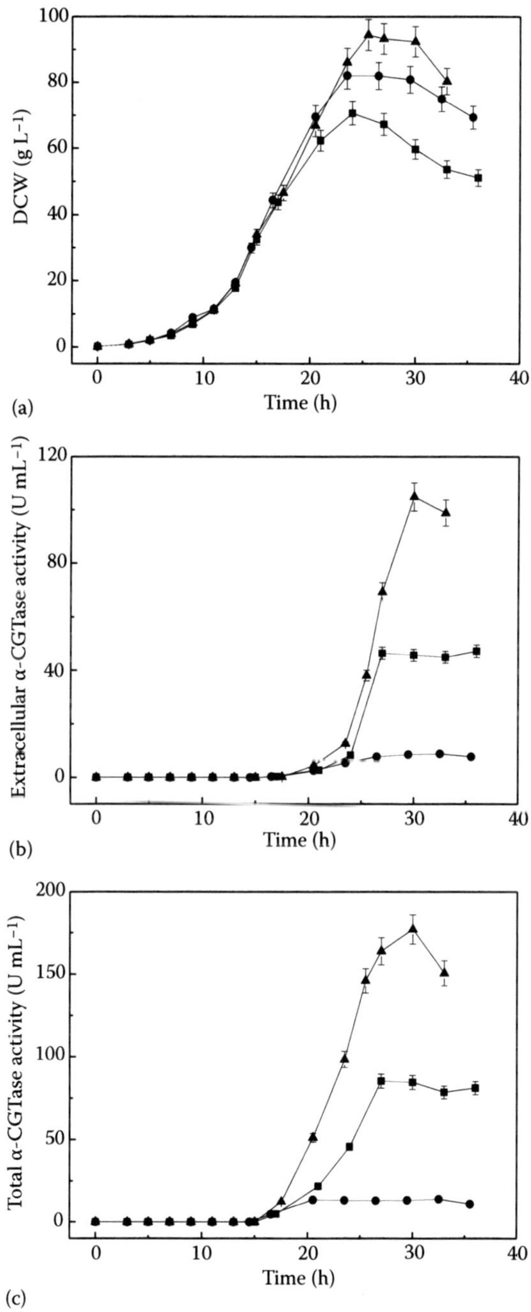

Figure 4.17 Comparison of the time profiles for cell concentration (a), extracellular α-CGTase activity (b), and total α-CGTase activity (c) obtained from cultivations with different nitrogen supplementation in feeding solution. ■: Control; ●: FS1; ▲: FS2. (From Cheng J et al., *Journal of Agricultural and Food Chemistry*, 59(8):3797–3802.)

Although there was significant influence of nitrogen supplementation in feeding solution on the total activity, no significant influence was detected on the capability of extracellular secretion, which was calculated as extracellular activity divided by total activity. Based on the above activity obtained in control, FS1, and FS2, the capability of extracellular secretion were 54.4%, 62.4%, and 59.3%, respectively.

4.6.3 Effect of Induction Temperature on Cell Growth and α-CGTase Production

Induction temperature is an important parameter for recombinant protein production in *E. coli*. Previously, flask experiments indicated that induction at 37°C had negative impact on α-CGTase production, possibly due to the excessive speed of α-CGTase synthesis, which leads to large amounts of inclusion bodies (data not shown). Based on this observation, a two-stage temperature fermentation strategy was developed in the present study, in which three different induction temperatures (20°C, 25°C, and 30°C) were investigated.

As seen from Figure 4.18a, the final biomass reached 94.4 $g{\cdot}L^{-1}$ at induction of 30°C, which was 1.6- and 1.1-fold of that at 20°C and 25°C, respectively. However, high biomass did not achieve high protein yield. The extracellular α-CGTase activity obtained at 30°C was lower than that of 25°C, but higher than that of 20°C. Thus, induction at 25°C was optimal for α-CGTase extracellular production, at which the total and extracellular α-CGTase activities reached 348.6 and 275.3 $U{\cdot}mL^{-1}$, respectively (Figure 4.18b,c). Besides, the capability of extracellular secretion at 25°C was also the highest, reaching 79.0%.

Lower temperature is believed to reduce protein synthesis and decrease fluidity of the cytoplasmic membrane, which influences membrane-associated cellular functions such as periplasmic transport and nonspecific periplasmic leakage. Thus, induction at 20°C resulted in the lowest total and extracellular enzyme. When induction occurred at a higher temperature, even though the fluidity of the cell membrane was enhanced, the rate of protein synthesis was also increased, and the extra higher protein synthesis rate caused the accumulation of large amounts of inclusion bodies, which may have affected the cell physiology and protein transport. In our studies, it

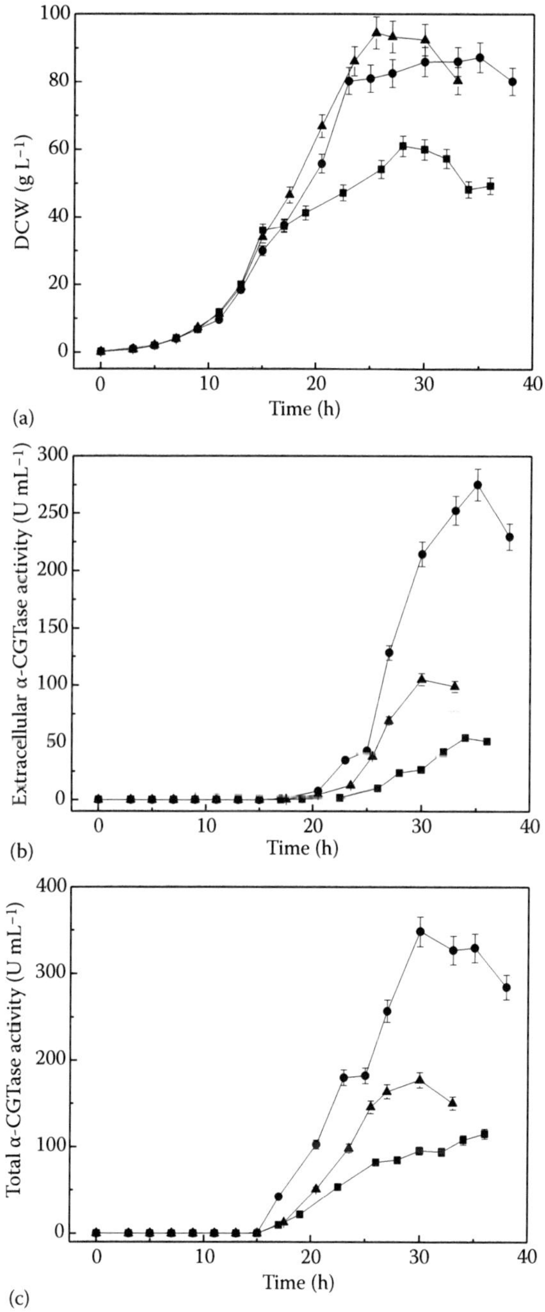

Figure 4.18 Comparison of the time profiles for cell concentration (a), total α-CGTase activity (b), and extracellular α-CGTase activity (c) obtained from cultivations with different induction temperature. ■: 20°C; ●: 25°C; ▲: 30°C. (From Cheng J et al., *Journal of Agricultural and Food Chemistry*, 59(8):3797–3802.)

seems like the induction at 25°C led to a better coordination between protein synthesis and translocation, and obtained the highest yield of α-CGTase in the culture medium.

4.6.4 Effect of Induction Point on Cell Growth and α-CGTase Production

Recombinant protein expression generally imposes certain metabolic burden on the cell, which reduces cellular growth rate, cell yield, protein expression, and plasmid stability. In order to adjust the extent of metabolic burden imposed on the cell, the optimal point for α-CGTase induction was investigated in the present study, in which lactose was fed at DCW of 15, 30, and 60 $g \cdot L^{-1}$, respectively.

As seen from Figure 4.19a, cell growth was seriously inhibited when induced at low cell density, under which the maximum DCW was only 23.4 $g \cdot L^{-1}$. In addition, due to the serious inhibition of cell growth, the extracellular α-CGTase activity was only 23.3 $U \cdot mL^{-1}$ (Figure 4.19b). Although no significant inhibition was observed at high-cell-density induction, the extracellular α-CGTase activity was also quite low, only 1.9 $U \cdot mL^{-1}$. Thus, induction at intermediate cell density was the optimal condition, and the maximum α-CGTase activity in the medium reached 275.3 $U \cdot mL^{-1}$ (Figure 4.19b). As seen from the SDS-PAGE analysis of extracellular fraction (Figure 4.20), the concentration of recombinant proteins in the medium increased with the time of induction, and recombinant proteins turned into the main component of fermentation supernatant at the end of cultivation.

Our above results showed that the extracellular production of recombinant α-CGTase was seriously affected by the time of induction, which was similar to that of production of human leptin and glycine oxidase. In order to analyze the effect of induction points on extracellular production of α-CGTase, the total activities under three conditions were compared. It was found that induction at high cell density rendered the highest total activity of 425.7 $U \cdot mL^{-1}$, but the capability of extracellular secretion was only 0.5%, leading 99.5% of total activity accumulated in the periplasm. In addition, it was observed that, in this induction condition, most of the enzymes were produced during the stationary phase (Figure 4.19a and c). Previously, it was reported that when the stationary phase approached, the membrane became more rigid. Increased rigidity

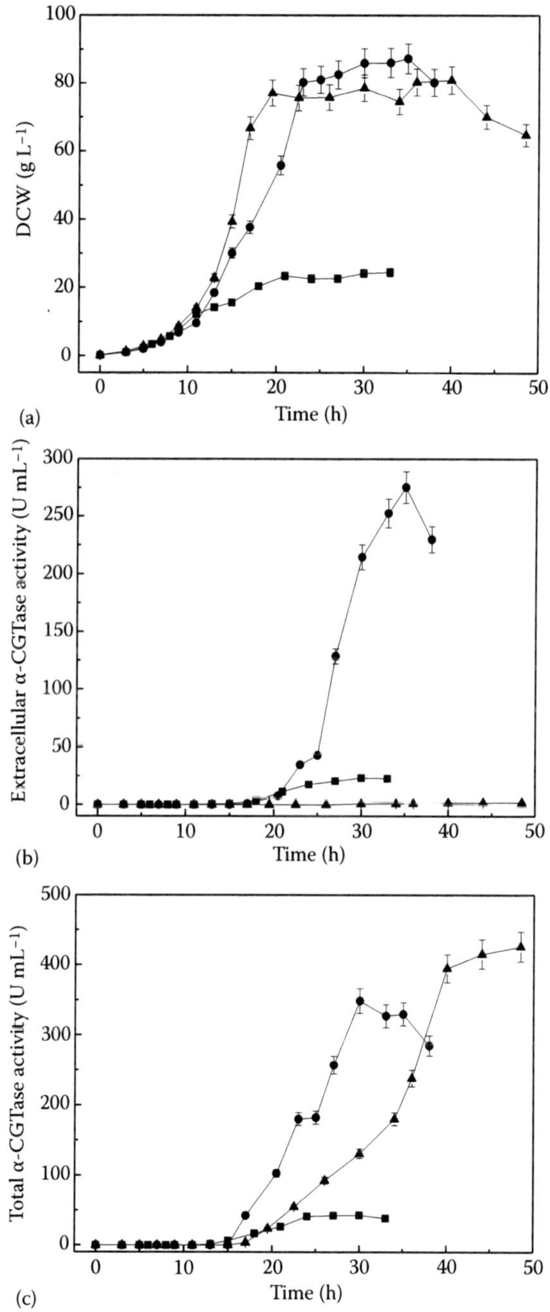

Figure 4.19 Comparison of the time profiles for cell concentration (a), total α-CGTase activity (b), and extracellular α-CGTase activity (c) obtained from cultivations induced at low cell density (■: 15 g·L^{-1}), intermediate cell density (●: 30 g·L^{-1}), and high cell density (▲: 60 g·L^{-1}). (From Cheng J et al., *Journal of Agricultural and Food Chemistry*, 59(8):3797–3802.)

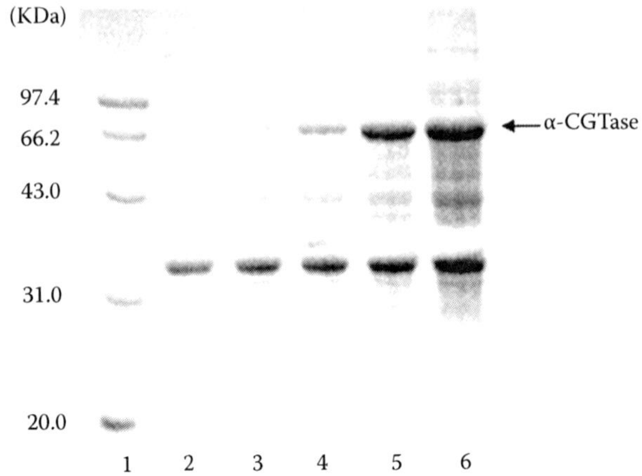

Figure 4.20 SDS-PAGE analysis of extracellular fraction in the optimal fermentation condition (2 μL). Lane 1: Marker; Lanes 2–6: culture time of 15, 20, 25, 30, and 35 h. (From Cheng J et al., *Journal of Agricultural and Food Chemistry*, 59(8):3797–3802.)

of cell membrane is believed to not be suitable for nonspecific periplasmic leakage. Thus, this may be the reason it led to the decreased secretion of periplasmic proteins into the culture medium when induction is done at high cell density.

4.7 Effect of Organic Solvents on the Yield and Specificity of Cyclodextrins by Recombinant CGTase

Despite the toxicity especially in human use products, and possibilities of environmental pollution, various organic solvents have been used as an option to increase the CD yield and specificity. Starch is the starting material for CD synthesis, and according to their amylopectin and amylose content, various starches provide different ratios and types of CDs. Due to its cost benefit and availability advantage, potato starch is preferably reported for industrial production. In bioconversion of starch by CGTases, starch is usually liquefied at high temperature by adding liquefying enzyme. Liquefaction increases starch vulnerability to CGTases. Beside amylase, some other thermophilic enzyme has also been reported. In this report, we investigated the effect of various organic solvents individually and in mixture (two organic solvents added at the same time) with a common complexant, decanol, on the biochemical activity of the recombinant CGTase. To the best of our knowledge, this is the first attempt to use two organic solvents added together to enhance the yield and specificity of CDs in the solvent process system (Rendleman 1997).

4.7.1 Optimum CGTase Reaction Conditions in the Absence/Presence of Organic Solvents

In the present study, it was also found that, at normal reaction condition, i.e., in the absence of organic solvents, the optimum temperature was 65°C in which a maximal 23% total conversion to CDs was achieved (Table 4.5), but in the presence of most organic solvents, a comparatively increased yield was achieved at a lower temperature of 40°C in which a maximal conversion of 45% was achieved in the presence of ethanol (Tables 4.6 and 4.7). Lower temperature could slightly reduce the burden of the organic solvents on the biochemical activity of the enzyme.

4.7.2 Effect of Organic Solvents on the Biochemical Activity of Recombinant CGTase

The effect of various organic solvents was investigated to minimize main factors such as product inhibition, intermolecular transglycosylation reaction, and/or hydrolysis reaction, which limit the increase in CD production by CGTases. A significant increase in both total

Table 4.5 Effect of Different Temperature on Starch Bioconversion by CGTase in the Absence of Organic Solvents

		PRODUCT RATIO (%)		
TEMPERATURE (°C)	TOTAL CONVERSION TO CDS (%)	α	β	γ
40	15 ± 3.5	34	53	13
50	18 ± 2.0	22	56	22
60	18 ± 0.5	22	56	22
65	23 ± 3.2	22	65	13
70	10 ± 0.5	40	50	10

Source: With kind permission from Springer Science+Business Media: *Journal of Inclusion Phenomena* Effect of organic solvents on the yield and specificity of cyclodextrins by recombinant cyclodextrin glucanotransferase (CGTase) from Anaerobranca gottschalkii, 77(1), 2012, Tesfai, B.T., Wu, D., Chen, S., Chen, J., and Wu, J., Tables 1 and 4–6.

Note: Statistical values of mean ± standard deviation are shown for total conversion to CDs (%), where $n = 3$.

Table 4.6 Bioconversion of Starch by the Recombinant CGTase at Different Temperatures in the Presence of Organic Solvents

	TEMPERATURE (°C)				
SOLVENT	40	50	60	65	70
Decanol (dec)	33 ± 2.0	32 ± 2.5	17 ± 0.6	14 ± 0.0	12 ± 1.7
Cyclohexane	33 ± 2.5	30 ± 2.5	26 ± 2.08	23 ± 2.1	17 ± 4.0
Butanol (but)	28 ± 2.0	33 ± 2.0	33 ± 5.3	34 ± 4.2	20 ± 0.6
Ethanol (eth)	45 ± 2.5	28 ± 0.6	24 ± 2.1	22 ± 0.6	18 ± 3.0
Dec-eth	36 ± 1.5	36 ± 0.6	35 ± 1.2	31 ± 0.0	30 ± 0.6
Dec-but	39 ± 2.5	44 ± 0.6	37 ± 3.8	25 ± 4.0	20 ± 4.2

Source: With kind permission from Springer Science+Business Media: *Journal of Inclusion Phenomena* Effect of organic solvents on the yield and specificity of cyclodextrins by recombinant cyclodextrin glucanotransferase (CGTase) from Anaerobranca gottschalkii, 77(1), 2012, Tesfai, B.T., Wu, D., Chen, S., Chen, J., and Wu, J., Tables 1 and 4–6.

Note: Statistical values of mean ± standard deviation are shown for total conversion to CDs (%), where $n = 3$. Organic solvents coupled with hyphen (-) indicate simultaneous addition of the two organic solvents.

yield and specificity of CD was achieved with all the tested organic solvents. A maximum conversion of starch into cyclodextrins (45%) was achieved in the presence of ethanol, while the maximum specific product yield of 91% α- and 64% β-CD was achieved in the presence of decanol and cyclohexane, respectively (Table 4.7). Bioconversion of

Table 4.7 Total and Product Ratios of the CDs from Starch Bioconversion at 40°C in the Presence of Organic Solvents

		PRODUCT RATIO (%)		
SOLVENT	TOTAL CONVERSION TO CDS (%)	α	β	γ
No solvent	15 ± 3.5	34	53	13
Butanol	28 ± 2.5	71	18	11
Decanol	33 ± 2.0	91	9	0
Cyclohexane	33 ± 2.5	30	64	6
Dec-but	36 ± 1.5	58	36	6
Dec-eth	39 ± 0.6	77	18	5
Ethanol	45 ± 2.5	58	38	4

Source: With kind permission from Springer Science+Business Media: *Journal of Inclusion Phenomena* Effect of organic solvents on the yield and specificity of cyclodextrins by recombinant cyclodextrin glucanotransferase (CGTase) from Anaerobranca gottschalkii, 77(1), 2012, Tesfai, B.T., Wu, D., Chen, S., Chen, J., and Wu, J., Tables 1 and 4–6.

Note: Statistical values of mean ± standard deviation are shown for total conversion to CDs (%), where $n = 3$. Solvents coupled with hyphen (-) indicate simultaneous addition of the two organic solvents.

starch by CGTase in the presence of organic solvents was affected by cultivation temperature (Table 4.6).

The use of decanol (complexing agent) was thoroughly investigated, and it was found that it minimized the product inhibition by forming a "strong" complex with the product α-CD and thus reduced its availability in the solution by forming a precipitate. As Blackwood and Bucke (2000) explained for their case of β-CD, use of organic solvents can affect the conformational structure of cyclodextrin, which, in turn, may affect the binding site or catalysis of the CGTase. Similarly, in this case, it can be proposed that an increase in α-CD product and specificity in the presence of decanol might mainly be attributed to the same phenomenon. A maximum α-CD product ratio of 91% was achieved using this complexant at a temperature of 40°C (Table 4.6). Although the enzyme stability only reduced slightly with increasing temperature, the influence of decanol at higher temperatures was found to dramatically decrease the overall yield and specificity. But the other organic solvents such as ethanol and cyclohexane and their simultaneous addition with decanol did not decrease the yield so dramatically as compared to decanol (Table 4.6).

The special effect of decanol in correlation to temperature though remains vague. It can in part be due to the fact that a higher temperature facilitates the interaction and susceptibility of enzymes, and thus, makes it vulnerable to organic solvents like decanol, which may have an effect on the enzyme's biochemical stability and activity. On the other hand, organic solvents also affect the conformational structure of CDs by forming a complex with them, which is assumed to prevent the CD from binding to the CGTase, thus reducing its activity.

Similar to the effect of decanol on its specific α-CD formation, cyclohexane also affected its specific β-CD yield by forming a complex with it and thus reducing its product inhibition. Additionally, similar to decanol, cyclohexane showed a small effect on the transglycosylation reaction. A total yield of 33% and a maximum specific yield of 64% β-CD (product ratio) were achieved in the presence of cyclohexane (Table 4.7). Compared to the other individual and/or simultaneous organic solvents, butanol showed a lower total conversion yield (Table 4.6).

As shown in Table 4.8, the use of organic solvents decreased the transglycosylation reaction of the recombinant CGTase. CGTases can cause degradation of cyclodextrins when there is availability of small linear dextrins such as lactose, which can be used as acceptors for cyclodextrin molecules. When maltose was added to act as a glucosyl acceptor in the cultivation of β-CD only with the CGTase, a significant reduction in the relative remaining concentration was observed. While in the presence of all the tested organic solvents the CD degradation by the CGTase was reduced significantly and reached its highest peak in the presence of ethanol (Table 4.8).

As far as we are aware, a simultaneous effect of two organic solvents has not yet been reported, and here, in this work, we have tried to investigate a special effect of an alternative organic solvent with a common complexant, decanol, aimed at specifically increasing the α-CD product ratio. The two different organic solvents were added together at the same time with the same volume. This special investigation was meant to sum up the two main effects of the different organic solvents-the product inhibition and transglycosylation (coupling) effects.

Table 4.8 Percentage Cyclodextrin Remaining after 4 h Incubation in the Presence of 3 $g \cdot L^{-1}$ Maltose and/or 10% (v/v) Organic Solvent

ORGANIC SOLVENT	RELATIVE CONC. OF REMAINING β-CD (%)
Control	40 ± 0.6
Decanol	61 ± 2.0
Cyclohexane	62 ± 0.0
Butanol (but)	96 ± 3.5
Ethanol	100 ± 2.52
Dec-but	84 ± 3.0
Dec-eth	90 ± 2.0

Source: With kind permission from Springer Science+Business Media: *Journal of Inclusion Phenomena* Effect of organic solvents on the yield and specificity of cyclodextrins by recombinant cyclodextrin glucanotransferase (CGTase) from Anaerobranca gottschalkii, 77(1), 2012, Tesfai, B.T., Wu, D., Chen, S., Chen, J., and Wu, J., Tables 1 and 4–6.

Note: The control was the cultivation of β-CD with the recombinant CGTase alone. The amount of the remaining β-CD was calculated in a relative concentration left in each sample compared to the initially used amount. Statistical values of mean ± standard deviation are shown for relative concentration of remaining β-CD (%), where $n = 3$. Solvents coupled with hyphen (-) indicate simultaneous addition of the two organic solvents.

The fact that decanol, as a complexant, maximized the α-CD yield by minimizing the product inhibition, and organic solvents like butanol and ethanol favored the α-CD production by affecting the transglycosylation activity of the enzyme, led us to believe that reacting these two solvents simultaneously could sum up into an increased total yield by taking advantage of the individual effect. The more α-CD produced by the presence of these organic solvents would mean the more product inhibition, but the presence of specific complexant like decanol for α-CD would minimize this problem by specifically complexing with it and thus avoiding its self-inhibition (product inhibition).

The increased CD yield and specificity in the presence of one organic solvent with decanol as complexant was attributed to a combined effect on product inhibition and intermolecular transglycosylation reaction. As speculated, the simultaneous effect of ethanol and butanol with common complexant, decanol, resulted in a significant increase in CD yield and specificity when compared to control (Table 4.7). This increased CD yield was confirmed by the counteractive results of minimized coupling activity (Table 4.8) and increased complex formation of decanol–ethanol and decanol–butanol. Although there was not much significant difference in the effect of individual or simultaneous use of organic solvents, in general, the simultaneous effect was comparatively less effective. This might be explained by the increased burden of the two organic solvents on the biochemical activity of the enzyme.

4.8 Applications of α-Cyclodextrin in the Food Industry

4.8.1 Technological Uses in Food

The acceptable daily intake of α-cyclodextrin has been allocated as "not specified." This acceptable daily intake was based on the known current uses of α-cyclodextrin under good manufacturing practices as a carrier and stabilizer for flavors, colors, and sweeteners; as a water-solubilizer for fatty acids and certain vitamins; as a flavor modifier in soya milk; and as an absorbent in confectionery products (WHO 2002).

For technological uses in food, α-cyclodextrin has usually been used as a carrier and stabilizer for bulky guests. For example, cinnamic acid (CA), a naturally occurring organic acid found in fruits and spices that has antimicrobial activity against spoilage and pathogenic bacteria, has low aqueous solubility that limits its use. Solubility-enhancing α-cyclodextrin–CA inclusion complexes were able to significantly reduce populations of *E. coli* O157:H7 and *Salmonella enterica* serovars suspended in apple cider or orange juice (Truong et al. 2010). McGowan et al. (1983) also found that α-cyclodextrin was very effective at solubilizing free fatty acids. Cyclodextrins are widely used as browning inhibitors in different fruit juices. The addition of α-cyclodextrin at 90 mM could prevent oxidation of the volatile precursors present in freshly squeezed pear juices. This resulted in juice with the best color, but with low aromatic intensity and low sensory quality. Addition of 15 mM α-cyclodextrin, in contrast, could lead to a pear juice that also had an acceptable color, but that retained a high intensity of fruity and pear-like odors/aromas, making it the best appreciated juice by the panel (Lopez-Nicolas et al. 2009). α-Cyclodextrin is also the most suitable agent for encapsulating flavors extracted from dried shiitake, including lenthionine. These flavors are encapsulated in powder form by spray-drying with α-cyclodextrin. The retention of flavor was markedly increased by using a combination of α-cyclodextrin and maltodextrin as the encapsulant (Shiga et al. 2004).

Because it has the smallest internal cavity of the common cyclodextrins, the application of α-cyclodextrin-assisted molecular encapsulation in foods might be limited. Nevertheless, we are convinced that there are many uses of α-cyclodextrin in food technology that can still be developed.

4.8.2 α-Cyclodextrin as Dietary Fiber

Dietary fiber is a medically important component of a healthy human diet. It provides a gastrointestinal health benefit and may reduce the risk of coronary heart disease and other lifestyle-related ailments (Ferrari et al. 2013; Threapleton et al. 2013; Zhang et al. 2013a,b). Soluble fiber can dissolve in water to form a gel-like material that may help lower blood cholesterol and glucose levels. Indigestible α-cyclodextrin has proven to be a natural, soluble dietary fiber. Its use as added soluble fiber has become the most important application of α-cyclodextrin in the food industry. As is generally known, although triglycerides are important to human life and are the main form of fat in the body, high triglyceride levels raise your risk of heart disease and may be a sign of metabolic syndrome (Hadaegh et al. 2009; Kisfali et al. 2010; Kasai et al. 2013). Due to its small cavity size, it had been generally believed that α-cyclodextrin could not form a complex with triglycerides (Waalkens-Berendsen et al. 2004). However, in a very interesting report, Artiss et al. (2006) proved that α-cyclodextrin could form a complex with triglyceride at a ratio that is distinctly different from the 1:1 that is typical for dietary fibers.

Among known dietary fibers, α-cyclodextrin has the unique ability to bind nine times its own weight in fat (Artiss et al. 2006; Grunberger et al. 2007; Jen et al. 2013). Furthermore, α-cyclodextrin might form a layer on the surface of the fat droplets; this was demonstrated by the fact that it was possible to produce "beads" with oily compartments by mixing an aqueous solution of α-cyclodextrin with soybean oil (Bochot et al. 2007).

α-Cyclodextrin has been shown to form a stable complex with dietary fat at a high ratio. Moreover, the α-cyclodextrin–fat complex was proved to be resistant to normal lipolytic hydrolysis by lipases. As a result, α-cyclodextrin reduces the absorption and bioavailability of dietary fat (Artiss et al. 2006), which makes it practical as a weight loss supplement (Suzuki and Sato 1985). Animal research has shown that α-cyclodextrin, marketed under the trade name FBCx (Wacker Biochem, Adrian, MI), significantly reduces weight gain in rats (Artiss et al. 2006). For obese patients with type 2 diabetes, α-cyclodextrin is also effective in reducing and/or maintaining body weight despite increasing their energy intake (Tonkova 1998). Comerford et al. (2011) observed that 1 month of α-cyclodextrin supplementation, without any diet or lifestyle changes, led to significant weight loss in healthy overweight non-obese individuals in the absence of any change in energy intake. Most importantly, since the α-cyclodextrin–fat complex is not accessible to the human gut flora and is commonly excreted in the stool intact, it does not lead to the gastrointestinal side effects associated with weight loss products that cause fat malabsorption (Penninga et al. 1995; Sabadini et al. 2006; Gallaher et al. 2007). By comparison, most weight loss products that inhibit lipase secretion allow free, uncomplexed dietary fats to pass through the digestive system, which may lead to steatorrhea and bowel incontinence (Sjostrom et al. 1998; Comerford et al. 2011).

Besides weight control, α-cyclodextrin also provides other health benefits. Artiss et al. (2006) demonstrated that α-cyclodextrin reduces serum triglyceride and leptin levels and increases insulin sensitivity and fecal fat excretion in rats, indicating that α-cyclodextrin might be effective in improving metabolic syndrome. In a hyperlipidemic experimental animal model, α-cyclodextrin lowered low-density lipoprotein cholesterol and altered the plasma fatty acid profile (Wagner et al. 2008); both saturated and *trans* fatty acids were decreased in the plasma, perhaps resulting from preferential binding of α-cyclodextrin with saturated fats in the intestine and thus selective increase in fecal excretion of saturated fats (Gallaher et al. 2007). In human models, when their diet was supplemented with α-cyclodextrin, obese type 2 diabetic individuals with hypertriglyceridemia showed significant reductions in blood lipid levels and increases in adiponectin levels (Grunberger et al. 2007), suggesting that α-cyclodextrin may potentially be helpful for the treatment of type 2 diabetes. The beneficial health effects of α-cyclodextrin on the blood lipid profile could also be found in healthy non-obese individuals; the effects in individuals with hyperlipidemia may be more significant than in those with normolipidemia (Comerford et al. 2011). In addition, Buckley et al. (2006) demonstrated that α-cyclodextrin reduced the postprandial glycemic response of healthy human subjects to a standard carbohydrate meal without affecting the insulin response, indicating that α-cyclodextrin may be useful as an ingredient for reducing the glycemic impact of such foods. Gentilcore et al. (2011) concluded that, at a dose of 10 g, α-cyclodextrin had modest effects to slow gastric emptying and modify the glycemic response to sucrose in healthy older adults, probably due to delayed intestinal carbohydrate absorption.

Since most people do not achieve the recommended daily intake of dietary fiber (25–30 g), food rich in dietary fiber has become a growing market. Thus, the market share of α-cyclodextrin as a natural, soluble dietary fiber will increase significantly in the next decade.

References

Ahmed, E. M. and El-Refai, H. A. (2010). Cyclodextrin glucosyltransferase production by *Bacillus megaterium* NCR: Evaluation and optimization of culture conditions using factorial design. *Indian Journal of Microbiology*. **50**(3), 303–308.

Ahuja, S. (1991). Chiral separations: An overview. In Ahuja, A. (ed.), *Chiral Separations by Liquid Chromatography.* Washington, D.C.: American Chemical Society, pp. 1–26.

Alvira, E., Cativiela, C., García, J. and Mayoral, J. (1995). Diels–Alder reactions in β-cyclodextrin cavities. A molecular modelling study. *Tetrahedron Letters.* **36**, 2129–2132.

Alvira, E., Mayoral, J. A. and García, J. I. (1995). A model for the interaction between β-cyclodextrin and some acrylic esters. *Chemical Physics Letters.* **245**, 335–342.

Armbruster, F. C. and Jacaway Jr., W. A., Inventors (1972). Procedure for production of alpha-cyclodextrin. United States Patent US 3,640,847.

Artiss, J. D., Brogan, K., Brucal, M., Moghaddam, M. and Jen, K. L. C. (2006). The effects of a new soluble dietary fiber on weight gain and selected blood parameters in rats. *Metabolism—Clinical and Experimental.* **55**(2), 195–202.

Bar, A., Krul, C. A., Jonker, D. and de Vogel, N. (2004). Safety evaluation of an alpha-cyclodextrin glycosyltranferase preparation. *Regulatory Toxicology and Pharmacology,* **39** (Suppl 1), S47–S56.

Bautista, V., Esclapez, J., Perez-Pomares, F., Martinez-Espinosa, R. M., Camacho, M. and Bonete, M. J. (2012). Cyclodextrin glycosyltrans ferase: A key enzyme in the assimilation of starch by the halophilic archaeon *Haloferax mediterranei. Extremophiles,* **16**, 147–159.

Binder, F., Huber, O. and Bock, A. (1986). Cyclodextrin-glycosyltransferase from *Klebsiella pneumoniae* M5a1: Cloning, nucleotide sequence and expression. *Gene,* **47**, 269–277.

Blackwood, A. D. and Bucke, C. (2000). Addition of polar organic solvents can improve the product selectivity of cyclodextrin glycosyltransferase. Solvent effects on cgtase. *Enzyme and Microbial Technology.* **27**, 704–708.

Bochot, A., Trichard, L., Le Bas, G., Alphandary, H., Grossiord, J. L., Duchene, D. and Fattal, E. (2007). alpha-Cyclodextrin/oil beads: An innovative self-assembling system. *International Journal of Pharmaceutics.* **339**(1–2), 121–129.

Buckley, J. D., Thorp, A. A., Murphy, K. J. and Howe, P. R. C. (2006). Dose-dependent inhibition of the post-prandial glycaemic response to a standard carbohydrate meal following incorporation of alpha-cyclodextrin. *Annals of Nutrition and Metabolism.* **50**(2), 108–114.

Cheng, J., Wu, D., Chen, S., Chen, J. and Wu, J. (2011). High-level extracellular production of alpha-cyclodextrin glycosyltransferase with recombinant *Escherichia coli* BL21 (DE3). *Journal of Agricultural and Food Chemistry.* **59**(8), 3797–3802.

Comerford, K. B., Artiss, J. D., Jen, K. L. and Karakas, S. E. (2011). The beneficial effects of alpha-cyclodextrin on blood lipids and weight loss in healthy humans. *Obesity.* **19**(6), 1200–1204.

Davis, M. E. and Brewster, M. E. (2004). Cyclodextrin-based pharmaceutics: Past, present and future. *Dressnature Reviews Drug Discovery.* **3**(12), 1023–1035.

Del Valle, E. M. M. (2004). Cyclodextrins and their uses: A review. *Process Biochemistry*. **39**(9), 1033–1046.

Ermolinsky, B., Peredelchuk, M. and Provenzano, D. (2013). alpha-Cyclodextrin decreases cholera toxin binding to GM1-gangliosides. *Journal of Medical Microbiology*. **62**(Pt 7), 1011–1014.

FDA (2004). GRAS Notice GRN 155: alpha-Cyclodextrin. Available from http://www.accessdata.fda.gov/scripts/fcn/gras_notices/grn0155.pdf. Last accessed July 10, 2013.

Ferrari, P., Rinaldi, S., Jenab, M. et al. (2013). Dietary fiber intake and risk of hormonal receptor-defined breast cancer in the European Prospective Investigation into Cancer and Nutrition study. *American Journal of Clinical Nutrition*. **97**(2), 344–353.

Flaschel, E., Landert, J. P., Spiesser, D. and Renken, A. (1984). The production of alpha-cyclodextrin by enzymatic degradation of starch. *Annals of the New York Academy of Science*. **434**, 70–77.

FSANZ. (2004). Final Assessment Report, Application A494, alpha-cyclodextrin as a novel food. Available at http://www.foodstandards.gov.au/code/applications/documents/A494_Alpha_cylclodextrin_FAR_FINAL.pdf. Last accessed July 10, 2013.

Fujiwara, S., Kakihara, H., Woo, K. B., Lejeune, A., Kanemoto, M., Sakaguchi, K. and Imanaka, T. (1992). Cyclization characteristics of cyclodextrin glucanotransferase are conferred by the NH2-terminal region of the enzyme. *Applied and Environmental Microbiology*. **58**(12), 4016–4025.

Gallaher, D. D., Gallaher, C. M. and Plank, D. W. (2007). Alpha-cyclodextrin selectively increases fecal excretion of saturated fats. *Faseb Journal*. **21**(5), A730–A730.

Gawande, B. N. and Patkar, A. Y. (1999). Application of factorial designs for optimization of cyclodextrin glycosyltransferase production from *Klebsiella pneumoniae pneumoniae* AS-22. *Biotechnology and Bioengineering*. **64**(2), 168–173.

Gawande, B. N. and Patkar, A. Y. (2001). Alpha-Cyclodextrin production using cyclodextrin glycosyltransferase from *Klebsiella pneumoniae* AS-22. *Starch—Stärke*. **53**(2), 75–83.

Gawande, B. N., Sonawane, A. M., Jogdand, V. V. and Patkar, A. Y. (2003). Optimization of cyclodextrin glycosyltransferase production from *Klebsiella pneumoniae* AS-22 in batch, fed-batch, and continuous cultures. *Biotechnology Progress*. **19**(6), 1697–1702.

Gentilcore, D., Vanis, L., Teng, J. C., Wishart, J. M., Buckley, J. D., Rayner, C. K., Horowitz, M. and Jones, K. L. (2011). The oligosaccharide α-cyclodextrin has modest effects to slow gastric emptying and modify the glycaemic response to sucrose in healthy older adults. *British Journal of Nutrition*. **106**, 583–587.

Grunberger, G., Jen, K. L. and Artiss, J. D. (2007). The benefits of early intervention in obese diabetic patients with FBCx: A new dietary fibre. *Diabetes/Metabolism Research and Reviews*. **23**(1), 56–62.

Hadaegh, F., Khalili, D., Ghasemi, A., Tohidi, M., Sheikholeslami, F. and Azizi, F. (2009). Triglyceride/HDL-cholesterol ratio is an independent predictor for coronary heart disease in a population of Iranian men. *Nutrition Metabolism and Cardiovascular Diseases.* **19**(6), 401–408.

Hashimoto, Y., Yamamoto, T., Fujiwara, S., Takagi, M. and Imanaka, T. (2001). Extracellular synthesis, specific recognition, and intracellular degradation of cyclomaltodextrins by the hyperthermophilic archaeon *Thermococcus sp.* strain B1001. *Journal of Bacteriology*, **183**, 5050–5057.

Hill, D. E., Aldape, R. and Rozzell, J. D. (1990). Nucleotide sequence of a cyclodextrin glucosyltransferase gene, cgtA, from *Bacillus licheniformis. Nucleic Acids Research.* **18**, 199.

Huang, H., Ridgway, D., Gu, T. and Moo-Young, M. (2003). A segregated model for heterologous amylase production by *Bacillus subtilis. Enzyme and Microbial Technology*, **32**, 407–413.

Ilk, N., Schumi, C. T., Bohle, B., Egelseer, E. M. and Sleytr, U. B. (2011). Expression of an endotoxin-free S-layer/allergen fusion protein in gram-positive *Bacillus subtilis* 1012 for the potential application as vaccines for immunotherapy of atopic allergy. *Microbial Cell Factories.* **10**, 6.

Jeang, C. L., Lin, D. G. and Hsieh, S. H. (2005). Characterization of cyclodextrin glycosyltransferase of the same gene expressed from *Bacillus macerans, Bacillus subtilis*, and *Escherichia coli. Journal of Agricultural and Food Chemistry.* **53**(16), 6301–6304.

Jen, K. L., Grunberger, G. and Artiss, J. D. (2013). On the binding ratio of α-cyclodextrin to dietary fat in humans. *Nutrition and Dietary Supplements.* **5**, 9–15.

Jiang, L. X., Yan, Y. and Huang, J. B. (2011). Versatility of cyclodextrins in self-assembly systems of amphiphiles. *Advances in Colloid and Interface Science.* **169**(1), 13–25.

Jodái, I., Kandra, L., Harangi, J., Nánási, P., Debrecen and Szejtli, J. (1984). Hydrolysis of cyclodextrin by *Aspergillus oryzae* alpha-amylase. *Starch—Stärke.* **36**(4), 140–143.

Jørgensen, S. T., Tangney, M., Starnes, R. L., Amemiya, K. and Jørgensen, P. L. (1997). Cloning and nucleotide sequence of a thermostable cyclodextrin glycosyltransferase gene from *Thermoanaerobacter sp.* ATCC 53627 and its expression in *Escherichia coli. Biotechnology Letters.* **19**, 1027–1031.

Kasai, T., Miyauchi, K., Yanagisawa, N., Kajimoto, K., Kubota, N., Ogita, M., Tsuboi, S., Amano, A. and Daida, H. (2013). Mortality risk of triglyceride levels in patients with coronary artery disease. *Heart.* **99**(1), 22–29.

Kim, J. E., Cho, H. J. and Kim, D. D. (2014). Budesonide/cyclodextrin complex-loaded lyophilized microparticles for intranasal application. *Drug Development and Industrial Pharmacy.* **40**(6), 743–748.

Kim, S. G., Kweon, D. H., Lee, D. H., Park, Y. C. and Seo, J. H. (2005). Coexpression of folding accessory proteins for production of active cyclodextrin glycosyltransferase of *Bacillus macerans* in recombinant *Escherichia coli. Protein Expression and Purification.* **41**(2), 426–432.

Kisfali, P., Polgar, N., Safrany, E., Sumegi, K., Melegh, B. I., Bene, J., Weber, A., Hetyesy, K. and Melegh, B. (2010). Triglyceride level affecting shared susceptibility genes in metabolic syndrome and coronary artery disease. *Current Medicinal Chemistry.* **17**(30), 3533–3541.

Kitahata, S., Tsuyama, N. and Okada, S. (1974). Purification and some properties of cyclodextrin glycosyltransferase from a strain of *Bacillus* species. *Agricultural and Biological Chemistry.* **38**, 387–393.

Kondo, H., Nakatani, H. and Hiromi, K. (1990). In vitro action of human and porcine alpha-amylases on cyclomalto-oligosaccharides. *Carbohydrate Research.* **204**, 207–213.

Kulkarni, A., DeFrees, K., Schuldt, R. A., Hyun, S. H., Wright, K. J., Yerneni, C. K., VerHeul, R. and Thompson, D. H. (2013). Cationic alpha-cyclodextrin:poly(ethylene glycol) polyrotaxanes for siRNA delivery. *Molecular Pharmaceutics.* **10**(4), 1299–1305.

Kuo, C. C., Lin, C. A., Chen, J. Y., Lin, M. T. and Duan, K. J. (2009). Production of cyclodextrin glucanotransferase from an alkalophilic *Bacillus* sp. by pH-stat fed-batch fermentation. *Biotechnology Letters.* **31**(11), 1723–1727.

Kwon, M. J., Park, S. L., Kim, S. K. and Nam, S. W. (2002). Overproduction of *Bacillus macerans* cyclodextrin glucanotransferase in *E. coli* by coexpression of GroEL/ES chaperone. *Journal of Microbiology and Biotechnology.* **12**(6), 1002–1005.

Li, B., Wang, L., Su, L., Chen, S., Li, Z., Chen, J. and Wu, J. (2012). Glycine and Triton X-100 enhanced secretion of recombinant α-CGTase mediated by OmpA signal peptide in *Escherichia coli*. *Biotechnology and Bioprocess Engineering.* **17**, 1128–1134.

Li, Z., Gu, Z., Wang, M., Du, G., Wu, J. and Chen, J. (2010). Delayed supplementation of glycine enhances extracellular secretion of the recombinant alpha-cyclodextrin glycosyltransferase in *Escherichia coli*. *Applied Microbiology and Biotechnology.* **85**, 553–561.

Li, Z., Li, B., Gu, Z., Du, G., Wu, J. and Chen, J. (2010). Extracellular expression and biochemical characterization of alpha-cyclodextrin glycosyltransferase from *Paenibacillus macerans*. *Carbohydrate Research.* **345**, 886–892.

Li, Z., Li, B., Liu, Z., Wang, M., Gu, Z., Du, G., Wu, J. and Chen, J. (2009). Calcium leads to further increase in glycine-enhanced extracellular secretion of recombinant alpha-cyclodextrin glycosyltransferase in *Escherichia coli*. *Journal of Agricultural and Food Chemistry.* **57**, 6231–6237.

Li, Z. F., Wang, M., Wang, F., Gu, Z. B., Du, G. C., Wu, J. and Chen, J. (2007). gamma-Cyclodextrin: A review on enzymatic production and applications. *Applied Microbiology and Biotechnology.* **77**(2), 245–255.

Li, Z., Zhang, J., Wang, M., Gu, Z., Du, G., Li, J., Wu, J. and Chen, J. (2009). Mutations at subsite -3 in cyclodextrin glycosyltransferase from *Paenibacillus macerans* enhancing alpha-cyclodextrin specificity. *Applied Microbiology and Biotechnology.* **83**, 483–490.

Lin, D. G. and Jeang, C. L. (1998). Extracellular production of *Bacillus macerans* authentic cyclodextrin glucanotransferase in *Bacillus subtilis*. *Journal of the Chinese Agricultural Chemical Society.* **36**, 640–651.

Linert, W., Margl, P. and Renz, F. (1992). Solute solvent interactions between cyclodextrin and water—A molecular mechanical study. *Chemical Physics*. **161**(3), 327–338.

Liu, B., Zhang, J., Li, B., Liao, X., Du, G. and Chen, J. (2013). Expression and characterization of extreme alkaline, oxidation-resistant keratinase from *Bacillus licheniformis* in recombinant *Bacillus subtilis* WB600 expression system and its application in wool fiber processing. *World Journal of Microbiology and Biotechnology*. **29**, 825–832.

Lopez-Nicolas, J. M., Andreu-Sevilla, A. J., Carbonell-Barrachina, A. A. and Garcia-Carmona, F. (2009). Effects of addition of alpha-cyclodextrin on the sensory quality, volatile compounds, and color parameters of fresh pear juice. *Journal of Agricultural and Food Chemistry*. **57**(20), 9668–9675.

McGowan, M. W., Artiss, J. D., Strandbergh, D. R. and Zak, B. (1983). A peroxidase-coupled method for the colorimetric determination of serum triglycerides. *Clinical Chemistry*. **29**(3), 538–542.

Messner, M., Kurkov, S. V., Flavia-Piera, R., Brewster, M. E. and Loftsson, T. (2011). Self-assembly of cyclodextrins: The effect of the guest molecule. *International Journal of Pharmaceutics*. **408**(1–2), 235–247.

Pavlov, G. M., Korneeva, E. V. and Smolina, N. A. (2009). Molecular characteristics of cyclodextrins as revealed by studies of their solutions. *Biophysics*. **54**, 450–454.

Penninga, D., Strokopytov, B., Rozeboom, H. J., Lawson, C. L., Dijkstra, B. W., Bergsma, J. and Dijkhuizen, L. (1995). Site-directed mutations in tyrosine-195 of cyclodextrin glycosyltransferase from *Bacillus–Circulans* strain-251 affect activity and product specificity. *Biochemistry*. **34**(10), 3368–3376.

Pinto, F. S. T., Flores, S. H., Ayub, M. A. Z. and Hertz, P. F. (2007). Production of cyclodextrin glycosyltransferase by alkaliphilic *Bacillus circulans* in submerged and solid-state cultivation. *Bioprocess and Biosystems Engineering*. **30**(5), 377–382.

Qi, Q. and Zimmermann, W. (2005). Cyclodextrin glucanotransferase: From gene to applications. *Applied Microbiology and Biotechnology*. **66**(5), 475–485.

Rashid, N., Cornista, J., Ezaki, S., Fukui, T., Atomi, H. and Imanaka, T. (2002). Characterization of an archaeal cyclodextrin glucanotransferase with a novel C-terminal domain. *Journal of Bacteriology*. **184**, 777–784.

Rendleman, J. A. J. (1997). Enhancement of cyclodextrin production through use of debranching enzymes. *Applied Biochemistry and Biotechnology*. **26**, 51–61.

Rosso, A. M., Ferrarotti, S. A., Krymkiewicz, N. and Nudel, B. C. (2002). Optimisation of batch culture conditions for cyclodextrin glucanotransferase production from *Bacillus circulans* DF 9R. *Microbial Cell Factories*. **1**(1), 1–9.

Sabadini, E., Cosgrove, T. and Egidio, F. C. (2006). Solubility of cyclomalto-oligosaccharides (cyclodextrins) in H2O and D2O: A comparative study. *Carbohydrate Research*. **341**(2), 270–274.

Saha, B. C. and Zeikus, J. G. (1992). Cyclodextrin degrading enzymes. *Starch—Starke*. **44**(8), 312–315.

Schmid, G. (2009). *Comprehensive Supramolecular Chemistry* (1st ed.). New York: Pergamon, pp. 41–55.

Shieh, W. J. and Hedges, A. R. (1994). Process for producing alpha-cyclodextrin using cyclomaltodextrin glucanotransferase in presence of cyclohexane. United States Patent. US 5,326,701.

Shiga, H., Yoshii, H., Ohe, H., Yasuda, M., Furuta, T., Kuwahara, H., Ohkawara, M. and Linko, P. (2004). Encapsulation of shiitake (*Lenthinus edodes*) flavors by spray drying. *Bioscience, Biotechnology and Biochemistry*. **68**(1), 66–71.

Sjostrom, L., Rissanen, A., Andersen, T., Boldrin, M., Golay, A., Koppeschaar, H. P. and Krempf, M. (1998). Randomised placebo-controlled trial of orlistat for weight loss and prevention of weight regain in obese patients. European Multicentre Orlistat Study Group. *Lancet*. **352**(9123), 167–172.

Strokopytov, B., Knegtel, R. M. A., Penninga, D., Rozeboom, H. J., Kalk, K. H., Dijkhuizen, L. and Dijkstra, B. W. (1996). Structure of cyclodextrin glycosyltransferase complexed with a maltononaose inhibitor at 2.6 angstrom resolution. Implications for product specificity. *Biochemistry*. **35**(13), 4241–4249.

Suzuki, M. and Sato, A. (1985). Nutritional significance of cyclodextrins: Indigestibility and hypolipemic effect of alpha-cyclodextrin. *Journal of Nutritional Science and Vitaminology* **31**(2), 209–223.

Szejtli, J. (1982). Cyclodextrins and their inclusion complexes. *Akadémiai Kiadó* **62**(43):1–18.

Szejtli, J. (1988). *Cyclodextrin Technology*. Dordrecht: Kluwer Academic Publishers.

Szente, L. and Szejtli, J. (2004). Cyclodextrins as food ingredients. *Trends in Food Science and Technology*. **15**(3–4), 137–142.

Tahir, M. N. and Lee, Y. (2013). Immobilisation of beta-cyclodextrin on glass: Characterisation and application for cholesterol reduction from milk. *Food Chemistry*. **139**(1–4), 475–481.

Takano, T., Fukuda, M., Monma, M., Kobayashi, S., Kainuma, K. and Yamane, K. (1986). Molecular cloning, DNA nucleotide sequencing, and expression in *Bacillus subtilis* cells of the *Bacillus macerans* cyclodextrin glucanotransferase gene. *Journal of Bacteriology*. **166**, 1118–1122.

Terada, Y., Yanase, M., Takata, H., Takaha, T. and Okada, S. (1997). Cyclodextrins are not the major cyclic alpha-1,4-glucans produced by the initial action of cyclodextrin glucanotransferase on amylose. *Journal of Biological Chemistry*. **272**, 15729–15733.

Tesfai, B. T., Wu, D., Chen, S., Chen, J. and Wu, J. (2012). Strategies for enhancing extracellular secretion of recombinant cyclodextrin glucanotransferase in *E. coli*. *Applied Biochemistry and Biotechnology*. **167**, 897–908.

Threapleton, D. E., Greenwood, D. C., Evans, C. E. L., Cleghorn, C. L., Nykjaer, C., Woodhead, C., Cade, J. E., Gale, C. P. and Burley, V. J. (2013). Dietary fiber intake and risk of first stroke a systematic review and meta-analysis. *Stroke*. **44**(5), 1360–1368.

Tonkova, A. (1998). Bacterial cyclodextrin glucanotransferase. *Enzyme and Microbial Technology*. **22**(8), 678–686.

Truong, V. T., Boyer, R. R., McKinney, J. M., O'Keefe, S. F. and Williams, R. C. (2010). Effect of alpha-cyclodextrin-cinnamic acid inclusion complexes on populations of *Escherichia coli* O157:H7 and *Salmonella enterica* in fruit juices. *Journal of Food Protection*. **73**(1), 92–96.

van der Veen, B. A., Uitdehaag, J. C. M., Dijkstra, B. W. and Dijkhuizen, L. (2000a). Engineering of cyclodextrin glycosyltransferase reaction and product specificity. *Biochimica et Biophysica Acta—Proteins and Proteomics*. **1543**(2), 336–360.

van der Veen, B. A., Uitdehaag, J. C. M., Dijkstra, B. W. and Dijkhuizen, L. (2000b). The role of arginine 47 in the cyclization and coupling reactions of cyclodextrin glycosyltransferase from *Bacillus circulans* strain 251—Implications for product inhibition and product specificity. *European Journal of Biochemistry*. **267**(12), 3432–3441.

van der Veen, B. A., Uitdehaag, J. C. M., Penninga, D., van Alebeek, G. J. W. M., Smith, L. M., Dijkstra, B. W. and Dijkhuizen, L. (2000c). Rational design of cyclodextrin glycosyltransferase from *Bacillus circulans* strain 251 to increase alpha-cyclodextrin production. *Journal of Molecular Biology*. **296**(4), 1027–1038.

van der Veen, B. A., van Alebeek, G. J., Uitdehaag, J. C., Dijkstra, B. W. and Dijkhuizen, L. (2000). The three transglycosylation reactions catalyzed by cyclodextrin glycosyltransferase from *Bacillus circulans* (strain 251) proceed via different kinetic mechanisms. *European Journal of Biochemistry*. **267**(3), 658–665.

Vollu, R. E., da Mota, F. F., Gomes, E. A. and Seldin, L. (2008). Cyclodextrin production and genetic characterization of cyclodextrin glucanotransferase of *Paenibacillus graminis*. *Biotechnology Letters*. **30**(5), 929–935.

Waalkens-Berendsen, D. H. and Bar, A. (2004). Embryotoxicity and teratogenicity study with neohesperidin dihydrochalcone in rats. *Regulatory Toxicology and Pharmacology*. **40**(1), 74–79.

Waalkens-Berendsen, D. H., Smits-Van Prooije, A. E. and Bar, A. (2004). Embryotoxicity and teratogenicity study with alpha-cyclodextrin in rabbits. *Regulatory Toxicology and Pharmacology*, **39**(Suppl. 1), S40–S46.

Wagner, E. M. A., Jen, K. L. C., Artiss, J. D. and Remaley, A. T. (2008). Dietary alpha-cyclodextrin lowers low-density lipoprotein cholesterol and alters plasma fatty acid profile in low-density lipoprotein receptor knockout mice on a high-fat diet. *Metabolism—Clinical and Experimental*. **57**(8), 1046–1051.

WHO. (2002). Safety evaluation of certain food additives and contaminants, α-cyclodextrin. In *WHO Food Additives Series*. **48**, 111–127.

Wind, R. D., Liebl, W., Buitelaar, R. M., Penninga, D., Spreinat, A., Dijkhuizen, L. and Bahl, H. (1995). Cyclodextrin formation by the thermostable alpha-amylase of *Thermoanaerobacterium thermosulfurigenes* EM1 and reclassification of the enzyme as a cyclodextrin glycosyltransferase. *Applied and Environmental Microbiology*. **61**(4), 1257–1265.

Wind, R. D., Uitdehaag, J. C. M., Buitelaar, R. M., Dijkstra, B. W. and Dijkhuizen, L. (1998). Engineering of cyclodextrin product specificity and pH optima of the thermostable cyclodextrin glycosyltransferase from *Thermoanaerobacterium thermosulfurigenes* EM1. *Journal of Biological Chemistry*. **273**(10), 5771–5779.

Zhang, Z. Z., Xu, G. L., Liu, D. Z., Zhu, W. S., Fan, X. Y. and Liu, X. F. (2013a). Dietary fiber consumption and risk of stroke. *European Journal of Epidemiology*. **28**(2), 119–130.

Zhang, Z. Z., Xu, G. L., Ma, M. M., Yang, J. and Liu, X. F. (2013b). Dietary fiber intake reduces risk for gastric cancer: A meta-analysis. *Gastroenterology*. **145**(1), 113–120.

Zhou, J., Liu, H., Du, G., Li, J. and Chen, J. (2012). Production of alpha-cyclodextrin glycosyltransferase in *Bacillus megaterium* MS941 by systematic codon usage optimization. *Journal of Agricultural and Food Chemistry*. **60**(41), 10285–10292.

Zhu, W., Zhang, K., Chen, Y. M. and Xi, F. (2013). Simple, clean preparation method for cross-linked alpha-cyclodextrin nanoparticles via inclusion complexation. *Langmuir*. **29**(20), 5939–5943.

5

Recent Progress on Galactooligosaccharides Synthesis by Microbial β-Galactosidase

LILI LU AND MIN XIAO

Contents

5.1 Introduction

Galactooligosaccharides (GOS) occur naturally in trace amounts in human milk (Lamsal 2012; Rodriguez-Colinas et al. 2013). *Bifidobacterial* intestinal microbiota appears to be established by the use of human milk oligosaccharides (Newburg 2000; Boehm et al. 2005; Bode 2012; Barile and Rastall 2013). Commercial GOS resembling the natural oligosaccharides present in human milk can be produced from lactose by enzymes (Mahoney 1998; Niittynen et al. 2007). These GOS products have been assessed to be one of the most attractive prebiotics that selectively stimulate the proliferation of intestinal probiotics (Sako et al. 1999; Intanon et al. 2014). They have GRAS (generally recognized as safe) status in the United States and FOSHU (food for specified health use) status in Japan (Torres et al. 2010).

GOS can result in substantially high growth of total bifidobacteria in in vitro cultures of infant fecal samples, and much of this growth is attributed to growth of *Bifidobacterium longum* (Stiverson et al. 2014). Additionally, GOS cause the largest decrease in harmful Clostridia, higher short-chain fatty acid generation, and lower gas production when compared with Raftilose P95, Raftiline LS, lactulose, xylo-, isomalto-, and soybean oligosaccharides (Rycroft et al. 2001). Moreover, GOS can act as decoy agents for the pathogen instead of host cell surface oligosaccharides and help flush the pathogen from the gastrointestinal tract (Searle et al. 2009, 2010; Sinclair et al. 2009). In short, GOS has emerged with strong clinical support for both digestive and immune health (Sangwan et al. 2011). As a result, a great deal of attention has been devoted to GOS synthesis, especially via enzymatic transglycosylation since chemical synthesis of GOS is very tedious (Gosling et al. 2010; Park and Oh 2010; Torres et al. 2010; Diez-Municio et al. 2014). This chapter describes the mechanism and development of GOS synthesis by microbial enzymes in recent years.

5.2 Mechanism of GOS Synthesis by β-Galactosidase

GOS are mainly synthesized by microbial β-galactosidases (EC 3.2.1.23) from lactose (Gosling et al. 2010; Torres et al. 2010). These enzymes are classified into four distinct glycosyl hydrolase (GH) families, GH1, GH2, GH35, and GH42, based on sequence similarities.* Besides β-galactosidases, other GH1 glycosidases such as β-glucosidases possessing broad substrate specificities can also catalyze transglycosylation from lactose (Akiyama et al. 2001; Hansson et al. 2001; Hassan et al. 2015, 2016).

The enzymes for GOS synthesis are usually retaining glycosidases from which the product retains the same stereochemistry as the starting state (Juers et al. 2001; Brás et al. 2010). In particular, retaining glycosidases proceed through a double displacement mechanism in which two key carboxylic acid residues act as nucleophile and acid–base catalysts (McCarter and Withers 1994; Rye and Withers 2000). In the first step, the acid–base residue serves as a general acid and provides protonic assistance to departure of glycosidic oxygen when the nucleophile residue attacks the substrate at its anomeric center, leading

* The section of EC Activities found in GH families in glycoside hydrolase family classification shown at the CAZy database (http://www.cazy.org/Glycoside-Hydrolases.html).

to formation of a covalent glycosyl-enzyme intermediate. In the second step, the acid–base residue, now functioning as a general base, activates an acceptor molecule that attacks the intermediate at the anomeric center, generating a product with the same anomeric configuration as the substrate. The acceptor can be either water or an alcohol, e.g., a sugar, resulting in hydrolysis or transglycosylation (Figure 5.1). Nevertheless, hydrolysis of transglycosylation products is inevitable since they are normal substrates for glycosidases (Perugino et al. 2004).

For β-galactosidases, they catalyze both lactose hydrolysis and GOS synthesis. Transglycosylation occurs by galactosyl transfer from lactose to lactose itself, or its hydrolysis products, galactose or glucose, or the newly produced oligosaccharides. The resulting reaction products are mixtures of monosaccharides, residual lactose, and GOS.

Currently, a series of crystallographic structures of β-galactosidases from different origins have been investigated for probing their catalytic mechanism (Table 5.1). The β-galactosidase from *Escherichia coli* (LacZ), which belongs to GH2, is one of the most clearly elucidated glycosidases with sufficient crystallographic structures. High-resolution structures of this enzyme are also available by using cryoelectron microscopy (Bartesaghi et al. 2014, 2015).

The LacZ enzyme is a homotetramer with 222 point symmetry. Each subunit consists of five domains, and the active center is a deep pocket built around the central (α/β)8 barrel with other domains apparently recruited to confer specificity for a disaccharide substrate (Juers et al. 2000, 2001). The key catalytic residues E461 (acid–base) and E537 (nucleophile) are approximately 5.5 Å apart (Juers et al. 2001; Bartesaghi et al. 2015). During the reaction, a complex network of hydrogen interactions with the N102, D201, H391, N460, E461, and E537 residues determines the binding pose of lactose. The

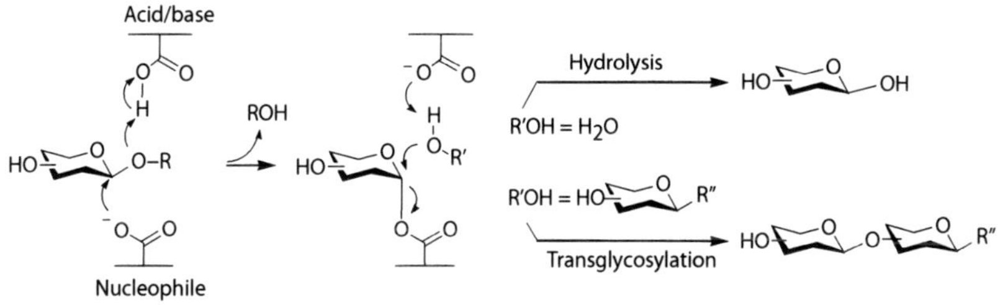

Figure 5.1 General mechanism for retaining glycosidases.

Table 5.1 Examples of β-Galactosidases with Crystal Structures

MICROBIAL SOURCE	GH	ENZYME	LIGAND IN ENZYME COMPLEX	PDB	REFERENCE
Arthrobacter sp. C2-2	2	C221-β-Gal	Free	1YQ2	Skálová et al. 2005
Bacillus circulans ATCC 31382	35	BgaC	Free	4MAD	Henze et al. 2014
Bacillus circulans sp. *alkalophilus*	42	Bca-β-gal	Free	3TTS	Maksimainen et al. 2012
			Galactose	3TTY	
Bifidobacterium bifidum	–	β-galactosidase	Free	4UZS	Unpublished
Escherichia coli	2	LacZ	2-Phenylethyl β-thiogalactoside	5A1A*	Bartesaghi et al. 2015
			Free	3J7H*	Bartesaghi et al. 2014
			Free	1DP0	Juers et al. 2000
			IPTG	1JYX	Juers et al. 2001
			2- Deoxy-2-fluoro-β-galactose	1JZ2	
			2-Deoxy-β-galactose	1JZ4	
			Galactonolactone	1JZ5	
			Galactotetrazole	1JZ6	
			Galactose	1JZ7	
		LacZ E537Q	Lactose	1JYN	
			Allolactose	1JZ8	
		LacZ N460D	Free	3VD3	Wheatley et al. 2012
		LacZ N460S	Free	3VD5	
		LacZ N460T	Free	3VDA	
		LacZ M542A	Free	3I3E	Dugdale et al. 2010a
		LacZ R599A	Free	3MUY	Dugdale et al. 2010b

(*Continued*)

Table 5.1 (Continued) Examples of β-Galactosidases with Crystal Structures

MICROBIAL SOURCE	GH	ENZYME	LIGAND IN ENZYME COMPLEX	PDB	REFERENCE
		LacZ G794A	Allolactose	4DUW	Wheatley et al. 2013
		LacZ S796A	Free	3SEP	Jancewicz et al. 2012
		LacZ S796T	Free	3T0A	
		LacZ S796D	Free	3T2O	
Geobacillus stearothermophilus T-6	42	GanB	Free	4OIF	Solomon et al. 2013
		GanB E323A	Free	4OJY	
Kluyveromyces lactis	2	KL-β-Gal	Free	3OBA	Pereira-Rodríguez et al. 2012
			Galactose	3OB8	
Penicillium sp.	35	LacA	Free	1TG7	Rojas et al. 2004
			Galactose	1XC6	
Sulfolobus solfataricus	1	LacS	Free	1UWQ	Gloster et al. 2004
			2-Deoxy-2-fluoro-galactose	1UWR	
			Galactohydroximo-1,5-lactam	1UWT	
Thermus thermophilus A4	42	A4-β-Gal	Free	1KWG	Hidaka et al. 2002
			Galactose	1KWK	
Trichoderma reesei	35	Tr-β-Gal	Free	3OG2	Maksimainen et al. 2011
			Galactose	3OGR	
			IPTG	3OGS	
			2-Phenylethyl β-thiogalactoside	3OGV	

*Structures obtained by cryo-electron microscopy.

lactose initially binds in a shallow mode of the enzyme, characterized by stacking on W999, and then it moves into a deep mode, further into the active site. The side chain of F601 rotates and the 794–804 loop also rearranges to allow this transition (Brás et al. 2010).

The LacZ predominantly produced allolactose (6′-galactosylglucose) by greater than 97% over other disaccharides (Huber et al. 1976). A possible explanation for this specificity is that 6-hydroxy of glucose, which has an extra atom linking it to the galactose ring, can reach further into the active site pocket and more easily attack the galactosyl-enzyme intermediate, when compared to the 1-, 2-, 3-, or 4-positions that appear sterically difficult (Juers et al. 2001). Also the formation of the β-1,6-glycosidic bond has the most negative value of the Gibbs energy, and this transglycosylating reaction is thermodynamically more favorable than others (Brás et al. 2010). The mechanism of the transglycosylation preference for β-1,3 or β-1,4 linkages by other β-galactosidases are still unknown.

5.3 Enzyme Source and GOS Structures Produced

β-Galactosidase naturally occurs in a wide variety of organisms including microorganisms, plants, and animals (Oliveira et al. 2011). Microbial sources of β-galactosidases are of great technological interest due to various advantages such as easy handling, high multiplication rate, and good production yield (Panesar et al. 2010). A large number of microorganisms from archaea, bacteria, and eukaryota have been assessed as potential enzyme sources (Gosling et al. 2010; Park and Oh 2010; Torres et al. 2010). *Bifidobacteria* and *Lactobacilli* have attracted particularly intense interest these years, possibly for the reason that GOS produced by their β-galactosidases might have better selectivity for growth and metabolic activity of these bacterial genera in the gut and thus will lead to improved prebiotic effects (Rabiu et al. 2001; Tzortzis et al. 2005a,b; Splechtna et al. 2006, 2007; Hsu et al. 2007; Tzortzis and Vulevic 2009; Lu et al. 2010; Nguyen et al. 2012; Arreola et al. 2014; Intanon et al. 2014).

The β-galactosidases from different sources vary in the regioselectivity for transgalactosylation from lactose (Table 5.2). Among commercially important β-galactosidases, the enzymes from *Aspergillus oryzae*, *Kluyveromyces fragilis*, *Kluyveromyces lactis*, and *Streptococcus thermophilus* predominantly form β-1,6 linkage, whereas the enzyme

Table 5.2 Examples of Identified Structures of GOS Produced by Microbial Enzymes from Lactose

MICROORGANISM	DP	KNOWN STRUCTURES OF GOS	REFERENCES
Aspergillus oryzae	DP3	Gal-β-1,6-β-Gal-1,6-Glc; Gal-β-1,4-β-Gal-1,6-Glc; Gal-β-1,6-Gal-β-1,4-Glc; Gal-β-1,4-Gal-β-1,4-Glc; Gal-β-1,3-Gal-β-1,4-Glc	Nakakuki 1993
	DP4	Gal-β-1,6-β-Gal-β-1,6-Gal-β-1,4-Glc; Gal-β-1,3-Gal-β-1,6-Gal-β-1,4-Glc	
	DP5	Gal-β-1,6-Gal-β-1,6-Gal-β-1,6-Gal-β-1,4-Glc	
Bacillus circulans	DP2	Gal-β-1,3-Glc; Gal-β-1,6-Glc; Gal-β-1,2-Glc	Yanahira et al. 1995
	DP3	Gal-β-1,4-Gal-β-1,4-Glc; Gal-β-1,6-[Gal-β-1,2]-Glc; Gal-β-1,6-[Gal-β-1,4]-Glc;Gal-β-1,4-Gal-β-1,3-Glc; Gal-β-1,4-Gal-β-1,2-Glc; Gal-β-1,4-[Gal-β-1,2]-Glc; Gal-β-1,4-Gal-β-1,6-Glc; Gal-β-1,6-[Gal-β-1,3]-Glc	
Bifidobacterium bifidum (pH 4.2~4.5)	DP2	Gal-β-1,3 Glc; Gal-β-1,6-Glc; Gal-β-1,6-Gal	Dumortier et al. 1990
	DP3	Gal-β-1,3-Gal-β-1,4-Glc; Gal-β-1,6-[Gal-β-1,4]-Glc; Gal-β-1,2-[Gal-1,6]-β-Glc	
	DP4	Gal-β-1,3-Gal-β-1,3-Gal-β-1,4-Glc	
	DP5	Gal-β-1,3-Gal-β-1,3-Gal-β-1,3-Gal-β-1,4-Clc	
	DP6	Gal-β-1,3-Gal-β-1,3-Gal-β-1,3-Gal-β-1,3-Gal-β-1,4-Glc	
	DP7	Gal-β-1,3-Gal-β-1,3-Gal-β-1,3-Gal-β-1,3-Gal-β-1,3-Gal-β-1,4-Glc	
Bifidobacterium bifidum NCIMB41171 (pH 6.8)	DP2	Gal-β-1,6-Gal; Gal-β-1,3-Glc; Gal-β-1,3-Gal	Tzortzis et al. 2005a
	DP3	Gal-β-1,6-Gal-β-1,4-Glc; Gal-β-1,3-Gal-β-1,4-Glc	
	DP4	Gal-β-1,6-Gal-β-1,6-Gal-β-1,4-Glc	
	DP5	Gal-β-1,6-Gal-β-1,6-Gal-β 1,6-Gal-β-1,4-Glc	
Cryptococcus laurentii OKN-4	DP3	Gal-β-1,4-Gal-β-1,4-Glc	Ohtsuka et al. 1990
Lactobacillusreuteri L103	DP2	Gal-β-1,6-Glc; Gal-β-1,6-Gal; Gal-β-1,3-Gal	Splechtna et al. 2006
	DP3	Gal-β-1,6-Gal-β-1,4-Glc; Gal-β-1,3-Gal-β-1,4-Glc	
Sterigmatomyces elviae CBS8119	DP3	Gal-β-1,4-Gal-β-1,4-Glc	Onishi et al. 1995
Streptococcus thermophillus	DP2	Gal-β-1,6-Glc; Gal-β-1,6-Gal; Gal-β-1,3-Glc;Gal-β-1,2-Glc	Nakakuki 1993

from *Bacillus circulans* and *Cryptococcus laurentii* prefer β-1,4 linkage (Ohtsuka et al. 1990; Nakakuki 1993; Yanahira et al. 1995; Gosling et al. 2010). Interestingly, β-galactosidases from strains of *B. bifidum* differ in preferences for β-1,3 or β-1,6 linkages under different reaction conditions (Dumortier et al. 1990; Tzortzis et al. 2005a). Correspondingly, famous commercial GOS products produced from various origins, such as Oigomate® 55 from *S. thermophilus* and *A. oryzae* (Yakult Honsha, Tokyo, Japan), CUP-Oligo® from *C. laurentii* (Nissin Sugar Manufacturing Company, Japan), Vivinal® GOS from *B. circulans* (Borcolo Domo Ingredients, Zwolle, The Netherlands), and Bimuno® from *B. bifidum* (Clasado Ltd, Milton, Keynes, UK), show different glycosidic linkages. Additionally, the proportion of monosaccharides, lactose, and GOS varies among these products. The GOS contents are 50–60% in Oigomate® 55 syrup, 70% in CUP-Oligo® syrup, 57–59% in Vivinal® GOS, and 48–55% in Bimuno® syrup (Torres et al. 2010; Lamsal 2012).

It is noteworthy that the composition and structure of pure GOS are always complex and vary according to enzyme source. Even in the case of a reaction with one enzyme, there are many geometrical isobaric oligomers formed (Table 5.2). Pure GOS synthesized by enzymes are still mixtures of linear and branched oligosaccharides with DPs of 2–8 and different types of glycosidic linkages. As physicochemical properties of these isomers are greatly similar, it is rather difficult to isolate them in large scale. For most GOS products, only major components (low DP) or linkage patterns have been described (van Leeuwen et al. 2014b). Structural characterization of a complex mixture of highly similar oligosaccharides with different linkage types often poses a challenge (van Leeuwen et al. 2014a,b).

Vivinal® GOS has been identified to comprise over 40 structures, covering over 99% of transglycosylated saccharides. This product contains 18.5% glucose, 1.7% galactose, 42.5% DP2, 23.6% DP3, 10.2% DP4, 3.0% DP5, and 0.5% DP6 and higher fractions. It was first subjected to DP analysis using HPLC and ^{1}H NMR analysis, and then fractionated into different DPs using size exclusion chromatography (SEC). Fractions of single DP, as well as their subfractions obtained by high-performance anion-exchange chromatography with pulsed amperometric detection (HPAEC-PAD), were analyzed by NMR spectroscopy. Detailed structural identifications of Vivinal® GOS involve 4 of transglycosylated DP2, 9 of DP3, 15 of DP4, and 5 of DP5. The library of NMR data established will facilitate the fast characterization of GOS of different origin (van Leeuwen et al. 2014a).

HPAEC-PAD is an analytical method currently used to measure GOS in food and feed products (De Slegte 2002), allowing mixtures of simple sugars and oligosaccharides to be separated with high resolution in a single run (Corradini et al. 2012). Recently, the capillary electrophoresis system together with a fused silica capillary column is attractive to distinguish GOS varying in DPs (Splechtna et al. 2006; Martínez-Villaluenga et al. 2008). Introducing capillary electrophoresis with laser-induced fluorescence detection can further facilitate sensitively qualitative and quantitative determination of the GOS contents in different food matrices (Albrecht et al. 2010). More strikingly, a recent mass spectrometry approach involving matrix-assisted laser desorption ionization–Fourier transform ion cyclotron resonance (MALDI-FTICR) has been developed for comprehensive profiling of oligosaccharides in GOS mixtures, with high resolution of detection of GOS with DP up to 15 (Barboza et al. 2009).

5.4 Influence of Reaction Conditions on GOS Production

Microbial β-galactosidases for GOS production include native crude or purified enzymes, recombinant enzymes, immobilized enzymes, and whole-cell biotransformations (Tables 5.3 and 5.4). Reactions can start from lactose or lactose-containing products such as milk and whey (Yi et al. 2011; Jovanovic-Malinovska et al. 2012; Rodriguez-Colinas et al. 2012, 2014; Golowczyc et al. 2013). The properties of the enzymes for GOS synthesis differ markedly according to their sources (Table 5.3). The yield of GOS (usually defined as a percentage of the initial lactose) varies widely, typically between 20% and 50% (w/w), whereas sometimes it exceeds 50%. The effects of the enzyme sources as well as the process parameters on the yield and the productivity of GOS production have been summarized in detail in recent reviews (Gosling et al. 2010; Torres et al. 2010; Sangwan et al. 2011; Intanon et al. 2014).

The maximal GOS yield can be influenced by reaction conditions such as lactose concentration, temperature, pH, and the reaction time. The initial lactose concentration is considered to be one of the key

Table 5.3 Microbial β-Galactosidases for GOS Production

ENZYME SOURCE	LACTOSE (g/L)	T (°C)	pH	YIELD (%)	PRODUCTIVITY (g/L/h)	REFERENCE
NATIVE ENZYMES						
Bacillus circulans	500	60	6.0	38.3	462.2	Urrutia et al. 2014
Bifidobacterium adolescentis ANB-7	300	55	7.5	43.1	18.5	Rabiu et al. 2001
Bifidobacterium infantis DSM-20088	300	55	7.5	47.6	20.4	Rabiu et al. 2001
Bifidobacterium angulatum	300	55	7.5	43.8	18.8	Rabiu et al. 2001
Bifidobacterium longum BCRC 15708	400	45	6.8	32.5	13.0	Hsu et al. 2007
Bifidobacterium pseudolongum DSM-20099	300	55	7.5	26.8	11.5	Rabiu et al. 2001
Bullera singularis KCTC 7534	180	50	5.0	50.0	3.9	Cho et al. 2003
Enterobacter agglomerans B1	125	50	7.5	38.0	4.0	Lu et al. 2007
Lactobacillus acidophilus R22	205	30	6.5	38.5	7.9	Nguyen et al. 2007
Lactobacillus plantarum WCFS1	205	37	6.5	41.0	10.5	Iqbal et al. 2010
Lactobacillus reuteri L103	205	30	6.5	37.8	3.9	Splechtna et al. 2006
Penicillium simplicissimum	600	50	6.5	30.5	11.4	Cruz et al. 1999
Saccharopolyspora rectivirgula	600	70	7.0	41.0	11.2	Nakao et al 1994
RECOMBINANT ENZYMES IN *E. COLI*						
Bifidobacterium breve DSM 20213	200	30	6.5	33.0	11.0	Arreola et al. 2014
Bacillus licheniformis DSM 13	200	60	6.5	12.0	–	Juajun et al. 2011
Bifidobacterium infantis HL96	300	60	7.5	63.3	12.7	Hung and Lee 2002
*Halothermothrix orenii**	400	70	7.0	46.3	61.7	Hassan et al. 2015b
Lactobacillus bulgaricus DSM 20081**	205	50	6.5	48.2	19.8	Nguyen et al. 2012

(*Continued*)

Table 5.3 (Continued) Microbial β-Galactosidases for GOS Production

ENZYME SOURCE	LACTOSE (g/L)	T (°C)	pH	YIELD (%)	PRODUCTIVITY (g/L/h)	REFERENCE
Lactobacillus sakei Lb790	215	37	6.5	41.0	29.7	Iqbal et al. 2011
Sulfolobus solfataricus	600	80	6.0	52.5	5.6	Park et al. 2008
Thermotoga maritime ATCC43589	500	80	6.0	18.2	18.2	Ji et al. 2005
Thermus sp. Z-1*	300	70	7.0	30.0	–	Akiyama et al. 2001
CELLS						
Bifidobacterium bifidum NCIMB 41171 (cells)	500	39	6.8	19.8	66.0	Tzortzis et al. 2005b
Enterobacter cloacae B5 (freeze-thawed cells)	275	50	7.0	55.0	12.6	Lu et al. 2009
Kluyveromyces marxianus NCIM 3551 (ethanol permeabilized cells)	200	40	6.5	36.0	24.0	Srivastava et al. 2015
Lactococcus lactis (ethanol permeabilized cells)	400	85	6.0	49.3	3.5	Yu and O'Sullivan 2014
Sirobasidium magnum CBS6803 (toluene treated cells)	360	50	6.0	37.8	3.2	Onishi et al. 1996
Sterigmatomyces elviae CBS8119 (toluene treated cells)	360	60	6.0	37.5	6.8	Onishi et al. 1995
S. elviae CBS8119 (in cell culture)	360	30	6.0	60.0	3.2	Onishi and Tanaka 1998

* β-Glucosidase.

** Recombinant enzyme in *Lactobacillus plantarum* WCFS1.

Table 5.4 Immobilized β-Galactosidases for GOS Synthesis

MICROORGANISM	β-GALACTOSIDASE SOURCE	IMMOBILIZATION MATRIX	LACTOSE (g/L)	T (°C)	pH	YIELD (%)	PRODUCTIVITY (g/L/h)	REUSE (CYCLES)*	REFERENCE
Aspergillus oryzae	Genencor international (US) enzyme	Cotton cloth	400	40	4.5	26.0	106.0	2 weeks	Albayrak and Yang 2002
	Bio-Cat (US) enzyme	Cotton cloth	270	40	4.5	20.0	190.6	–	Matella et al. 2006
	Sigma (US) enzyme	Chitosan	200	40	4.0	17.3	17.3	3	Gaur et al. 2006
	Not mentioned	Magnetic polysiloxane-polyvinyl alcohol	500	40	4.5	26.0	130	10	Neri et al. 2009
	Sigma (US) enzyme	Magnetic Fe_3O_4–chitosan	500	40	4.6	15.5	38.8	>15	Pan et al. 2009
	Enzyme Development Corporation (US) enzyme	Glyoxyl-agarose	546	50	4.5	28.7	13.6	5	Huerta et al. 2011
Bacillus circulans	Biolacta FN5, Japan	PVDF membrane	200	40	6.0	28	2.8	1	Palai and Bhattacharya 2013
	Biolacta FN5, Japan	Poly-*N*-isopropyl acrylamide	50	40	6.0	29.4	–	1	Palai et al. 2015
	Biolactasa-NTL CONC X2, Spain	Chitosan	500	60	6.0	39.1	473.7	>10	Urrutia et al. 2014
Bifidobacterium bifidum NCIMB 41171	Recombinant BbgIV in *E. coli*	DEAE-Cellulose	430	55	6.5	50.7	35.1	–	Osmana et al. 2014
		Q-Sepharose	430	55	6.5	51.9	143.5	>6	
		Amino-ethyl agarose	430	55	6.5	44.1	30.5	–	

(*Continued*)

Table 5.4 (Continued) Immobilized β-Galactosidases for GOS Synthesis

MICROORGANISM	β-GALACTOSIDASE SOURCE	IMMOBILIZATION MATRIX	LACTOSE (g/L)	T (°C)	pH	YIELD (%)	PRODUCTIVITY (g/L/h)	REUSE (CYCLES)*	REFERENCE
Bullera singularis ATCC 24193	Partially purified enzyme	Chitosan beads	300	45	3.7	54.0	4.8	15 days	Shin et al. 1998
Lactobacillus bulgaricus L3	CBD fusing enzyme in *E. coli*	Microcrystalline cellulose	400	45	7.6	49.0	156.8	20	Lu et al. 2012
Kluyveromyces fragilis	Partially purified enzyme	Magnetic nanosized epoxy support	200	30	6.5	32	16.0	10	Liu et al. 2012
Kluyveromyces lactis	Novo Nordisk A/S (Denmark) enzyme	Anion exchange resin Duolite A-568	159	40	6.5	27.7	22.0	–	Maugard et al. 2003
	Sigma enzyme	Polystyrene nanofiber	400	37	7	28.0	56.0	2	Misson et al. 2015
Penicillium expansum F3	Crude enzyme	Calcium alginate	380	50	5.4	28.7	3.0	7	Li et al. 2008
Thermus aquaticus YT-1	Crude enzyme	Agarose beads	160	70	4.6	34.8	2.3	–	Berger et al. 1995
Talaromyces thermophilus CBS 236.58	Purified enzyme	Acrylic polymer Eupergit C	200	40	6.5	50	13		Nakkharat et al. 2006

* With above 80% GOS yield or enzyme activity of the first cycle.

factors affecting the maximal GOS concentration achieved (Mahoney 1998). The increase in lactose can increase availability of saccharide acceptors, and decrease availability of water, which would be expected to increase the rate of GOS synthesis, and decrease both the rates of GOS degradation and lactose hydrolysis (Gosling et al. 2010).

Raising the temperature can increase the solubility of lactose, which in turn can improve GOS yield (Gosling et al. 2010). High temperatures also seem to favor transgalactosylation relative to hydrolysis (Torres et al. 2010). Recent research has therefore been directed toward the development of thermostable enzymes for GOS synthesis, including the enzymes from *Halothermothrix orenii* (Hassan et al. 2015), *Lactococcus lactis* (Yu and O'Sullivan 2014), *Pyrococcus furiosus* (Bruins et al. 2003), *Saccharopolyspora rectivirgula* (Nakao et al. 1994), *Sulfolobus solfataricus* (Park et al. 2008), *Thermotoga maritime* (Ji et al. 2005), and *Thermus* sp. (Akiyama et al. 2001). These enzymes are usually prepared in recombinant form from *E. coli* and used for GOS synthesis at over 70°C. Microwave irradiation, a nonconventional heating energy source, has also been successfully attempted for GOS synthesis by using the immobilized β-galactosidase from *K. lactis* (Maugard et al. 2003).

During the reaction catalyzed by the enzymes, hydrolysis of the oligosaccharide product competes with the transglycosylation reaction. The concentration of GOS at its peak is determined by the relative kinetics of synthesis and degradation. Hence, a knowledge of the time course of the reaction is required to determine the point of maximum yield of the desired product (Gosling et al. 2010).

It also should be noted that a large amount of water existing in the reaction system is unfavorable for GOS synthesis. The use of organic solvents such as cyclohexane in aqueous–organic reactions can control water activity to some extent and shift thermodynamic equilibrium to favor synthesis over hydrolysis (Shin and Yang 1994; Wang et al. 2012). Reverse micelles have also been used as an effective low-water media for GOS synthesis. Their structures are self-assembled water pools in organic solution, in which surfactant molecules organize with the polar part to the inner side able to solubilize water and the apolar part in contact with the organic solvent (Melo et al. 2001). Lactose is condensed in micro-aqueous core of the reverse micelles, and GOS formation is enhanced as more lactose competes with water as an acceptor for glycosylation (Chen S. X. et al. 2001;

Chen C. W. et al. 2003). Nevertheless, the use of nonaqueous system for GOS synthesis has disadvantages considering that some organic solvents have limits in food applications.

For either aqueous or biphasic reactions, optimization of the reaction conditions is required to reach the maximal GOS yield. Conventional optimization usually investigates one factor at a time. Nowadays, statistical experimental design, such as response surface methodology, has been used for optimizing reaction parameters with overall consideration of the interactive effects among various factors (Roy et al. 2002; Ghaffari-Moghaddam et al. 2014; Palai et al. 2015). An example is the model development for the formation of GOS by whole cells of *Bifidobacterium infantis* RW-8120. The most appropriate reaction conditions have been established by estimation of the effects of multiple factors (such as lactose concentration, cell concentration, reaction time, and temperature) on GOS production using central composite designs (Roy et al. 2002).

5.5 Immobilization of β-Galactosidases for GOS Production

Enzyme immobilization has always received a great deal of attention for use in industrial-scale processes. The advantages of immobilization include easy separation of the enzyme from the reaction mixture, reuse of the enzyme, continuous product formation, and improved stability (Gosling et al. 2010; Panesar et al. 2010; Park and Oh 2010).

Various strategies for β-galactosidase immobilization have been applied in GOS synthesis by utilizing a series of support materials, such as agarose (Berger et al. 1995; Huerta et al. 2011; Benavente et al. 2015), chitosan (Shin et al. 1998; Gaur et al. 2006; Urrutia et al. 2014), anion exchange resin (Maugard et al. 2003), cotton cloth (Albayrak and Yang 2002; Matella et al. 2006), magnetic particles (Neri et al. 2009; Pan et al. 2009; Liu et al. 2012), cellulose (Lu et al. 2012; Osman et al. 2014), responsive polymers (Palai et al. 2015), polystyrene nanofiber (Misson et al. 2015), and membranes (Gulec 2013; Palai and Bhattacharya 2013) (Table 5.4).

Inexpensive fibrous matrix, such as cotton cloth, has the advantages of high porosity, large specific surface area, and excellent mechanical strength for enzyme immobilization (Albayrak and Yang 2002; Matella et al. 2006). The β-galactosidase from *A. oryzae* has been immobilized on cotton cloth for continuous production of GOS from lactose in a plug-flow reactor. The maximal GOS production reached 26% yield from 400 g/L of lactose, corresponding to a high reactor productivity of 106 g/L/h. Stable, continuous operation in a plug-flow reactor was demonstrated for 2 weeks without any apparent problem (Albayrak and Yang 2002).

Chitosan is also an outstanding biopolymer for enzyme immobilization thanks to its relatively low cost, innocuousness, biocompatibility, antimicrobial activity, and the presence of functional groups for chemical modification (Shin et al. 1998; Gaur et al. 2006; Urrutia et al. 2014). Partially purified β-galactosidase from *Bullera singularis* ATCC 24193 has been simply absorbed with chitosan bead and used for production of GOS in a packed bed reactor; 55% yield of GOS was obtained continuously with a productivity of 4.4 g/L/h from 100 g/L lactose during a 15-day operation (Shin et al. 1998).

Chemical derivation of chitosan can further improve the properties of immobilized enzymes. The immobilization of β-galactosidase from *B. circulans* on the epichlorohydrin activated chitosan results in a high thermal stability, with a half-life of 449 times the value of the soluble enzyme. The immobilized enzyme can run 10 batches without any reduction of GOS yield. In recent years, the magnetic nanoparticles for the immobilization of enzymes have been paid increasing attention as they can be used repeatedly and easily separated from the reaction medium by a magnetic field (Neri et al. 2009; Pan et al. 2009; Liu et al. 2012). The β-galactosidase from *A. oryzae* immobilized on magnetic Fe_3O_4–chitosan nanoparticles can retain 92% of its initial activity after successive utilization for 15 cycles (Pan et al. 2009).

Besides being immobilized by physical or chemical attachment to a carrier, the enzyme can also be rendered immobilized by keeping it inside an ultrafiltration (UF)-membrane (molecular weight cutoff of ~10 kDa) reactor (Foda and Lopez-Leiva 2000; Chockchaisawasdee et al. 2005; Splechtna et al. 2007). Due to the difference in molecular sizes, the enzyme remains in the UF unit while the sugars permeate the membrane. The β-galactosidase from *Kluyveromyces lactis* with molecular weight of 135 kDa has been applied in the production of GOS at a pilot plant-scale membrane reactor using a Romicon hollow fiber cartridge. The largest yield of GOS reached 31% from a whey UF permeate containing initially 200 g/L lactose (Foda and Lopez-Leiva 2000). The β-galactosidase from *Lactobacillus reuteri* has been used for GOS synthesis in a reactor employing polyethersulfone membrane. The reactor can run for 6 days using a 600 mM lactose solution, with stable GOS yield at 24% (Splechtna et al. 2007).

Membranes have also been used as supports for enzyme immobilization these years (Gulec 2013; Palai and Bhattacharya 2013). The β-galactosidase from *K. lactis* has been immobilized onto the chemically activated cellulose acetate membrane. Although high enzyme loading was achieved (65–83%), the GOS yield (34%) by the immobilized enzyme decreased by 6% as compared with the free enzyme (40%). The immobilized enzyme could be successively reutilized for five to eight cycles with enzyme activity retaining above 75% (Gulec 2013).

Recently, the inclusion of an affinity tag fused to the target protein by genetic engineering has proved useful to purify and immobilize the protein in one simple step. The use of biospecific affinity of cellulose binding domain (CBD) to cellulose offers industrially attractive low-cost advantages, as well as the spontaneous and specific binding property (Boraston et al. 2002; Shoseyov et al. 2006). The β-galactosidase from *Lactobacillus bulgaricus* L3 has been subjected to gene cloning and fused with CBD using *E. coli* expression system. The recombinant CBD fusion protein was immobilized onto microcrystalline cellulose by absorption. Using the immobilized enzyme, the yield of GOS reached a maximum of 49% from 400 g/L lactose, with a high productivity of 156.8 g/L/h. The immobilized enzyme can retain over 85% activity after 20 batches with the GOS yields all above 40% (Lu et al. 2012).

The use of molecular biotechnology not only contributes to easily generating tag-bearing enzymes for immobilization but also directly creates immobilized enzymes. One promising strategy is to anchor the enzyme on the cell surface of engineered microorganisms (Lee et al. 2003), such as yeast that is generally regarded as safe. The β-galactosidase from *Penicillium expansum* F3 has been successfully anchored on the cell surface of *Saccharomyces cerevisiae* EBY-100 as an immobilized catalyst. The β-galactosidase-anchored yeast can directly utilize lactose to produce GOS as well as byproducts of glucose and galactose. The glucose that would inhibit GOS synthesis was consumed by the yeast for cell growth and the galactose was used for enzyme induction, thus greatly facilitating GOS synthesis. The GOS yield reached 43.6% when the recombinant yeast was cultivated in a lactose-containing medium (100 g/L). The yield can remain at high levels over 40% after seven batches detected (Li et al. 2009).

Currently, carrier-free immobilized enzymes, particularly cross-linked enzyme aggregates (CLEAs), have aroused extensive attention (Gaur et al. 2006; Li et al. 2015). CLEAs are insoluble biocatalysts prepared by precipitating enzyme with appropriate precipitating agents, usually with highly concentrated enzyme, good stability, and low production. A recombinant novel β-galactosidase from a marine metagenomic library has been prepared to CLEAs for GOS synthesis. High GOS yield of about 59.4% was achieved from 360 g/L of lactose. This catalyst can maintain around 82.1% activity after 10 cycles of reuse (Li et al. 2015).

5.6 Engineering of Microbial Enzymes for GOS Synthesis

Nowadays, the utility of glycosidases in oligosaccharide synthesis also benefits from various protein engineering strategies. Modulation of the molecular structure of β-galactosidases/β-glucosidases enhances the GOS yields produced from lactose (Hansson et al. 2001; Jorgensen et al. 2001; Placier et al. 2009; Wu et al. 2013; Hassan et al. 2016).

Deletion mutagenesis has been employed to create an efficient catalyst from *B. bifidum* DSM20215 β-galactosidase bearing a C-terminal galactose-binding motif. Truncation of approximately C-terminal 580 amino acid residues converts the enzyme from a normal, hydrolytic β-galactosidase into a highly efficient transgalactosylating enzyme. The truncated enzyme utilized approximately 90% of the reacted lactose for the production of GOS (Jorgensen et al. 2001).

Site-directed mutagenesis has been directed to improve the transgalactosylation properties of a series of glycosidases. An exchange of one phenylalanine to tyrosine (F426Y) of β-glucosidase CelB from *P. furiosus* increased the oligosaccharide yield (45%) compared with the wild-type CelB (40%). Incorporation of a positively charged group in the active site (M424K) increased the pH optimum of transglycosylation reaction of CelB. The double mutant, M424K/F426Y, showed much better transglycosylation properties at low lactose concentrations (10–20%) compared to the wild type. At a lactose concentration of 10%, the oligosaccharide yield for the mutant was 40% compared to 18% for the wild type (Hansson et al. 2001).

Rational design of β-glucosidase from *Halothermothrix orenii* generated variants that targeted the glycone (-1) and aglycone (+1) subsites. Three variants Y296F, F417S, and F417Y all produced increased GOS yield from 39.3% to above 50% (Hassan et al. 2016). Site-directed mutagenesis of β-galactosidase from *S. solfataricus* P2 generated two mutants, F359Q and F441Y, with increased GOS yields from the wild type 50.9% to 58.3% and 61.7%, respectively (Wu et al. 2013).

Thermostable mutants have also been successfully constructed from β-galactosidase from *B. circulans* ATCC 31382. Combinations of the mutations (K166P, G307P, and A833P) produced a triple mutant with the most thermostability. Despite a nonsignificant improvement for the transglycosylation, the thermostable mutants exhibited the potential for the industrial use at a higher temperature (Ishikawa et al. 2015).

Semirational site-saturation mutagenesis has exhibited promising prospects for converting poorly transglycosylating glycosidases into an efficient synthetic catalyst. A change of arginine to lysine (R109K) of a thermophilic GH 42 β-galactosidase from *Geobacillus stearothermophilus* increased the 3′-galactosyllactose yield from lactose to 11.5% compared to 2% for the wild type. Subsequent saturation mutagenesis at this position demonstrated that valine and tryptophan further increased the transglycosylation performance to 21% and 23%. Surprisingly, nonhydrolysis activity of the R109W toward trisaccharide was observed after it reached the maximal yield. The residue R109 was predicted to be one of the best targets to improve the transgalactosylation yield of GH 42 glycoside hydrolases (Placier et al. 2009).

5.7 GOS Purification

GOS produced by glycosidases are always mixtures of transgalactosylated saccharides, unreacted lactose, glucose, and galactose. The removal of non-GOS impurities will expand the applications of GOS in the food and pharmaceutical industries, such as the inclusion of ingredients in diabetic, lactose-intolerant, and low-calorie foods (Hernandez et al. 2009; Michelon et al. 2014). Currently, there are more and more applications for production of high-purity GOS with decreased amounts of monosaccharides and lactose.

Large-scale removal of monosaccharides is usually conducted by a chromatographic process with ion-exchange resins or activated charcoal (Hernandez et al. 2009; Nobre et al. 2009; Moravčík et al. 2012). An early patent has been registered to eliminate monosaccharides from GOS by using a preparative column with a strong acid cation-exchange resin such as Unibead UBK-530 in Na form (Keisuke et al. 1988). This method has recently been adapted for purification of Vivinal® GOS by exchange of Na^+ ion with a K^+ ion (Sinclair et al. 2009).

Selective enzymatic oxidation of lactose into lactobionic acid combined with the ion-exchange chromatography can help produce lactose-free GOS syrup (Splechtna et al. 2001; Maischberger et al. 2008). The GOS produced by recombinant β-glycosidase from *Sulfolobus solfataricus*, with a purity of 41%, has been subjected to a series of ion-exchange columns after being treated with cellobiose dehydrogenase, finally resulting in a high purity of 97% (Splechtna et al. 2001). However, large loss of total GOS (above 25%) occurs during the process.

Early studies for separation of GOS also include the use of activated charcoal column to remove monosaccharides and lactose (Prakash et al. 1989; Onishi et al. 1996; Kim et al. 2001). Activated charcoal has a higher affinity for oligosaccharides compared to mono- and disaccharides, which makes their operation at the industrial level more advantageous, since regeneration can take place offline without large substrate losses (Torres et al. 2010). Interestingly, GOS syrup can also be simply mixed with charcoal particles for purification (Hernandez et al. 2009; Rodriguez-Colinas et al. 2013). The charcoal loaded with sugars can be filtered through filter paper and washed with ethanolic solution. The selectivity in GOS recovery depends on the concentration of ethanolic water solutions used (Hernandez et al. 2009).

SEC is also widely used for GOS purification with high purity and recovery. Oigomate® 55 has been purified from a purity of 55% to above 99% by SEC via Sephadex G-10 column (Shoaf et al. 2006; Huebner et al. 2007). Mono- and disaccharides in the commercial Vivinal® GOS have been removed using Bio-Gel P-2 column for in vitro fermentation assays with human fecal slurries (Cardelle-Cobas et al. 2009). Additionally, SEC can fractionate saccharides based on their DPs, enabling the purification of GOS with different DPs. Oligosaccharide fractions with DPs 2 to 5 have been separated from Bimuno® through Biogel P2 column to test their functionality, respectively (Tzortzis et al. 2005a; Searle et al. 2010).

Nanofiltration through membranes with small pore sizes (nominal molecular weight cutoff, MWCO, <1000 Da) appears as an alternative step to chromatographic procedures for the separation of GOS, due to advantages such as energy saving, low cost of implementation, simplicity of operation, and scale-up (Goulas et al. 2003; Catarino et al. 2008). Vivinal® GOS has been purified by polyethersulfone and cellulose acetate membranes with MWCO of 1000 Da, respectively (Goulas et al. 2003). The former membrane gave 80% removal of monosaccharides, whereas the latter gave better removal of the monosaccharides but a larger loss of oligosaccharides at the same time. Similarly, a GOS mixture from Yantai Beer Limited Co. (China) has been separated through cellulose acetate membrane NF-3 with MWCO of 800–1000 Da. The oligosaccharides purity improved from 36.4% to 54.5%, with GOS recovery of 70.0% (Feng et al. 2009).

Nanofiltration of GOS produced by β-galactosidase from *Kluyveromyces marxianus* CCT 7082 has been accessed by various membranes. A polyethersulfone membrane NP030 with MWCO of 400 Da shows better property for purification of GOS, with oligosaccharide purity increased by 1.8-fold along with ~61% recovery yield (Michelon et al. 2014). Using the membrane NP030, the purity of GOS produced by the recombinant β-galactosidase from *L. bulgaricus* was increased from 58.5% to 85% at 5°C, with a GOS recovery yield of 82% (Pruksasri et al. 2015). Operation at a low temperature not only reduced the microorganism contamination but also improved the GOS recovery.

Nonchromatographic strategies for GOS purification also involve yeast selective fermentation of digestible sugars including glucose, galactose, and lactose. *S. cerevisiae* is usually used to eliminate monosaccharides (Goulas et al. 2007; Hernandez et al. 2009), while *K. marxianus* and *K. lactis* are applied to remove lactose in addition to monosaccharides in the GOS syrup (Cheng et al. 2006; Li et al. 2008). Notably, immobilized yeast cells can be recycled for sequential batch purification of GOS. The GOS syrups produced by β-galactosidase from *P. expansum* F3 has been treated by fermentation with *S. cerevisiae* L1 and *K. lactis* L3 immobilized in the calcium alginate, resulting in increased GOS purity from 28.7% to 39.4% and 97.5%, respectively.

Purity greater than 37% of nonmonosaccharide GOS was maintained by *S. cerevisiae* L1 for 19 batches, and purity greater than 97% of high-purity GOS was produced using *K. lactis* L3 for two batches (Li et al. 2008).

It is also noteworthy to mention that large-scale separation of GOS isomers with identical DPs is currently available by chemical acetylation and deacetylation method. GOS synthesized by *L. bulgaricus* L3 β-galactosidase were first fermented by yeast cells, giving a nearly 100% purified GOS mainly composed of 6′-galactosyllactose and 3′-galactosyllactose. The purified GOS was acetylated and then subjected to silica gel column chromatography. The trisaccharide derivatives were easily separated and could be prepared in gram scale by this method. Acetylation differentiated between the isobaric oligosaccharides and allowed the separation in large scale. In addition, it was quite simple to remove the acetyl protecting groups from the isolated GOS derivatives and prepare the free oligosaccharides (Lu et al. 2010).

5.8 Outlook

GOS produced by microbial enzymes have attracted worldwide attention owing to their similarity to human milk oligosaccharides (Sangwan et al. 2011). As a stable, soluble ingredient, GOS can be incorporated in different types of food, feed, and pharmaceutical products for digestive and immune health (Torres et al. 2010). For instance, the introduction of GOS into baby foods has been very successful (Sangwan et al. 2011).

Future efforts on GOS synthesis might focus on decreased production costs and an increase in prebiotic potency (Gosling et al. 2010). There is still a need to find good microbial sources of β-galactosidases with high transgalactosylation activities. Currently, screening of the β-galactosidases from the metagenomic library provides an alternative method to find novel enzymes with good properties (Wang et al. 2012; Li et al. 2015). The recombinant production systems are also crucial as the microbial sources of β-galactosidases are usually not GRAS and may present slow growth and low productivities (Oliveira et al. 2011). Excitingly, recombinant GRAS hosts such as *S. cerevisiae* and *L. plantarum* have already been reported for GOS synthesis (Li et al. 2009; Nguyen et al. 2012). More importantly, using the recombinant system, enzyme engineering can be conducted to improve transglycosylation activity and stability, which will facilitate increased GOS yields as well as recyclability for immobilization (Gosling et al. 2010). Elucidation of enzymes structures will further disclose their catalytic mechanisms, which is definitely helpful for rational design and engineering.

Production of high-content GOS is still important for delivering purer and more efficient mixtures, with narrow and specific target ranges. With the rapid development of sugar separation and identification methods, the overall maps of GOS produced by various enzymes are likely uncovered, and large-scale separation of various GOS isomers might become possible in the near future. This would facilitate the evaluation of elaborate functions of single GOS structures. It is worth noting that the studies on the mechanisms of GOS utilization and functionalization in vivo have started (Bhatia et al. 2015; Shigehisa et al. 2015), which will help to find the GOS structures with high prebiotic efficacy.

References

Akiyama, K., Takase, M., Horikoshi, K. and Okonogi, S. (2001). Production of galactooligosaccharides from lactose using a β-glucosidase from *Thermus* sp. Z-1. *Bioscience, Biotechnology, and Biochemistry.* **65**(2), 438–441.

Albayrak, N. and Yang, S. T. (2002). Production of galacto-oligosaccharides from lactose by *Aspergillus oryzae* beta-galactosidase immobilized on cotton cloth. *Biotechnology and Bioengineering.* **77**(1), 8–19.

Albrecht, S., Schols, H. A., Klarenbeek, B., Voragen, A. G. and Gruppen, H. (2010). Introducing capillary electrophoresis with laser-induced fluorescence (CE-LIF) as a potential analysis and quantification tool for galactooligosaccharides extracted from complex food matrices. *Journal of Agricultural and Food Chemistry.* **58**(5), 2787–2794.

Arreola, S. L., Intanon, M., Suljic, J., Kittl, R., Pham, N. H., Kosma, P., Haltrich, D. and Nguyen, T. H. (2014). Two beta-galactosidases from the human isolate *Bifidobacterium breve* DSM 20213: Molecular cloning and expression, biochemical characterization and synthesis of galacto-oligosaccharides. *PLoS One.* **9**(8), e104056.

Barboza, M., Sela, D. A., Pirim, C., Locascio, R. G., Freeman, S. L., German, J. B., Mills, D. A. and Lebrilla, C. B. (2009). Glycoprofiling bifidobacterial consumption of galacto-oligosaccharides by mass spectrometry reveals strain-specific, preferential consumption of glycans. *Applied and Environmental Microbiology.* **75**(23), 7319–7325.

Barile, D. and Rastall, R. A. (2013). Human milk and related oligosaccharides as prebiotics. *Current Opinion in Biotechnology*. **24**(2), 214–219.

Bartesaghi, A., Matthies, D., Banerjee, S., Merk, A. and Subramaniam, S. (2014). Structure of β-galactosidase at 3.2-Å resolution obtained by cryo-electron microscopy. *Proceedings of the National Academy of Sciences*. **111**(32), 11709–11714.

Bartesaghi, A., Merk, A., Banerjee, S., Matthies, D., Wu, X., Milne, J. L. and Subramaniam, S. (2015). 2.2 Å resolution cryo-EM structure of β-galactosidase in complex with a cell-permeant inhibitor. *Science*. **348**(6239), 1147–1151.

Benavente, R., Pessela, B. C., Curiel, J. A., de las Rivas, B., Munoz, R., Guisan, J. M., Mancheno, J. M., Cardelle-Cobas, A., Ruiz-Matute, A. I. and Corzo, N. (2015). Improving properties of a novel beta-galactosidase from *Lactobacillus plantarum* by covalent immobilization. *Molecules*. **20**(5), 7874–7889.

Berger, J., Lee, B. and Lacroix, C. (1995). Oligosaccharides synthesis by free and immobilized β-galactosidases from *Thermus aquaticus* YT-1. *Biotechnology Letters*. **17**(10), 1077–1080.

Bhatia, S., Prabhu, P. N., Benefiel, A. C., Miller, M. J., Chow, J., Davis, S. R. and Gaskins, H. R. (2015). Galacto-oligosaccharides may directly enhance intestinal barrier function through the modulation of goblet cells. *Molecular Nutrition and Food Research*. **59**(3), 566–573.

Bode, L. (2012). Human milk oligosaccharides: Every baby needs a sugar mama. *Glycobiology*. **22**(9), 1147–1162.

Boehm, G., Stahl, B., Jelinek, J., Knol, J., Miniello, V. and Moro, G. E. (2005). Prebiotic carbohydrates in human milk and formulas. *Acta Paediatrica*. **94**(449), 18–21.

Boraston, A. B., McLean, B. W., Kavoosi, M., Haynes, C. A. and Kilburn, D. G. (2002). Cellulose-binding fusion proteins. In Gupta, M. N. (ed.), *Methods for Affinity-Based Separations of Enzymes and Proteins. Methods and Tools in Biosciences and Medicine*. Basel: Birkhäuser, pp. 148–162.

Brás, N. F., Fernandes, P. A. and Ramos, M. J. (2010). QM/MM studies on the β-galactosidase catalytic mechanism: Hydrolysis and transglycosylation reactions. *Journal of Chemical Theory and Computation*. **6**(2), 421–433.

Bruins, M., Strubel, M., Van Lieshout, J., Janssen, A. and Boom, R. (2003). Oligosaccharide synthesis by the hyperthermostable β-glucosidase from Pyrococcus furiosus: Kinetics and modelling. *Enzyme and Microbial Technology*. **33**(1), 3–11.

Cardelle-Cobas, A., Fernández, M., Salazar, N., Martínez-Villaluenga, C., Villamiel, M., Ruas-Madiedo, P. and de los Reyes-Gavilán, C. G. (2009). Bifidogenic effect and stimulation of short chain fatty acid production in human faecal slurry cultures by oligosaccharides derived from lactose and lactulose. *Journal of Dairy Research*. **76**(3), 317–325.

Catarino, I., Minhalma, M., Beal, L. L., Mateus, M. and de Pinho, M. N. (2008). Assessment of saccharide fractionation by ultrafiltration and nanofiltration. *Journal of Membrane Science*. **312**(1–2), 34–40.

Chen, C. W., Ou-Yang, C. C. and Yeh, C. W. (2003). Synthesis of galactooligosaccharides and transgalactosylation modeling in reverse micelles. *Enzyme and Microbial Technology*. **33**(4), 497–507.

Chen, S. X., Wei, D. Z. and Hu, Z. H. (2001). Synthesis of galacto-oligosaccharides in AOT/isooctane reverse micelles by beta-galactosidase. *Journal of Molecular Catalysis B—Enzymatic*. **16**(2), 109–114.

Cheng, C. C., Yu, M. C., Cheng, T. C., Sheu, D. C., Duan, K. J. and Tai, W. L. (2006). Production of high-content galacto-oligosaccharide by enzyme catalysis and fermentation with *Kluyveromyces marxianus*. *Biotechnology Letters*. **28**(22), 793–797.

Cho, Y. J., Shin, H. J. and Bucke, C. (2003). Purification and biochemical properties of a galactooligosaccharide producing beta-galactosidase from *Bullera singularis*. *Biotechnology Letters*. **25**(24), 2107–2111.

Chockchaisawasdee, S., Athanasopoulos, V. I., Niranjan, K. and Rastall, R. A. (2005). Synthesis of galacto-oligosaccharide from lactose using β-galactosidase from *Kluyveromyces lactis*: Studies on batch and continuous UF membrane-fitted bioreactors. *Biotechnology and Bioengineering*. **89**(4), 434–443.

Corradini, C., Cavazza, A. and Bignardi, C. (2012). High-performance anion-exchange chromatography coupled with pulsed electrochemical detection as a powerful tool to evaluate carbohydrates of food interest: Principles and applications. *International Journal of Carbohydrate Chemistry*. **2012**, 1687–9341.

Cruz, R., Cruz, V. d'A., Belote, J. G., de Oliveira Khenayfes, M., Dorta, C., dos Santos Oliveira, L. H., Ardiles, E. and Galli, A. (1999). Production of transgalactosylated oligosaccharides (TOS) by galactosyltransferase activity from *Penicillium simplicissimum*. *Bioresource Technology*. **70**(2), 165–171.

De Slegte, J. (2002). Determination of trans-galactooligosaccharides in selected food products by ion-exchange chromatography: Collaborative study. *Journal of AOAC International*. **85**(2), 417–423.

Diez-Municio, M., Herrero, M., Olano, A. and Moreno, F. J. (2014). Synthesis of novel bioactive lactose-derived oligosaccharides by microbial glycoside hydrolases. *Microbial Biotechnology*. **7**(4), 315–331.

Dugdale, M. L., Dymianiw, D. L., Minhas, B. K., D'Angelo, I. and Huber, R. E. (2010a). Role of Met-542 as a guide for the conformational changes of Phe-601 that occur during the reaction of β-galactosidase (*Escherichia coli*). *Biochemistry and Cell Biology*. **88**(5), 861–869.

Dugdale, M. L., Vance, M. L., Wheatley, R. W., Driedger, M. R., Nibber, A., Tran, A. and Huber, R. E. (2010b). Importance of Arg-599 of β-galactosidase (*Escherichia coli*) as an anchor for the open conformations of Phe-601 and the active-site loop. *Biochemistry and Cell Biology*. **88**(6), 969–979.

Dumortier, V., Montreuil, J. and Bouquelet, S. (1990). Primary structure of ten galactosides formed by transglycosylation during lactose hydrolysis by *Bifidobacterium bifidum*. *Carbohydrate Research*. **201**(1), 115–123.

Feng, Y., Chang, X., Wang, W. and Ma, R. (2009). Separation of galacto-oligosaccharides mixture by nanofiltration. *Journal of the Taiwan Institute of Chemical Engineers*. **40**(3), 326–332.

Foda, M. I. and Lopez-Leiva, M. (2000). Continuous production of oligosaccharides from whey using a membrane reactor. *Process Biochemistry*. **35**(6), 581–587.

Gaur, R., Pant, H., Jain, R. and Khare, S. (2006). Galacto-oligosaccharide synthesis by immobilized *Aspergillus oryzae* β-galactosidase. *Food Chemistry*. **97**(3), 426–430.

Ghaffari-Moghaddam, M., Yekke-Ghasemi, Z., Khajeh, M., Rakhshanipour, M. and Yasin, Y. (2014). Application of response surface methodology in enzymatic synthesis: A review. *Russian Journal of Bioorganic Chemistry*. **40**(3), 275–285.

Gloster, T. M., Roberts, S., Ducros, V. M., Perugino, G., Rossi, M., Hoos, R., Moracci, M., Vasella, A. and Davies, G. J. (2004). Structural studies of the β-glycosidase from *Sulfolobus solfataricus* in complex with covalently and noncovalently bound inhibitors. *Biochemistry*. **43**(20), 6101–6109.

Golowczyc, M., Vera, C., Santos, M., Guerrero, C., Carasi, P., Illanes, A., Gomez-Zavaglia, A. and Tymczyszyn, E. (2013). Use of whey permeate containing in situ synthesised galacto-oligosaccharides for the growth and preservation of *Lactobacillus plantarum*. *Journal of Dairy Research*. **80**(3), 374–381.

Gosling, A., Stevens, G. W., Barber, A. R., Kentish, S. E. and Gras, S. L. (2010). Recent advances refining galactooligosaccharide production from lactose. *Food Chemistry*. **121**(2), 307–318.

Goulas, A. K., Grandison, A. S. and Rastall, R. A. (2003). Fractionation of oligosaccharides by nanofiltration. *Journal of the Science of Food and Agriculture*. **83**(7), 675–680.

Goulas, A. K, Tzortzis, G. and Gibson, G. R. (2007). Development of a process for the production and purification of alpha- and beta-galactooligosaccharides from *Bifidobacterium bifidum* NCIMB 41171. *International Dairy Journal*. **17**(6), 648–656.

Gulec, H. A. (2013). Immobilization of beta-galactosidase from *Kluyveromyces lactis* onto polymeric membrane surfaces: Effect of surface characteristics. *Colloids and Surfaces B—Biointerfaces*. **104**, 83–90.

Hansson, T., Kaper, T., van der Oost, J., de Vos, W. M. and Adlercreutz, P. (2001). Improved oligosaccharide synthesis by protein engineering of beta-glucosidase CelB from hyperthermophilic *Pyrococcus furiosus*. *Biotechnology and Bioengineering*. **73**(3), 203–210.

Hassan, N., Geiger, B., Gandini, R., Patel, B. K., Kittl, R., Haltrich, D., Nguyen, T. H., Divne, C. and Tan, T. C. (2016). Engineering a thermostable *Halothermothrix orenii* beta-glucosidase for improved galacto-oligosaccharide synthesis. *Applied Microbiology and Biotechnology*. **100**(8), 3533–3543.

Hassan, N., Nguyen, T. H., Intanon, M., Kori, L. D., Patel, B. K., Haltrich, D., Divne, C. and Tan, T. C. (2015). Biochemical and structural characterization of a thermostable beta-glucosidase from *Halothermothrix orenii* for galacto-oligosaccharide synthesis. *Applied Microbiology and Biotechnology*. **99**(4), 1731–1744.

Henze, M., You, D.-J., Kamerke, C., Hoffmann, N., Angkawidjaja, C., Ernst, S., Pietruszka, J., Kanaya, S. and Elling, L. (2014). Rational design of a glycosynthase by the crystal structure of β-galactosidase from *Bacillus circulans* (BgaC) and its use for the synthesis of N-acetyllactosamine type 1 glycan structures. *Journal of Biotechnology.* **191**, 78–85.

Hernandez, O., Ruiz-Matute, A. I., Olano, A., Moreno, F. J. and Sanz, M. L. (2009). Comparison of fractionation techniques to obtain prebiotic galactooligosaccharides. *International Dairy Journal.* **19**(9), 531–536.

Hidaka, M., Fushinobu, S., Ohtsu, N., Motoshima, H., Matsuzawa, H., Shoun, H. and Wakagi, T. (2002). Trimeric crystal structure of the glycoside hydrolase family 42 beta-galactosidase from *Thermus thermophilus* A4 and the structure of its complex with galactose. *Journal of Agricultural and Food Chemistry.* **322**(1), 79–91.

Hsu, C. A., Lee, S. L. and Chou, C. C. (2007). Enzymatic production of galactooligosaccharides by beta-galactosidase from *Bifidobacterium longum* BCRC 15708. *Journal of Agricultural and Food Chemistry.* **55**(6), 2225–2230.

Huber, R., Kurz, G. and Wallenfels, K. (1976). A quantitation of the factors which affect the hydrolase and transgalactosylase activities of β-galactosidase (*E. coli*) on lactose. *Biochemistry.* **15**(9), 1994–2001.

Huebner, J., Wehling, R. L. and Hutkins, R. W. (2007). Functional activity of commercial prebiotics. *International Dairy Journal.* **17**(7), 770–775.

Huerta, L. M., Vera, C., Guerrero, C., Wilson, L. and Illanes, A. (2011). Synthesis of galacto-oligosaccharides at very high lactose concentrations with immobilized β-galactosidases from *Aspergillus oryzae. Process Biochemistry.* **46**(1), 245–252.

Hung, M.-N. and Lee, B. (2002). Purification and characterization of a recombinant β-galactosidase with transgalactosylation activity from Bifidobacterium infantis HL96. *Applied Microbiology and Biotechnology.* **58**(4), 439–445.

Intanon, M., Arreola, S. L., Pham, N. H., Kneifel, W., Haltrich, D. and Nguyen, T. H. (2014). Nature and biosynthesis of galacto-oligosaccharides related to oligosaccharides in human breast milk. *FEMS Microbiology Letters.* **353**(2), 89–97.

Iqbal, S., Nguyen, T. H., Nguyen, H. A., Nguyen, T. T., Maischberger, T., Kittl, R. and Haltrich, D. (2011). Characterization of a heterodimeric GH2 beta-galactosidase from *Lactobacillus sakei* Lb790 and formation of prebiotic galacto-oligosaccharides. *Journal of Agricultural and Food Chemistry.* **59**(8), 3803–3811.

Iqbal, S., Nguyen, T. H., Nguyen, T. T., Maischberger, T. and Haltrich, D. (2010). beta-Galactosidase from *Lactobacillus plantarum* WCFS1: Biochemical characterization and formation of prebiotic galacto-oligosaccharides. *Carbohydrate Research.* **345**(10), 1408–1416.

Ishikawa, K., Kataoka, M., Yanamoto, T., Nakabayashi, M., Watanabe, M., Ishihara, S. and Yamaguchi, S. (2015). Crystal structure of beta-galactosidase from *Bacillus circulans* ATCC 31382 (BgaD) and the construction of the thermophilic mutants. *FEBS Journal.* **282**(13), 2540–2552.

Jancewicz, L. J., Wheatley, R. W., Sutendra, G., Lee, M., Fraser, M. E. and Huber, R. E. (2012). Ser-796 of β-galactosidase (*Escherichia coli*) plays a key role in maintaining a balance between the opened and closed conformations of the catalytically important active site loop. *Archives of Biochemistry and Biophysics*. **517**(2), 111–122.

Ji, E.-S., Park, N.-H. and Oh, D.-K. (2005). Galacto-oligosaccharide production by a thermostable recombinant β-galactosidase from *Thermotoga maritima*. *World Journal of Microbiology and Biotechnology*. **21**(5), 759–764.

Jorgensen, F., Hansen, O. C. and Stougaard, P. (2001). High-efficiency synthesis of oligosaccharides with a truncated beta-galactosidase from *Bifidobacterium bifidum*. *Applied Microbiology and Biotechnology*. **57**(5-6), 647–652.

Jovanovic-Malinovska, R., Fernandes, P., Winkelhausen, E. and Fonseca, L. (2012). Galacto-oligosaccharides synthesis from lactose and whey by β-galactosidase immobilized in PVA. *Applied Biochemistry and Biotechnology*. **168**(5), 1197–1211.

Juajun, O., Nguyen, T. H., Maischberger, T., Iqbal, S., Haltrich, D. and Yamabhai, M. (2011). Cloning, purification, and characterization of beta-galactosidase from *Bacillus licheniformis* DSM 13. *Applied Microbiology and Biotechnology*. **89**(3), 645–654.

Juers, D. H., Heightman, T. D., Vasella, A., McCarter, J. D., Mackenzie, L., Withers, S. G. and Matthews, B. W. (2001). A structural view of the action of *Escherichia coli* (lacZ) beta-galactosidase. *Biochemistry*. **40**(49), 14781–14794.

Juers, D. H., Jacobson, R. H., Wigley, D., Zhang, X. J., Huber, R. E., Tronrud, D. E. and Matthews, B. W. (2000). High resolution refinement of beta-galactosidase in a new crystal form reveals multiple metal-binding sites and provides a structural basis for alpha-complementation. *Protein Science*. **9**(9), 1685–1699.

Keisuke, M., Tamura, N. and Watanabe, T. (1988). Method for producing galactooligosaccharide. Eur. Patent EP0272095.

Kim, J.-H., Lee, D.-H. and Lee, J.-S. (2001). Production of galactooligosaccharide by β-galactosidase from *Kluyveromyces maxianus* var *lactis* OE-20. *Biotechnology and Bioprocess Engineering*. **6**(5), 337–340.

Lamsal, B. P. (2012). Production, health aspects and potential food uses of dairy prebiotic galactooligosaccharides. *Journal of the Science of Food and Agriculture*. **92**(10), 2020–2028.

Lee, S. Y., Choi, J. H. and Xu, Z. (2003). Microbial cell-surface display. *Trends in Biotechnology*. **21**(1), 45–52.

Li, L., Li, G., Cao, L. C., Ren, G. H., Kong, W., Wang, S. D., Guo, G. S. and Liu, Y. H. (2015). Characterization of the cross-linked enzyme aggregates of a novel beta-galactosidase, a potential catalyst for the synthesis of galacto-oligosaccharides. *Journal of Agricultural and Food Chemistry*. **63**(3), 894–901.

Li, Y., Lu, L., Wang, H., Xu, X. and Xiao, M. (2009). Cell surface engineering of a beta-galactosidase for galactooligosaccharide synthesis. *Applied and Environmental Microbiology*. **75**(18), 5938–5942.

Li, Z., Xiao, M., Lu, L. and Li, Y. (2008). Production of non-monosaccharide and high-purity galactooligosaccharides by immobilized enzyme catalysis and fermentation with immobilized yeast cells. *Process Biochemistry.* **43**(8), 896–899.

Liu, H., Liu, J., Tan, B., Zhou, F., Qin, Y. and Yang, R. (2012). Covalent immobilization of *Kluyveromyces fragilis* beta-galactosidase on magnetic nanosized epoxy support for synthesis of galacto-oligosaccharide. *Bioprocess and Biosystems Engineering.* **35**(8), 1287–1295.

Lu, L., Gu, G., Xiao, M. and Wang, F. (2010). Separation and structure analysis of trisaccharide isomers produced from lactose by *Lactobacillus bulgaricus* L3 β-galactosidase. *Food Chemistry.* **121**(4), 1283–1288.

Lu, L., Xiao, M., Li, Z., Li, Y. and Wang, F. (2009). A novel transglycosylating β-galactosidase from *Enterobacter cloacae* B5. *Process Biochemistry.* **44**(2), 232–236.

Lu, L., Xiao, M., Xu, X., Li, Z. and Li, Y. (2007). A novel beta-galactosidase capable of glycosyl transfer from *Enterobacter agglomerans* B1. *Biochemical and Biophysical Research Communications.* **356**(1), 78–84.

Lu, L., Xu, S., Zhao, R., Zhang, D., Li, Z., Li, Y. and Xiao, M. (2012). Synthesis of galactooligosaccharides by CBD fusion beta-galactosidase immobilized on cellulose. *Bioresource Technology.* **116**, 327–333.

Mahoney, R. R. (1998). Galactosyl-oligosaccharide formation during lactose hydrolysis: A review. *Food Chemistry.* **63**(2), 147–154.

Maischberger, T., Nguyen, T. H., Sukyai, P., Kittl, R., Riva, S., Ludwig, R. and Haltrich, D. (2008). Production of lactose-free galacto-oligosaccharide mixtures: Comparison of two cellobiose dehydrogenases for the selective oxidation of lactose to lactobionic acid. *Carbohydrate Research.* **343**(12), 2140–2147.

Maksimainen, M., Hakulinen, N., Kallio, J. M., Timoharju, T., Turunen, O. and Rouvinen, J. (2011). Crystal structures of *Trichoderma reesei* beta-galactosidase reveal conformational changes in the active site. *Journal of Structural Biology.* **174**(1), 156–163.

Maksimainen, M., Paavilainen, S., Hakulinen, N. and Rouvinen, J. (2012). Structural analysis, enzymatic characterization, and catalytic mechanisms of beta-galactosidase from *Bacillus circulans* sp. *alkalophilus. FEBS Journal.* **279**(10), 1788–1798.

Martínez-Villaluenga, M., Cobas, A. C., Corzo, N., Olano, A. and Villamiel, M. (2008). Optimization of conditions for galactooligosaccharide synthesis during lactose hydrolysis by β-galactosidase from *Kluyveromyces lactis* (Lactozym 3000 L HP G). *Food Chemistry.* **107**, 258–264.

Matella, N., Dolan, K. and Lee, Y. (2006). Comparison of galactooligosaccharide production in free-enzyme ultrafiltration and in immobilized-enzyme systems. *Journal of Food Science.* **71**(7), C363–C368.

Maugard, T., Gaunt, D., Legoy, M. D. and Besson, T. (2003). Microwave-assisted synthesis of galacto-oligosaccharides from lactose with immobilized beta-galactosidase from *Kluyveromyces lactis. Biotechnology Letters.* **25**(8), 623–629.

McCarter, J. D. and Withers, S. G. (1994). Mechanisms of enzymatic glycoside hydrolysis. *Current Opinion in Structural Biology*. **4**(6), 885–892.

Melo, E. P., Aires-Barros, M. R. and Cabral, J. M. (2001). Reverse micelles and protein biotechnology. *Biotechnology Annual Review*. **7**, 87–129.

Michelon, M., Manera, A. P., Carvalho, A. L. and Maugeri Filho, F. (2014). Concentration and purification of galacto-oligosaccharides using nanofiltration membranes. *International Journal of Food Science and Technology*. **49**(8), 1953–1961.

Misson, M., Jin, B., Chen, B. and Zhang, H. (2015). Enhancing enzyme stability and metabolic functional ability of β-galactosidase through functionalized polymer nanofiber immobilization. *Bioprocess and Biosystems Engineering*. **38**(10), 1915–1923.

Moravčík, J., Gramblička, M., Wiśniewski, Ł., Vaňková, K. and Polakovič, M. (2012). Influence of the ionic form of a cation-exchange adsorbent on chromatographic separation of galactooligosaccharides. *Chemical Papers*. **66**(6), 583–588.

Nakakuki, T. (1993). *Oligosaccharides: Production, Properties and Applications*. Basel, Switzerland: Gordon and Breach, pp. 90–106.

Nakao, M., Harada, M., Kodama, Y., Nakayama, T., Shibano, Y. and Amachi, T. (1994). Purification and characterization of a thermostable β-galactosidase with high transgalactosylation activity from *Saccharopolyspora rectivirgula*. *Applied Microbiology and Biotechnology*. **40**(5), 657–663.

Nakkharat, P., Kulbe, K. D., Yamabhai, M. and Haltrich, D. (2006). Formation of galacto-oligosaccharides during lactose hydrolysis by a novel beta-galactosidase from the moderately thermophilic fungus *Talaromyces thermophilus*. *Biotechnology Journal*. **1**(6), 633–638.

Neri, D. F., Balcão, V. M., Costa, R. S., Rocha, I. C., Ferreira, E. M., Torres, D. P., Rodrigues, L. R., Carvalho, L. B. and Teixeira, J. A. (2009). Galacto-oligosaccharides production during lactose hydrolysis by free *Aspergillus oryzae* β-galactosidase and immobilized on magnetic polysiloxane-polyvinyl alcohol. *Food Chemistry*. **115**(1), 92–99.

Newburg, D. S. (2000). Oligosaccharides in human milk and bacterial colonization. *Journal of Pediatric Gastroenterology and Nutrition*. **30**, S8–S17.

Nguyen, T.-H., Splechtna, B., Krasteva, S., Kneifel, W., Kulbe, K. D., Divne, C. and Haltrich, D. (2007). Characterization and molecular cloning of a heterodimeric β-galactosidase from the probiotic strain *Lactobacillus acidophilus* R22. *FEMS Microbiology Letters*. **269**(1), 136–144.

Nguyen, T.-T., Nguyen, H. A., Arreola, S. L., Mlynek, G., Djinović-Carugo, K., Mathiesen, G., Nguyen, T.-H. and Haltrich, D. (2012). Homodimeric β-galactosidase from *Lactobacillus delbrueckii* subsp. *bulgaricus* DSM 20081: Expression in *Lactobacillus plantarum* and biochemical characterization. *Journal of Agricultural and Food Chemistry*. **60**(7), 1713–1721.

Niittynen, L., Kajander, K. and Korpela, R. (2007). Galacto-oligosaccharides and bowel function. *Scandinavian Journal of Food and Nutrition*. **51**(2), 62–66.

Nobre, C., Santos, M. J., Dominguez, A., Torres, D., Rocha, O., Peres, A. M., Rocha, I., Ferreira, E. C., Teixeira, J. A. and Rodrigues, L. R. (2009). Comparison of adsorption equilibrium of fructose, glucose and sucrose on potassium gel-type and macroporous sodium ion-exchange resins. *Analytica Chimica Acta*. **654**(1), 71–76.

Ohtsuka, K., Tanoh, A., Ozawa, O., Kanematsu, T., Uchida, T. and Shinke, R. (1990). Purification and properties of a β-galactosidase with high galactosyl transfer activity from *Cryptococcus laurentii* OKN-4. *Journal of Fermentation and Bioengineering*. **70**(5), 301–307.

Oliveira, C., Guimarães, P. M. and Domingues, L. (2011). Recombinant microbial systems for improved β-galactosidase production and biotechnological applications. *Biotechnology Advances*. **29**(6), 600–609.

Onishi, N., Kira, I. and Yokozeki, K. (1996). Galacto-oligosaccharide production from lactose by *Sirobasidium magnum* CBS6803. *Letters in Applied Microbiology*. **23**(4), 253–256.

Onishi, N. and Tanaka, T. (1998). Galacto-oligosaccharide production using a recycling cell culture of *Sterigmatomyces elviae* CBS8119. *Letters in Applied Microbiology*. **26**(2), 136–139.

Onishi, N., Yamashiro, A. and Yokozeki, K. (1995). Production of galacto-oligosaccharide from lactose by *Sterigmatomyces elviae* CBS8119. *Applied and Environmental Microbiology*. **61**(11), 4022–4025.

Osman, A., Symeou, S., Trisse, V., Watson, K. A., Tzortzis, G. and Charalampopoulos, D. (2014). Synthesis of prebiotic galactooligosaccharides from lactose using bifidobacterial β-galactosidase (BbgIV) immobilised on DEAE-Cellulose, Q-Sepharose and amino-ethyl aga rose. *Biochemical Engineering Journal*. **82**(15), 188–199.

Palai, T. and Bhattacharya, P. K. (2013). Kinetics of lactose conversion to galacto-oligosaccharides by β-galactosidase immobilized on PVDF membrane. *Journal of Bioscience and Bioengineering*. **115**(6), 668–673.

Palai, T., Kumar, A. and Bhattacharya, P. K. (2015). Kinetic studies and model development for the formation of galacto-oligosaccharides from lactose using synthesized thermo-responsive bioconjugate. *Enzyme and Microbial Technology*. **70**, 42–49.

Pan, C., Hu, B., Li, W., Sun, Y., Ye, H. and Zeng, X. (2009). Novel and efficient method for immobilization and stabilization of β-d-galactosidase by covalent attachment onto magnetic Fe_3O_4–chitosan nanoparticles. *Journal of Molecular Catalysis B: Enzymatic*. **61**(3), 208–215.

Panesar, P. S., Kumari, S. and Panesar, R. (2010). Potential applications of immobilized β-galactosidase in food processing industries. *Enzyme Research*. **2010**, 473137.

Park, A.-R. and Oh, D.-K. (2010). Galacto-oligosaccharide production using microbial β-galactosidase: Current state and perspectives. *Applied Microbiology and Biotechnology*. **85**(5), 1279–1286.

Park, H.-Y., Kim, H.-J., Lee, J.-K., Kim, D. and Oh, D.-K. (2008). Galactooligosaccharide production by a thermostable β-galactosidase from *Sulfolobus solfataricus*. *World Journal of Microbiology and Biotechnology*. **24**(8), 1553–1558.

Pereira-Rodríguez, Á., Fernández-Leiro, R., González-Siso, M. I., Cerdán, M. E., Becerra, M. and Sanz-Aparicio, J. (2012). Structural basis of specificity in tetrameric *Kluyveromyces lactis* β-galactosidase. *Journal of Structural Biology*. **177**(2), 392–401.

Perugino, G., Trincone, A., Rossi, M. and Moracci, M. (2004). Oligosaccharide synthesis by glycosynthases. *Trends in Biotechnology*. **22**(1), 31–37.

Placier, G., Watzlawick, H., Rabiller, C. and Mattes, R. (2009). Evolved β-galactosidases from *Geobacillus stearothermophilus* with improved transgalactosylation yield for galacto-oligosaccharide production. *Applied Environmental Microbiology*. **75**(19), 6312–6321.

Prakash, B. S., Suyama, K., Itoh, T. and Adachi, S. (1989). Structure elucidation of major galactooligosaccharides formed by growing cultures of *Trichoderma harzianum*. *Journal of Agricultural and Food Chemistry*. **37**(2), 334–337.

Pruksasri, S., Nguyen, T. H., Haltrich, D. and Novalin, S. (2015). Fractionation of a galacto-oligosaccharides solution at low and high temperature using nanofiltration. *Separation and Purification Technology*. **151**, 124–130.

Rabiu, B. A., Jay, A. J., Gibson, G. R. and Rastall, R. A. (2001). Synthesis and fermentation properties of novel galacto-oligosaccharides by β-galactosidases from *Bifidobacterium* species. *Applied Environmental Microbiology*. **67**(6), 2526–2530.

Rodriguez-Colinas, B., Fernandez-Arrojo, L., Ballesteros, A. O. and Plou, F. J. (2014). Galactooligosaccharides formation during enzymatic hydrolysis of lactose: Towards a prebiotic-enriched milk. *Food Chemistry*. **145**, 388–394.

Rodriguez-Colinas, B., Kolida, S., Baran, M., Ballesteros, A. O., Rastall, R. A. and Plou, F. J. (2013). Analysis of fermentation selectivity of purified galacto-oligosaccharides by in vitro human faecal fermentation. *Applied Microbiology Biotechnology*. **97**(13), 5743–5752.

Rodriguez-Colinas, B., Poveda, A., Jimenez-Barbero, J., Ballesteros, A. O. and Plou, F. J. (2012). Galacto-oligosaccharide synthesis from lactose solution or skim milk using the β-galactosidase from *Bacillus circulans*. *Journal of Agricultural and Food Chemistry*. **60**(25), 6391–6398.

Rojas, A., Nagem, R., Neustroev, K., Arand, M., Adamska, M., Eneyskaya, E., Kulminskaya, A., Garratt, R., Golubev, A. and Polikarpov, I. (2004). Crystal structures of β-galactosidase from *Penicillium* sp. and its complex with galactose. *Journal of Molecular Biology*. **343**(5), 1281–1292.

Roy, D., Daoudi, L. and Azaola, A. (2002). Optimization of galacto-oligosaccharide production by *Bifidobacterium infantis* RW-8120 using response surface methodology. *Journal of Industrial Microbiology and Biotechnology*. **29**(5), 281–285.

Rycroft, C., Jones, M., Gibson, G. and Rastall, R. (2001). A comparative in vitro evaluation of the fermentation properties of prebiotic oligosaccharides. *Journal of Applied Microbiology*. **91**(5), 878–887.

Rye, C. S. and Withers, S. G. (2000). Glycosidase mechanisms. *Current Opinion in Chemical Biology*. **4**(5), 573–580.

Sako, T., Matsumoto, K. and Tanaka, R. (1999). Recent progress on research and applications of non-digestible galacto-oligosaccharides. *International Dairy Journal.* **9**(1), 69–80.

Sangwan, V., Tomar, S., Singh, R., Singh, A. and Ali, B. (2011). Galactooligosaccharides: Novel components of designer foods. *Journal of Food Science.* **76**(4), R103–R111.

Searle, L. E., Best, A., Nunez, A., Salguero, F. J., Johnson, L., Weyer, U., Dugdale, A. H., Cooley, W. A., Carter, B. and Jones, G. (2009). A mixture containing galactooligosaccharide, produced by the enzymic activity of *Bifidobacterium bifidum*, reduces *Salmonella enterica* serovar *Typhimurium* infection in mice. *Journal of Medical Microbiology.* **58**(1), 37–48.

Searle, L. E., Cooley, W. A., Jones, G., Nunez, A., Crudgington, B., Weyer, U., Dugdale, A. H., Tzortzis, G., Collins, J. W. and Woodward, M. J. (2010). Purified galactooligosaccharide, derived from a mixture produced by the enzymic activity of *Bifidobacterium bifidum*, reduces *Salmonella enterica* serovar *Typhimurium* adhesion and invasion in vitro and in vivo. *Journal of Medical Microbiology.* **59**(12), 1428–1439.

Shigehisa, A., Sotoya, H., Sato, T., Hara, T., Matsumoto, H. and Matsuki, T. (2015). Characterization of a bifidobacterial system that utilizes galacto-oligosaccharides. *Microbiology.* **161**(7), 1463–1470.

Shin, H.-J., Park, J.-M. and Yang, J.-W. (1998). Continuous production of galacto-oligosaccharides from lactose by *Bullera singularis* β-galactosidase immobilized in chitosan beads. *Process Biochemistry.* **33**(8), 787–792.

Shin, H.-J. and Yang, J.-W. (1994). Galacto-oligosaccharide production by β-galactosidase in hydrophobic organic media. *Biotechnology Letters.* **16**(11), 1157–1162.

Shoaf, K., Mulvey, G. L., Armstrong, G. D. and Hutkins, R. W. (2006). Prebiotic galactooligosaccharides reduce adherence of enteropathogenic *Escherichia coli* to tissue culture cells. *Infection and Immunity.* **74**(12), 6920–6928.

Shoseyov O, Shani Z, Levy I. (2006). Carbohydrate binding modules: Biochemical properties and novel applications. *Microbiology and Molecular Biology Reviews.* **70**(2), 283–295.

Sinclair, H. R., de Slegte, J., Gibson, G. R. and Rastall, R. A. (2009). Galactooligosaccharides (GOS) inhibit *Vibrio cholerae* toxin binding to its GM1 receptor. *Journal of Agricultural and Food Chemistry.* **57**(8), 3113–3119.

Skálová, T., Dohnálek, J., Spiwok, V., Lipovová, P., Vondráčková, E., Petroková, H., Dušková, J., Strnad, H., Kralova, B. and Hašek, J. (2005). Cold-active β-galactosidase from *Arthrobacter* sp. C2-2 forms compact 660kDa hexamers: Crystal structure at 1.9 Å resolution. *Journal of Molecular Biology.* **353**(2), 282–294.

Solomon, H. V., Tabachnikov, O., Feinberg, H., Govada, L., Chayen, N. E., Shoham, Y. and Shoham, G. (2013). Crystallization and preliminary crystallographic analysis of GanB, a GH42 intracellular beta-galactosidase from *Geobacillus stearothermophilus. Acta Crystallographica Section F—Structural Biology Communications.* **69**(10), 1114–1119.

Splechtna, B., Nguyen, T. H. and Haltrich, D. (2007). Comparison between discontinuous and continuous lactose conversion processes for the production of prebiotic galacto-oligosaccharides using beta-galactosidase from *Lactobacillus reuteri*. *Journal of Agricultural and Food Chemistry*. **55**(16), 6772–6777.

Splechtna, B., Nguyen, T. H., Steinbock, M., Kulbe, K. D., Lorenz, W. and Haltrich, D. (2006). Production of prebiotic galacto-oligosaccharides from lactose using beta-galactosidases from *Lactobacillus reuteri*. *Journal of Agricultural and Food Chemistry*. **54**(14), 4999–5006.

Splechtna, B., Petzelbauer, I., Baminger, U., Haltrich, D., Kulbe, K. D. and Nidetzky, B. (2001). Production of a lactose-free galacto-oligosaccharide mixture by using selective enzymatic oxidation of lactose into lactobionic acid. *Enzyme and Microbial Technology*. **29**(6–7), 434–440.

Srivastava, A., Mishra, S. and Chand, S. (2015). Transgalactosylation of lactose for synthesis of galacto-oligosaccharides using *Kluyveromyces marxianus* NCIM 3551. *New Biotechnology*. **32**(4), 412–418.

Stiverson, J., Williams, T., Chen, J., Adams, S., Hustead, D., Price, P., Guerrieri, J., Deacon, J. and Yu, Z. (2014). Prebiotic oligosaccharides: Comparative evaluation using in vitro cultures of infants' fecal microbiomes. *Applied and Environmental Microbiology*. **80**(23), 7388–7397.

Torres, D. P. M., Goncalves, M. P., Teixeira, J. A. and Rodrigues, L. R. (2010). Galacto-oligosaccharides: Production, properties, applications, and significance as prebiotics. *Comprehensive Reviews in Food Science and Food Safety*. **9**(5), 438–454.

Tzortzis, G., Goulas, A. K., Gee, J. M. and Gibson, G. R. (2005a). A novel galactooligosaccharide mixture increases the bifidobacterial population numbers in a continuous in vitro fermentation system and in the proximal colonic contents of pigs in vivo. *Journal of Nutrition*. **135**(7), 1726–1731.

Tzortzis, G., Goulas, A. K. and Gibson, G. R. (2005b). Synthesis of prebiotic galactooligosaccharides using whole cells of a novel strain, *Bifidobacterium bifidum* NCIMB 41171. *Applied Microbiology and Biotechnology*. **68**(3), 412–416.

Tzortzis, G. and Vulevic, J. (2009). Galacto-oligosaccharide prebiotics. In D. Charalampopoulos and R. A. Rastall (eds.), *Prebiotics and Probiotics Science and Technology*. New York: Springer, pp. 207–244.

Urrutia, P., Bernal, C., Wilson, L. and Illanes, A. s. (2014). Improvement of chitosan derivatization for the immobilization of *Bacillus circulans* β-galactosidase and its further application in galacto-oligosaccharide synthesis. *Journal of Agricultural and Food Chemistry*. **62**(41), 10126–10135.

van Leeuwen, S. S., Kuipers, B. J., Dijkhuizen, L. and Kamerling, J. P. (2014a). ^{1}H NMR analysis of the lactose/β-galactosidase-derived galacto-oligosaccharide components of Vivinal® GOS up to DP5. *Carbohydrate Research*. **400**, 59–73.

van Leeuwen, S. S., Kuipers, B. J., Dijkhuizen, L. and Kamerling, J. P. (2014b). Development of a ^{1}H NMR structural-reporter-group concept for the analysis of prebiotic galacto-oligosaccharides of the [β-d-Galp-(1→x)] n-d-Glcp type. *Carbohydrate Research.* **400**, 54–58.

Wang, K., Lu, Y., Liang, W. Q., Wang, S. D., Jiang, Y., Huang, R. and Liu, Y. H. (2012). Enzymatic synthesis of galacto-oligosaccharides in an organic-aqueous biphasic system by a novel beta-galactosidase from a metagenomic library. *Journal of Agricultural and Food Chemistry.* **60**(15), 3940–3946.

Wheatley, R. W., Kappelhoff, J. C., Hahn, J. N., Dugdale, M. L., Dutkoski, M. J., Tamman, S. D., Fraser, M. E. and Huber, R. E. (2012). Substitution for Asn460 cripples beta-galactosidase (*Escherichia coli*) by increasing substrate affinity and decreasing transition state stability. *Journal of Agricultural and Food Chemistry.* **521**(1–2), 51–61.

Wheatley, R. W., Lo, S., Jancewicz, L. J., Dugdale, M. L. and Huber, R. E. (2013). Structural explanation for allolactose (lac operon inducer) synthesis by lacZ β-galactosidase and the evolutionary relationship between allolactose synthesis and the lac repressor. *Journal of Biological Chemistry.* **288**(18), 12993–13005.

Wu, Y., Yuan, S., Chen, S., Wu, D., Chen, J. and Wu, J. (2013). Enhancing the production of galacto-oligosaccharides by mutagenesis of *Sulfolobus solfataricus* beta-galactosidase. *Food Chemistry.* **138**(2–3), 1588–1595.

Yanahira, S., Kobayashi, T., Suguri, T., Nakakoshi, M., Miura, S., Ishikawa, H. and Nakajima, I. (1995). Formation of oligosaccharides from lactose by *Bacillus circulans* beta-galactosidase. *Bioscience, Biotechnology and Biochemistry.* **59**(6), 1021–1026.

Yi, S. H., Alli, I., Park, K. H. and Lee, B. (2011). Overexpression and characterization of a novel transgalactosylic and hydrolytic beta-galactosidase from a human isolate *Bifidobacterium breve* B24. *New Biotechnology.* **28**(6), 806–813.

Yu, L. and O'Sullivan, D. (2014). Production of galactooligosaccharides using a hyperthermophilic β-galactosidase in permeabilized whole cells of *Lactococcus lactis. Journal of Dairy Science.* **97**(2), 694–703.

6

Biodegradation and Utilization of Hemicellulose

XIN ZHOU, HONGYU ZHANG, AND YONG XU

Contents

6.1 Introduction

The term "hemicellulose" was first introduced by Schulze in 1891 to describe the fractions isolated or extracted from plant materials with dilute alkali (Schulze 1891). Hemicellulose, which is regarded as the second most abundant polysaccharide in nature after cellulose, is made up of approximately 20–35% of lignocellulosic biomass, generated in large concentration through conversion of agricultural, forest, fruit, and vegetable processing operation as residual waste. Moreover, hemicellulose is composed of the heterogeneous polymer of pentoses (xylose and arabinose), hexoses (mannose, glucose, and galactose), and uronic acids. The sugars from hemicellulose can be converted to various valued products by chemical and biorefined processing. Therefore, hemicellulose is getting more and more attention as a potential source to provide a major feedstock for the production of biofuels, biochemicals, and biomaterials (Wyman et al. 2005).

6.1.1 Existence State and Distribution

Diverse polysaccharides are present in the majority of cell walls, the specific makeup varying in different species and tissues. The plant cell wall can usually be grouped into cellulose, hemicelluloses, and proteins, and in some cells of terrestrial plants, the cell wall is surrounded by phenolic compounds (Bauer et al. 1973). Therefore, cellulose is the main structural polysaccharide of plant cell walls, which is a linear β-(1,4)-D-glucopyranoside polymer. Pectins, known as heterogeneous polysaccharides, contain a large amount of galacturonic acid residues. In contrast with cellulose, hemicelluloses have branches with short lateral chains consisting of different sugars. They usually consist of xyloglucans, xylans, mannans, glucomannans, and β-(1,3/1,4)-glucans.

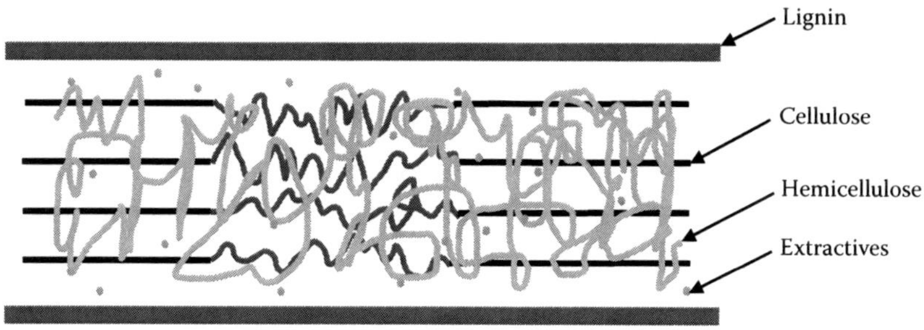

Figure 6.1 Schematic of cellulose, hemicellulose, lignin, and extractives in plants.

The hemicelluloses perform a very important biological function for the plants (Bauer et al. 1973; Jeffries 1994; Scheller and Ulvskov 2010). Generally, the accepted view for the main function of hemicelluloses is that they take part in building the cell wall structure and the regulation of the cell growth process. The main structure and distribution in the plant cell wall is described in Figure 6.1.

6.1.2 Applications and Potential

For the lignocellulosic biomass, estimated global annual production is about 1×10^{11} tons (Smeets et al. 2007) and is regarded as a significant renewable resource for bioproducts. Lignocellulosic biomass, in general, is composed of 40–50% cellulose and 25–30% hemicellulose, which can be hydrolyzed to various sugars. Therefore, lignocellulosic biomass represents an abundant, renewable, widespread, and low-cost source of sugar platforms for the bioproduction of special and value-added chemicals. Meanwhile, how to effectively and integrally utilize all of the lignocellulosic biomass to add eco-friendly value is currently the global challenge (Singhania et al. 2009).

Hemicellulose, as the second most abundant polysaccharide substance after cellulose (Menon and Rao 2012; Zhu and Zhuang 2012), consists of various monosaccharides that can be converted into ethanol, xylitol, lactic acid, functional oligosaccharides, and other valuable products by biochemical or chemical conversion. Nowadays, xylonic acid from the oxidization of xylose is currently generating considerable interest with significant economic potential as a valuable platform chemical (Karimi 2015; Zhou et al. 2015).

6.2 Complex Structure of Hemicellulose

The majority of the polysaccharides in the plant resources are the cellulose, hemicellulose, and pectin classes. The major function of cellulose is regarded as being the rigid, load-bearing component of the plant cell wall, with a linear polymer structure of β-D-glucopyranose units forming microfibril that strengthens the cell wall within hemicelluloses bonded to other plant cell wall components by covalent and hydrogen bonds, and ionic and hydrophobic interactions (Bauer et al. 1973; Spiridon and Popa 2008). Hemicelluloses consist of D-xylose, D-mannose, D-galactose, D-glucose, L-arabinose, 4-O-methylglucuronic, D-galacturonic, and D-glucuronic acids and are interspersed with the microfibrils of cellulose, conferring consistency and flexibility to the structure of the plant cell wall (Bauer et al. 1973; Held et al. 2015).

The chemical constitution and structure of hemicellulose are varied mainly from composition of main and branched chains in different plants. The main chain generally comprises one or more types of glycosyls; moreover, the connections between glycosyls are also different. Therefore, to illustrate the chemical structure of the hemicellulose, glycans are classified into different types. It is generally believed that hemicellulose is the xylan in the matrix of the cell, and the main components are xylan, xyloglucan, glucomannan, mannan, galactomannan, and so on (Sjöström 1981; Yin and Fan 1999). Table 6.1 lists main types of polysaccharides present in lignocellulosic biomass (Ebringerova et al. 2005; Gírio et al. 2010).

6.2.1 Xylan

Unlike cellulose, hemicelluloses are not homogeneous. The term "xylan" is a catchall for a diverse group of polysaccharides. Xylan is the major component present in hardwood and grass hemicellulose, and it occurs as a β-(1,4)-D-xylopyranose (xylose) backbone with a variety of branch chains, which consist of O-acetyl, α-L-arabinofuranosyl, α-1, 2-linked glucuronic, or 4-oxymethylglucuronic acid substituents, depending on the different kinds of plants (Kormelink and Voragen 1993; Xu et al. 2003; Yang 2008; Scheller and Ulvskov 2010; Peng et al. 2012).

Table 6.1 Main Types of Di-, Oligo-, and Polysaccharides Present in Biomass

SACCHARIDE TYPE	BIOLOGICAL ORIGIN	AMOUNT (%, DRY WEIGHT)	UNITS: BACKBONE	UNITS: SIDE CHAINS	UNITS: LINKAGE	DP
Arabinogalactan	Softwoods	1–3	β-D-Galp	β-D-Galp α-L-Araf β-L-Arap	β-(1→6) α-(1→3) β-(1→3)	100–600
Xyloglucan	Hardwoods, softwood, grasses	2–25	β-D-Glcp β-D-Xylp	β-D-Xylp β-D-Galp α-L-Araf α-L-Fucp Acetyl	β-(1→4) α-(1→3) β-(1→2) α-(1→2) α-(1→2)	
Galactoglucomannan	Softwoods	10–25	β-D-Manp β-D-Glcp	β-D-Galp Acetyl	α-(1→6)	40–100
Glucomannan	Hardwoods and softwoods	2–5	β-D-Manp β-D-Glcp		β-(1→4)	40–70
Glucuronoxylan	Hardwoods	15–30	β-D-Xylp	4-O-Me-α-D-GlcpA Acetyl	α-(1→2)	100–200
Arabinoglucuronoxylan	Grasses and cereals, softwoods	5–10	β-D-Xylp	4-O-Me-α-D-GlcpAb-L-Araf	α-(1→2) α-(1→3)	50–185
Arabinoxylans	Cereals	0.15–30	β-D-Xylp	α-L-Araf-Feruloy	α-(1→2) α-(1→3)	
Glucuronoarabinoxylans	Grasses and cereals	15–30	β-D-Xylp	α-L-Araf 4-O-Me-α-D-GlcpA Acetyl	α-(1→2) α-(1→3)	

Sources: Ebringerova, A. et al., 2005, *Hemicellulose*; Gírio, F. M. et al., 2010, *Bioresource Technology*. **101**(13), 4775–4800.

Almost all terrestrial plants contain xylan, which is the predominant hemicellulose in the majority of plant cell walls, generally comprising approximately one-third of the total plant biomass (Prade 1996). In the plant cell, xylans are primarily found in the secondary cell wall and accumulated on the surface of cellulose microfibril as the cellulose synthase complex extrudes the microfibril. This deposition continues as the additional cell wall material is laid down, resulting in heavier xylan deposition on the cellulose microfibrils in the outermost layers of the secondary cell wall (Awano et al. 2002).

The function of xylans is regarded as forming cross-links between the other cell wall components, such as cellulose, lignin, other hemicellulose, and pectin. These interactions are always connected by hydrogen bonding to the other polysaccharides and by covalent linkages through the arabinofuranosyl side chains to the ferulic and coumaric acids found in lignin (Scheller and Ulvskov 2010).

6.2.2 Xyloglucan

Xyloglucan is abundant in primary walls of higher plants except for grasses, responsible for bounding cellulose, which is usually found in the middle lamella, primary walls, and gelatinous wall layer of higher plants (de Vries and Visser 2001). The main chains of xyloglucan and cellulose are composed of a β-(1,4)-linked β-D-Glcp (β-D-glucopyranose) backbone having 50–75% of the β-D-Glcp residues covalently linked to xylose at the O-6 position (Powlowski et al. 2009). Xylose can be further substituted by galactose or arabinose, and the galactose may be fucosylated. The glucose, xylose, and galactose, and its residues, ratio about 2.8:2.25:1.0. According to differences in family, xyloglucan may also contain fucose and arabinose. In addition, xyloglucans may also contain O-linked acetyl groups (Maruyama et al. 1996; Sims et al. 1996; Chen 2014). Cellulose microfibrils interact with xyloglucans by hydrogen bonds, thus contributing to integrating the structure of the cellulose network (Carpita and Gibeaut 1993). Xyloglucan is also a storage polymer in the seed endosperm of many plants (Goyal et al. 2007).

6.2.3 Mannans

The polysaccharides in lignocellulosic biomass containing mannose are also widely distributed and act as the main hemicellulose in charophytes

(Popper 2008). Therefore, mannan compounds include some different types, such as mannan, galactomannan, glucomannan, glucuronic acid mannan, and so on (Petkowicz et al. 2001). Mannan is formed by mannose residues connected by a β-(1,4) bond, except for galactomannan composed of β-(1,6) bonds linked to mannose and galactose.

Mannans are the major component of the hemicellulose fraction in softwoods and show a widely spread distribution in plant tissues (Preston 1968; Ximenes 1998). In plants, mannans play a significant structural role as hemicelluloses bound with cellulose (Preston 1968; Brennan et al. 1996; Liepman et al. 2007). In addition, they have been much studied in their role as seed storage compounds (Meier and Reid 1982). In gymnosperms, galactoglucomannans are major components of the secondary walls (Ebringerova et al. 2005). Mannans appear to have been very abundant in early land plants and are still abundant in mosses and lycophytes (Moller et al. 2007; Harholt et al. 2010). However, results obtained by Liepman et al. (2007) give evidence of an additional mannan function as a signaling molecule in plant growth and development.

6.2.4 Complex Heteroxylans

Moreover, many kinds of complex heteroxylans are displayed in cereals, seeds, gum exudates, and lucilages with a more complex structure (Stephen 1983). The β-(1,4)-xylose backbone is substituted by single uronic acid and arabinosyl residues and also with various mono- and oligoglycosyl branch chains (Gírio et al. 2010).

6.3 Biodegradation and Hydrolysis of Hemicellulose

6.3.1 Degradation of Hemicellulose

Hemicellulose is a heteropolymer and is characterized by a random, amorphous, and complex structure with little strength. It can be depolymerized into a simple form of monomer sugars by methods of dilute acid or enzymatic hydrolysis (Chandel et al. 2010; Chaturvedi and Verma 2013). A mild pretreatment with dilute acid or base prior to enzymatic hydrolysis could be effective for depolymerization of hemicelluloses into the monomer sugars. Nowadays, the most promising method for hydrolysis of lignocellulosic biomass to monomer sugars is by use of enzymes, mainly including cellulases and hemicellulases (Chandel et al. 2010). Moreover, hemicellulases are conducive to cellulose hydrolysis by exposing the cellulosic fibers to some extent; that is to say, it can improve accessibility of cellulases, thereby increasing the efficiency of enzymatic hydrolysis (Shallom and Shoham 2003). Lignin interfering with hydrolysis is generally caused by blocking the access of cellulases to cellulose and the irreversible bind with these enzymes. Therefore, the preceded removal of the lignin and hemicellulose by pretreatment is necessary for maximizing enzymatic hydrolysis (McMillan 1994).

Various pretreatment processes can be used for fractionating, solubilizing, hydrolyzing, and separating cellulose, hemicellulose, and lignin. The common methods include acid, alkaline, steam explosion, wet oxidation, liquid hot water, organic solvent treatments, ionic liquids, and supercritical fluids (Dale 1982; Gould 1984; Clark and Mackie 1987; Chum et al. 1988; Goldstein and Easter 1992; Koullas et al. 1993; Dale et al. 1996; Schmidt and Thomsen 1998; Kaar and Holtzapple 2000; Fernandez-Bolanos et al. 2001; Laser et al. 2002; Miyafuji et al. 2005; Ohno and Fukaya 2009; Zavrel et al. 2009).

6.3.2 Main Hemicellulase

During pretreatment processing, hemicellulose is commonly decomposed and removed from lignocellulosic materials by chemical and microbial degradation, and then is further hydrolyzed into fermentable sugars by diverse hemicellulases (Shallom and Shoham 2003; Chandel et al. 2010; Chaturvedi and Verma 2013).

A wide array of enzymes are needed for the complete hydrolysis of hemicelluloses because of the complex structures composed of various building blocks and polymer chain structures. In general, hemicellulases are a broad group of enzymes, which can be classified into depolymerizing enzymes that act on the main chain of hemicellulose and debranching enzymes (also referred to as accessory enzymes) that act on the polymer of branches. Meanwhile, backbone-depolymerizing hemicellulases are hydrolases with different specificities for the various structures of hemicellulose (Decker et al. 2008). The main hemicellulases are described in Table 6.2.

Table 6.2 Lists of Some Enzymes Involved in the Degradation of Hemicellulose and Their Modes of Action

ENZYME	MODE OF ACTION
Endoxylanase	Hydrolyzes interior β-1,4-xylose linkages of the xylan backbone
Exoxylanase	Hydrolyzes β-1,4-xylose linkages
β-Xylosidase	Hydrolyzes xylobiose or short-chain XOS to xylose
α-Arabinofuranosidase	Hydrolyzes terminal nonreducing α-arabinofuranose from arabinoxylans
α-Glucuronidase	Releases glucuronic acid from glucuronoxylans
Acetylxylan esterase	Hydrolyzes acetylester bonds in acetyl xylans
Ferulic acid esterase	Hydrolyzes feruloylester bonds in xylans
Endo-β-1,4-mannanase	Hydrolyzes mannan-based hemicelluloses, liberates β-1,4-manno-oligomers
Endo-β-1,4-mannosidase	Hydrolyzes β-1,4-manno-oligomers to mannose
Acetyl mannan esterase	2- or 3-O-acetyl xylan

Source: Shallom, D. and Shoham, Y., 2003, *Current Opinion in Microbiology.* **6**(3), 219–228.

6.3.2.1 Xylanase Generally, degradation of xylans requires endo-β-xylanase, β-xylosidase, and several accessory enzymes. These enzyme components contain α-L-arabinofuranosidase, β-glucuronidase, acetylxylan esterase, ferulic acid esterase, and p-coumaric acid esterase, which are necessary for hydrolysis of diverse substituted xylans (Bakır 2002; Shallom and Shoham 2003).

Endo-1,4-β-xylanase is responsible for cleaving the glycosidic bond in the xylan backbone, and thus the degree of polymerization of the substrate can be reduced. Xylan cannot be attacked randomly by the endo-1,4-β-xylanase, as the bonds selected for hydrolysis depend on the nature of the substrate, i.e., degree of branching, chain length, and presence of substituents (Reilly 1981). Initially, the materials of hemicellulose are hydrolyzed into β-D-xylopyranosyl oligomers; later, they will continue to be converted into small molecules such as mono-, di-, and trisaccharides of β-D-xylopyranosyl (Saha 2003; Polizeli et al. 2005).

Compared to endoxylanases and β-xylosidases, exoxylanases also act as a significant variation in the degradation of xylan. Endoxylanases cleave the xylan backbone from the inside producing long-chain xylo-oligomers. Conversely, exoxylanases act on the xylan backbone from the reducing end yielding short-chain oligomers or hydrolyzing short-chain xylo-oligomers with a degree polymerization of ≥2–3 (Honda and Kitaoka 2004).

β-Xylosidase (1,4-β-D-xylan xylohydrolase) is mainly responsible for hydrolyzing short xylo-oligomers chains into the monosaccharide. The mode of enzyme action is prior to removing the monomers from the nonreducing end of the xylo-oligomer with affinity increasing and decreasing degree of polymerization (DP); (Biely 1993). In addition, the best substrate for β-xylosidase is xylobiose and their affinity for xylo-oligosaccharides (XOS), which are inversely proportional to its degree of polymerization. The main action of β-xylosidase is to act after the xylan has been hydrolyzed by exo- and endoxylanase. Due to the short oligomers that accumulated in the reaction, it can inhibit the activity of endoxylanase, and thus the action of β-xylosidase can decrease the concentration of short oligomers, which improves the performance of xylan hydrolysis. Moreover, an important consideration about using the β-xylosidase is also its susceptibility for xylose end product (Poutanen et al. 1991).

Due to hemicellulosic xylan complex constitutes, general completion of xylan hydrolysis requires the action of a complete enzyme system, which is usually composed of β-xylanase, β-xylosidase, and other accessory enzymes such as α-L-arabinofuranosidase, α-glucuronidase, acetylxylan esterase, and hydroxycinnamic acid esterases that are used for cleaving side-chain residues from the xylan backbone.

Figure 6.2 describes the mechanism of xylanolytic enzymes involved in the degradation of hemicellulose (Shallom and Shoham 2003). All these enzymes work cooperatively on converting xylan to its constituents (Biely 1993). Xylanases attack the backbone of xylan to produce both substituted and nonsubstituted shorter-chain oligomers, xylobiose, and xylose. Xylosidases are responsible for the breakdown of xylan as they hydrolyze XOS to xylose (Deshpande et al. 1986). The enzymes arabinosidase, α-glucuronidase, and acetylxylan esterase act synergistically with the xylanases and xylosidases by releasing the substituents on the xylan backbone to implement a complete hydrolysis of xylan to monosaccharides (Gilbert and Hazlewood 1993).

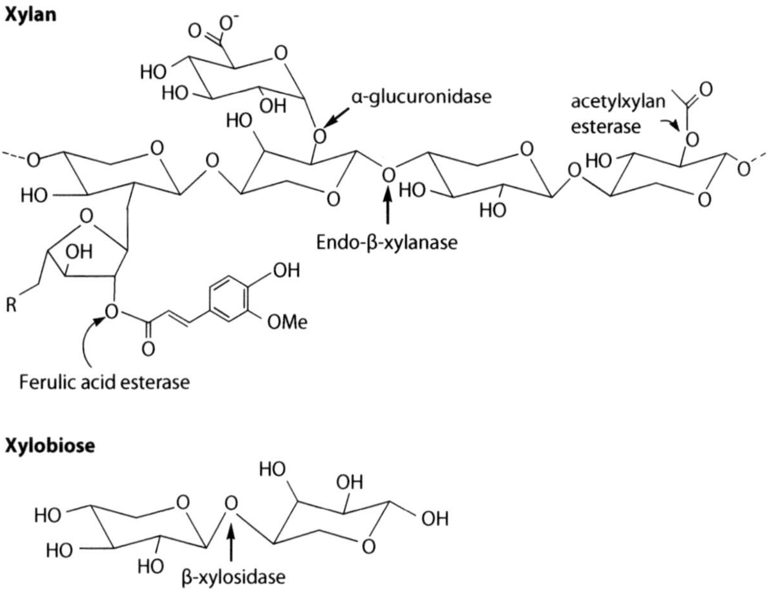

Figure 6.2 Xylan structure showing different substituent groups with sites of attacks by enzymes.

6.3.2.2 Mannase The main sugar moiety of galactoglucomannans is D-mannose, which is also a significant component of the hemicellulose family, and it can be grouped into four subfamilies: linear mannan, glucomannan, galactomannan, and galactoglucomannan (Petkowicz et al. 2001). Thus, hydrolysis of mannans also requires a wide array of enzymes (Moreira and Filho 2008). Endo-1,4-β-mannanase is essential for cleaving the glucomannan main chain to oligosaccharides leaving a mannose residue on the reducing end (Tenkanen et al. 1997). The oligosaccharides will be further hydrolyzed into mannose and glucose by β-mannosidase and β-glucosidase. The action of endo-1,4-β-mannanases is restricted by the substitution groups on the backbone, and the galactosyl side groups can be cleaved off by α-galactosidase (McCleary and Matheson 1983; Tenkanen et al. 1997). Acetylmannan esterase acts as catalyzing the deacetylation of the glucomannan backbone in softwood (Tenkanen et al. 1995).

6.4 Production of High-Valued Chemicals from Hemicellulose

In recent years, the conversion of hemicellulose has obtained more and more attention because of its potential wide spread applications in different processes for the production of value-added chemicals. Various materials from agriculture, forestry, and fruit and vegetable processing operations are valuable as the residual waste contains about 20–35% hemicellulose, which is composed of pentoses (xylose and arabinose), hexoses (mannose, glucose, and galactose), and sugar acids, which provide a renewable, widespread, and low-cost carbohydrate source for the production of special and high-valued products (Jeffries 1994; Wyman et al. 2005).

6.4.1 Bioconversion and Fermentation of Xylose

Xylose, the predominant sugar of hemicellulose, constitutes one-third of the total carbohydrate sugars in lignocellulosic biomass. Thus, cost-effective methods for the large-scale conversion of lignocellulosic materials require efficient utilization of xylose which is still a critical bottleneck for the commercialization of lignocellulose biorefinery (Rodriguez and O'Connell 2011) or not readily utilized like glucose bioprocessing. Some typical bioconversion applications are summarized as follows.

6.4.1.1 Xylitol D-Xylitol is a five-carbon sugar alcohol, which can be found in many fruits and vegetables as a natural composition in low amounts (Hyvönen et al. 1982). The industrial production of xylitol from hemicellulosic xylose has been used for decades as sweetener additives of food, oral product, pharmacy, and cosmetics (Emodi 1978; Saha and Bothast 1997; Edgar 1998). The current value of the xylitol market is about $340 million with a global consumption of 50,000 tons (Kadam et al. 2008).

Generally, both chemical and biological methods can be used for the production of xylitol from xylan-rich agricultural and wood residues (Winkelhausen and Kuzmanova 1998). The chemical method of xylose hydrogenation demands huge production costs in high temperatures and pressure input. It also forms some by-products that require expensive separation and purification. In addition, xylose should be purified before hydrogenation, which further increases the investment and costs for xylitol chemical process (Santos et al. 2005). Conversely, xylitol production by bioconversion is usually operated under mild conditions and could achieve higher yield that will lead to an eco-friendly process. Hence, the biological process for xylitol production is attracting wider attention (Saha and Bothast 1997). Several bacteria, such as filamentous fungi and *Candida* spp., can implement the biotransformation of xylose to xylitol. Different strategies such as entrapment of xylitol-producing microorganisms, cell recycling, fed-batch cultivation, and using metabolically engineered cells have been suggested for improvement of the yield and hence the economy of xylose production by fermentation (Granstrom et al. 2007; Sjoman et al. 2008; Cheng, Zhang et al. 2010; Chandel et al. 2012; Su et al. 2013; Wang et al. 2013).

Although thus far the chemical conversion of xylose is still used for production of xylitol, various bioconversion methods have been explored for the production of xylitol from hemicellulose using microorganisms or their enzymes (Nigam and Singh 1995). And some research-based companies try to commercialize the biotechnological production method such as the kinetics of xylitol crystallization from lignocellulosic hydrolysates (Martinez et al. 2007).

6.4.1.2 Ethanol Currently, ethanol biofuel is still the most strategic product for the economic success of lignocellulose biorefinery. Bioproduction of biofuels from lignocellulosic materials is an optimal alternative for the second generation of biofuels, which was proposed as a solution for food-energy conflict that resulted from the first-generation starch ethanol fuel (Menon and Rao 2012; Zhu and Zhuang 2012).

Xylose is not readily and efficiently utilized as glucose for the production of ethanol by microorganisms such as *Saccharomyces cerevisiae* and *Zymomonas mobilis*. As documented in recent reviews of xylose fermentation, several microorganisms have been recognized to be capable of fermenting lignocellulosic xylose to ethanol (McMillan 1993; Nigam 2001; Hahn-Hägerdal and Pamment 2004; Mohagheghi et al. 2004; O'Brien et al. 2004; Sedlak and Ho 2004; Saha et al. 2005). Table 6.3 lists microbiological genera that contain species capable of direct xylose fermentation. Several recombinant bacterial and yeast genera incapable of direct high-yield xylose fermentation are listed (McMillan 1993).

Table 6.3 Main Genera with the Fermentability of Converting of Xylose to Ethanol

EUKARYOTES	WILD TYPE
	Brettanomyces
	Candida
	Clavispora
	Fusarium
	Kluyveromyces
	Monilia
	Neurospora
	Pachysolen
	Paecilomyces
	Pichia
	Polyporus
	Rhizopus
	Schizosaccharomyces
	RECOMBINANT
	Saccharomyces
	Schizosaccharomyces
PROKARYOTES	WILD TYPE
	Aerobacter
	Bacillus
	Bacteroides
	Clostridium
	Erwinia
	Klebsiella
	Thermoanaerobacter
	RECOMBINANT
	Erwinia
	Escherichia
	Klebsiella
	E. coli

The conversion process from lignocellulose to ethanol, with the aid of microorganisms, always occurs through four events: production of saccharolytic enzymes, hydrolysis of the polysaccharides present in pretreated biomass, fermentation of hexose, and fermentation of pentose to ethanol (Hahn-Hägerdal et al. 1994). Commercial exploitation of these microorganisms for ethanol production from xylose is still restricted by their inefficient ethanol tolerance, low yield and productivity, and extra sensitivity to lignocellulose degraded inhibitors, although efforts had been carried out for years (Nigam 2001; Mohagheghi et al. 2004; Sedlak and Ho 2004; Saha et al. 2005). In general, the technological process of high ethanol yield still requires high productivity to be developed in order to reduce costs. To improve the microorganic ability of inhibitor tolerance, temperature tolerance for utilizing hemicellulose-sugars is also essential.

6.4.1.3 Xylonic Acid Thus far, the efficient utilization of xylose remains a critical bottleneck for the commercialization of lignocellulosic biorefinery products (Rodriguez and O'Connell 2011). As a result, new technologies and products must be developed in order to achieve economic feasibility. Xylonic acid, from xylose, has applications similar to gluconic acid and serves as its substitute. It has been identified as a valuable platform chemical, ranking in the top 30 high-value chemicals by NREL and PNNL. Xylonic acid and its derivated xylonates have a promising potential to be used in a wide variety of applications including as a chelator, a dispersant of concrete, a retardant for oil wells, a precursor of 1,2,4-butanetriol, an edible acidifier, a polyester, and so on (Chun et al. 2006). Xylonic acid can be produced in laboratories by enzymatic (Holscher et al. 2009), electrochemical or chemical oxidation. However, these methods are inaccessible for industrial production because of their pollution and cost. Fortunately, the bioconversion of xylose to xylonic acid has a promising future because of its high efficiency (Buchert and Viikari 1988; Xu, Wang et al. 2012; Yu, Yong et al. 2012).

Xylonic acid can be formed in the first step of oxidative metabolism of xylose by some archaea and bacteria via the action of xylose or glucose dehydrogenases. And xylonic acid has been produced from various bacteria, in particular *Gluconobacter oxydans* and *Pseudomonas putida* (Buchert et al. 1988). Moreover, high yields of xylonic acid from xylose make *G. oxydans* an attractive choice for xylonic acid production. Recently, xylonic acid has been produced by the genetically modified bacterium *Escherichia coli* and yeast *S. cerevisiae* and *Kluyveromyces lactis* (Toivari et al. 2010, 2012).

Recently, Zhou et al. established a novel method of pure oxygen supplied bioreactor for the efficient bioconversion of xylose to xylonic acid. By means of this fantastic system of compressed oxygen supply in a stirred and sealed tank reactor (COS-SSTR), 586.3 g/L of xylonic acid could be produced from pure xylose via fed-batch fermentation, and 143.9 g/L of xylonic acid could be obtained from an in situ and fed-batch fermentation of crude acidic lignocellulosic pre-hydrolysates without any detoxification step. These results greatly encourage industrial production and application of hemicellulosic xylose (Zhou et al. 2015).

6.4.1.4 2,3-Butanediol 2,3-Butanediol is an interesting chemical feedstock because of its large number of industrial applications, such as in the manufacture of inks, perfumes, fumigants, explosives, plasticizers, foods, pharmaceutical solvents, antifreeze agents, and use for synthetic polymers and resins (Syu 2001; Saha 2003). Due to the long-term prospects of rising petroleum prices, 2,3-butanediol from renewable biomass by the promising microbial route has attracted more and more attention worldwide for developing an eco-friendly economy and a portal to a more sustainable future (Li et al. 2010).

Microbial 2,3-butanediol production already has a long history that was first investigated in 1906 by Harden and Walpole and in 1912 by Harden and Norris (Magee and Kosaric 1987).

Hexose and pentose all can be converted to 2,3-butanediol. In bacterial metabolism, monosaccharides must first be converted to pyruvate before generation of major products. From glucose, pyruvate is formed in a relatively simple manner by the Embden–Meyerhof pathway (EMP). For xylose, the production of pyruvate must proceed by a combination of pentose phosphate pathway (PPP) and EMP (Jansen and Tsao 1983; Jansen et al. 1984). Species that are reported for this ability include *Aeromonas* (Willetts 1984), *Bacillus* (Groleau et al. 1985), *Paenibacillus* (Nakashimada et al. 2000), *Serratia*, *Aerobacter* (Kosaric and Velikonja 1995), *Enterobacter* (Saha and Bothast 1999), and also *Klebsiella* (Yu and Saddler 1982). Until now, *K. pneumoniae*, *Klebsiella oxytoca*, and *P. polymyxa* have been the most widely used for the efficient production of 2,3-butanediol.

So far, some research has been conducted to utilize the hemicellulose portion of forest residues for 2,3-butanediol production (Willetts 1984; Grover et al. 1990; Afschar et al. 1991; Motwani et al. 1993; Perego et al. 2003; Sun et al. 2009). Cheng et al. (2010) have reported production of 2,3-butanediol from corn cob hydrolysate by fed-batch fermentation using *K. oxytoca*, and the maximal 2,3-butanediol production was 35.7 g/L obtained after 60 h of fed-batch fermentation, giving a yield of 0.5 g/g reducing sugar and a productivity of 0.59 g/h/L.

6.4.1.5 Lactic Acid Lactic acid is one kind of organic acid that has many applications in food, cosmetics, pharmaceuticals, leather, textiles, and the production of polylactic acid is considered a promising substitute for synthetic biodegradable plastics (Datta et al. 1995; Brown 2003).

Nowadays, lactic acid can be produced via biological and chemical methods. However, the chemical method has a limited potential for expansion, and therefore, increased demand for lactic acid should be supplied through the fermentation process. Approximately 90% of lactic acid production is performed by biological routes. Higher production volume dictates the requirement for a low-cost and widely available feedstock for the substitute of current sugar and starchy substrates. Thus, lignocellulosic materials have been regarded as suitable alternative substrates for lactic acid production (Hofvendahl and Hahn-Hägerdal 2000).

Generally, lignocellulosic materials are mainly composed of hexose and pentose. Therefore, many strains of lactic acid bacteria such as *lactobacilli*, which is most widely used for fermentation of lactic acid, are not able to ferment pentose sugars. In recent years, the utilization of lignocellulose as raw material for lactic acid production has been developed for efficient utilization of xylose (Yoshida et al. 2011). For example, *L. xylosus* (Tyree et al. 1990) and *L. rhamnosus* (Iyer et al. 2000) have been successfully applied in media containing a mixture of xylose and glucose and acidic hemicellulosic hydrolysates of wood, respectively. *L. pentosus* allowed obtaining 33 g/L of lactic acid and 17 g/L of acetic acid from detoxified hemicellulosic liquor from reeds (Perttunen et al. 2001), and 44.8 g/L of lactic acid and 6.5 g/L of acetic acid from concentrated hemicellulosic hydrolysates of trimming vine shoots (Bustos et al. 2005). In addition, genetic modification of lactic acid bacteria has already been performed (Hofvendahl and Hahn-Hägerdal 2000; Ou et al. 2011).

6.4.2 Chemical Conversion of Hemicellulose

6.4.2.1 Furfural Recently, furfural, the sleeping beauty biorenewable building block chemical, has received renewed attention as a biobased alternative for the production of biofuels and biochemicals. Wide applications are industrially known for furfural including oil refining, plastics, and pharmaceutical and agrochemical industries.

Furfural usually can be derived by acid hydrolysis of pentosans. It is regarded as the only organic compound derived from biomass that can replace the crude oil-based organics used in industry. Furfural is used as a chemical feedstock for furfuryl alcohol (in the production of furan resins), in the manufacture of furan (as an intermediate in the synthesis of pharmaceuticals, agricultural chemicals, stabilizers, and fine chemicals), as well as a solvent for refining lubricating oils, as a decolorizing and wetting agent (Mamman et al. 2008; Marinova et al. 2010; Menon et al. 2010). In addition, world markets for furfural are currently estimated to be about 200,000 to 210,000 ton/year of which about 60–62% is used for the production of furfuryl alcohol (Win 2005).

The preferred alternative for furfural production are agricultural residues because of their homogeneity and regular availability in large quantities from food processing plants (Montane et al. 2002). Nowadays, furfural is mainly synthesized from pentoses available in hemicellulosic hydrolysate through the cyclic dehydration reactions with sulfuric acid. In the process, the pentosans or xylans present in the hemicelluloses are first acid-hydrolyzed into pentoses or xylose, and then successively, cyclodehydration of the latter forms furfural. Some mild conditions can also be used for dehydration of xylose into furfural followed by initial hydrolysis of the pentosans or xylans (Perego and Bianchi 2010; Serrano et al. 2010; Sahu and Dhepe 2012). Although the presence of a developed industrial process for furfural production exists, the use of conventional acid catalysts (such as H_2SO_4 or HCl) continue to cause serious undesirable side reactions such as environmental problems and the formation of various toxic wastes (Vazquez et al. 2007). Hence, eco-friendly catalytic processes also need to be developed to eliminate environmental corrosion as well as waste-disposal problems.

6.4.2.2 Functional Oligosaccharides Production from Hemicellulose XOS are oligosaccharides containing 2–10 xylose molecules linked by 1–4 bonds, as shown in Figure 6.3. Considered as nondigestible oligosaccharides, XOS can be used as dietary sweeteners in low-calorie diet foods. Furthermore, XOS can be used as a feed additive, which is beneficial for microorganisms in the gut such as *Bifidobacterium bifidum* and *Bifidobacterium adolescentis*. They are considered to improve bowel functions (Zhang et al. 2007; Carvalho et al. 2013). Xu et al. reported that XOS could effectively proliferate *B. bifidum* and *B. adolescentis* under anaerobic conditions in vitro. Cell density of *B. bifidum* and *B. adolescentis* could be increased respectively by 1.8 and 4.7 times when they grew in 5 g/L XOS after 30 h. Microbes in layers could also be proliferated with the XOS products, and as a result of the proliferation by XOS, the layers gain health protection and the yield of eggs rises remarkably (Yong et al. 2005). As substitutes for antibiotics in feed, XOS also have a huge market potential and economic benefits (Carvalho et al. 2013).

Nowadays, XOS are mainly produced from xylan-rich lignocellulosic material by the method of acid or enzymatic hydrolysis (Otieno and Ahring 2012; Samanta et al. 2012; Carvalho et al. 2013). Generally, in lignocellulosic biomass, xylan exists in xylan–lignin complex and becomes resistant to hydrolysis (Zhu et al. 2006). Therefore, XOS production is carried out in two treatment steps. The first step is the xylan extraction from lignocellulosic material, which includes a chemical pretreatment. The second step includes the xylanase enzymatic reaction or the acid hydrolysis of xylan. The most widely used methods for pretreatments are autohydrolysis and acid

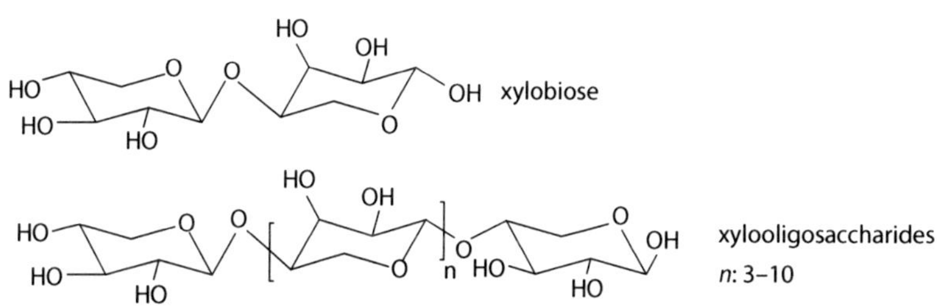

Figure 6.3 Schematic structure of xylose and XOS.

or alkaline pre-hydrolysis (Moure et al. 2006; Aachary and Prapulla 2009; Sabiha-Hanim et al. 2011; Carvalho et al. 2013).

The enzymatic production of XOS mainly depends on xylanase composed of endoxylanase (1,4-β-d-xylan xylohydrolases), β-d-xylosidases (1,4-β-xyloside xylanohydrolases), and debranching enzymes (esterases) (Vazquez et al. 2000). For production of XOS, the enzyme complex should have low exoxylanase or β-xylosidase activity. Enzymes with high exoxylanase or β-xylosidase activity will produce high amounts of xylose, and XOS production will be inhibited (Vazquez et al. 2002).

Xu et al. have reported the production of XOS using xylanase from *Trichoderma reesei* and corncob. Results showed that the maximum yields of XOS and total sugars were 35.2% and 41.9%, respectively, and oligomeric xylan of DP 2-5 constituted 25.9% of solid in the substrate product (Xu et al. 2002).

XOS are generally obtained from the separated xylan or pretreated xylan-rich biomass by enzymatic or chemical hydrolysis (Carvalho et al. 2013). Recently, Zhang et al. (2017) provided a method of XOS production by straight acidic hydrolysis of different agricultural residues at 120–150°C for 10–60 min using 10–30% acetic acid effectively, without pretreatment to the feedstock. Simple and fast kinetics are the major advantages of acetic acid hydrolysis. Moreover, the moderately acidic nature of acetic acid makes it a selective hydrolysis process. More importantly, the recovery of acetic acid is industrially feasible, which is more environmentally favorable (Ahsan et al. 2014). They investigated the impacts of each factor on the production of XOS by selective hydrolysis using concentrated acetic acid. It was found that reaction temperature was the most significant factor influencing the XOS yield. As shown in Table 6.4, the yield of XOS increases when the reaction temperature is in the range of 120–140°C. On the other hand, a sharp decrease in the XOS yield was observed at 160°C. This can be explained by the fact that acetic acid increases the breakdown of the glycosidic bond of xylan (within certain limits) with the temperature; excessively high temperatures can lead to the degradation of XOS (Samala et al. 2015), resulting in a considerable amount of xylose and then furfural (Table 6.5). An even distribution of XOS was found at 120°C. The XOS products obtained at 140°C were mainly

Table 6.4 HPAEC-PAD Analysis of XOS Profiles from Different Temperature Processing

	XOS YIELD (%)					
TEMPERATURE (°C)	XYLOBIOSE	XYLOTRIOSE	XYLOTETRAOSE	XYLOPENTAOSE	XYLOHEXAOSE	XOS (DP2-6)
120	3.23 ± 0.09	3.43 ± 0.11	4.44 ± 006	3.13 ± 0.07	3.28 ± 0.03	17.51 ± 0.36
140	12.04 ± 0.46	8.20 ± 0.28	6.42 ± 0.06	2.74 ± 0.02	1.66 ± 0.04	31.05 ± 0.52
160	1.89 ± 0.09	0.51 ± 0	0 ± 0	0 ± 0	0 ± 0	2.40 ± 0.09

Table 6.5 By-Products of Acetic Acid Catalysis at Different Temperatures

	BY-PRODUCTS YIELD (%)	
TEMPERATURE (°C)	XYLOSE	FURFURAL
120	4.33 ± 0.01	0 ± 0
140	32.71 ± 1.11	0.39 ± 0.05
160	58.81 ± 0.94	9.80 ± 0.14

xylobiose, xylotriose, and xylotetraose, along with small amounts of xylopentaose and xylohexaose.

The yield of XOS can reach 45.86% with desirable oligomer profiles under the optimal reaction conditions (140°C, 20% acetic acid, 20 min). Accurate HPAEC-PAD analysis of component profiles showed that the main components in the XOS product derived from xylan materials were xylobiose, xylotriose, and xylotetraose, with yields of 10.35%, 10.38%, and 11.81%, respectively. These products had higher yields than xylopentaose and xylohexaose, whose yields were 7.14% and 6.19%, respectively. A small amount of xylose and traces of furfural were produced. The yields were comparable to values reported by other researchers: 57.8% from the wastewater of a viscose fiber mill after acid precipitation and washing, as well as the subsequent dilute acid pretreatment (30 min) and enzymatic hydrolysis (12 h) (Zhang et al. 2016), and 52% from corncob after acid pretreatment (24 h) (Akpinar et al. 2009).

6.4.3 Other High-Valued Products from Hemicellulose

6.4.3.1 Biohydrogen Hydrogen, one of the potential energy carriers of the future, has a high potential of being renewable and an environmentally friendly alternative for future energy carriers (Koroneos and Rovas 2012). It has a high energy content (122 kJ/g) and yields 2.7 times higher than other hydrocarbon fuels, and has the potential to be used in fuel cells for the generation of electricity (Kapdan and Kargi 2006).

Various methods are used for hydrogen production such as solar gasification, thermochemical gasification, and super critical conversions. However, compared with the conventional hydrogen generation process (thermochemical and electrochemical), biohydrogen production processes are more environmentally friendly and less energy intensive. Dark fermentative biohydrogen production has received considerable attention because of the high conversion efficiency. Various biomasses, e.g., lignocellulosic wastes from wheat straw, corn stover, sugarcane bagasse, sweet sorghum, and corn straw, have been evaluated as feedstocks for biohydrogen production (Oh et al. 2003; Kapdan and Kargi 2006; Pattra et al. 2008; Cao et al. 2009; Ivanova et al. 2009; Kaparaju et al. 2009; Kongjan and Angelidaki 2010; Perera et al. 2010; Xu et al. 2010). Biological hydrogen provides a renewable, more sustainable alternative, but has not yet to reach a scale large enough for consideration in replacing a significant portion of the hydrogen supply (Brentner et al. 2010).

6.4.3.2 Chitosan Currently, chitosan is commercially produced from shellfish waste through chemical deacetylation even though there are environmental drawbacks of using chemical methods. Chitosan has found many applications in food, cosmetics, and pharmaceutical industries because of its unique properties such as nontoxicity, biodegradability, biocompatibility, film-forming, and chelating properties along with its antimicrobial activity (Rabea et al. 2003; Synowiecki and Al-Khateeb 2003; Goksungur 2004; Silva et al. 2006). Thus, fermentation processes receive more attention for production of polysaccharide. Biosynthesis of chitosan takes place in a family of fungi called zygomycetes, and chitosan, as an important kind of biomacromolecules, has protective and supportive functions in cell walls of zygomycetes (Goksungur 2004). Tai et al. (2010) have studied the potential of hemicellulose hydrolysate of corn straw for the production of chitosan by *Rhizopus oryzae*. It was shown that hemicellulose hydrolysate of corn straw is a good source for chitosan production (Tai et al. 2010).

6.4.3.3 Microbial Polyesters Microbial polyesters, classified as polyhydroxyalkanoates, are another large group of biopolymers that are produced by fermentation processes. Low concentrations of some nutrients together with excess amounts of carbon source lead to the formation of intracellular granules of microbial polyesters by several bacterial strains. There is a very high potential for the replacement of synthetic plastics by polyhydroxyalkanoates if these biopolymers can be broadly produced at reasonable cost (Pan et al. 2012).

Fortunately, many of polyester-producing microorganisms consume xylose even better than glucose, and therefore, lignocellulosic hydrolysates seem to be ideal substrates for production of these biopolymers. The production of microbial polyesters from lignocellulosic materials is usually performed in two steps of hydrolysis (chemical or enzymatic) and fermentation. Recombinant strains of *E. coli* have also been developed with the ability to hydrolyze hemicelluloses (namely xylan) and convert the obtained xylose to microbial polyesters. After fermentation, polyesters are separated and purified by extraction from microbial biomass using suitable solvents (Lee and Na 2013; Zhang et al. 2013; Salamanca-Cardona et al. 2014).

6.5 Future Prospects

Lignocelluloses are widely available, renewable, low-cost, and non-food-based materials considered a suitable substitute for a major part of fossil-derived fuels and chemicals in the future. So far, much research has focused on the production of valuable biofuels, biochemicals, and biomaterials from lignocelluloses. Currently, the conversion of hemicellulosic substrates to high-valued chemicals is still a challenging area. We need further research to develop eco-friendly pretreatment methods, cost-effective enzymes for hemicellulose conversion, and robust and efficient microbes to ferment hemicellulosic sugars. Above all, the new bioproducts and integrated solutions are the first task for commercial production systems.

References

Aachary, A. A. and Prapulla, S. G. (2009). Value addition to corncob: Production and characterization of xylooligosaccharides from alkali pretreated lignin-saccharide complex using *Aspergillus oryzae* MTCC 5154. *Bioresource Technology*. **100**(2), 991–995.

Afschar, A. S., Bellgardt, K. H., Rossell, C. E. V., Czok, A. and Schaller, K. (1991). The production of 2,3-butanediol by fermentation of high test molasses. *Applied Microbiology and Biotechnology*. **34**(5), 582–585.

Ahsan, L., Jahan, M. S. and Ni, Y. H. (2014). Recovering/concentrating of hemicellulosic sugars and acetic acid by nanofiltration and reverse osmosis from prehydrolysis liquor of kraft based hardwood dissolving pulp process. *Bioresource Technology*. **155**, 111–115.

Akpinar, O., Erdogan, K. and Bostanci, S. (2009). Production of xylooligosaccharides by controlled acid hydrolysis of lignocellulosic materials. *Carbohydrate Research*. **344**(5), 660–666.

Awano, T., Takabe, K. and Fujita, M. (2002). Xylan deposition on secondary wall of *Fagus crenata* fiber. *Protoplasma*. **219**(1–2), 106–115.

Bakır, U. (2002). Microbial xylanase production. In A. Pandey (ed.), *Encyclopedia of Bioresource Technology*, Pandey, A. (ed). Food Products Press, Binghamton, New York. 592–599.

Bauer, W. D., Talmadge, K. W., Keegstra, K. and Albersheim, P. (1973). The structure of plant cell walls II. The hemicellulose of the walls of suspension-cultured sycamore cells. *Plant Physiology*. **51**(1), 174–187.

Biely, P. (1993). *Biochemical Aspects of the Production of Microbial Hemicellulases*. Portland Press, London. 29–51.

Brennan, C. S., Blake, D. E., Ellis, P. R. and Schofield, J. D. (1996). Effects of guar galactomannan on wheat bread microstructure and on the in vitro and in vivo digestibility of starch in bread. *Journal of Cereal Science*. **24**(2), 151–160.

Brentner, L. B., Peccia, J. and Zimmerman, J. B. (2010). Challenges in developing biohydrogen as a sustainable energy source: Implications for a research agenda. *Environmental Science & Technology*. **44**(7), 2243–2254.

Brown, S. F. (2003). Bioplastic fantastic. *Fortune*. **148**(2), 92–94.

Buchert, J., Puls, J. and Poutanen, K. (1988). Comparison of *Pseudomonas fragi* and *Gluconobacter oxydans* for production of xylonic acid from hemicellulose hydrolyzates. *Applied Microbiology and Biotechnology*. **28**(4–5), 367–372.

Buchert, J. and Viikari, L. (1988). The role of xylonolactone in xylonic acid production by *Pseudomonas fragi*. *Applied Microbiology and Biotechnology*. **27**(4), 333–336.

Bustos, G., Moldes, A. B., Cruz, J. M. and Dominguez, J. M. (2005). Influence of the metabolism pathway on lactic acid production from hemicellulosic trimming vine shoots hydrolyzates using *Lactobacillus pentosus*. *Biotechnology Progress*. **21**(3), 793–798.

Cao, G., Ren, N., Wang, A., Lee, D., Guo, W., Liu, B., Feng, Y. and Zhao, Q. (2009). Acid hydrolysis of corn stover for biohydrogen production using *Thermoanaerobacterium thermosaccharolyticum* W16. *International Journal of Hydrogen Energy*. **34**(17), 7182–7188.

Carpita, N. C. and Gibeaut, D. M. (1993). Structural models of primary cell walls in flowering plants: Consistency of molecular structure with the physical properties of the walls during growth. *The Plant Journal*. **3**(1), 1–30.

Carvalho, A. F. A., Neto, P. D., Da Silva, D. F. and Pastore, G. M. (2013). Xylo-oligosaccharides from lignocellulosic materials: Chemical structure, health benefits and production by chemical and enzymatic hydrolysis. *Food Research International*. **51**(1), 75–85.

Chandel, A. K., da Silva, S. S., Carvalho, W. and Singh, O. V. (2012). Sugarcane bagasse and leaves: Foreseeable biomass of biofuel and bio-products. *Journal of Chemical Technology and Biotechnology*. **87**(1), 11–20.

Chandel, A. K., Singh, O. V. and Rao, L. V. (2010). Biotechnological applications of hemicellulosic derived sugars: State-of-the-art. In O. V. Singh and S. P. Harvey (eds.), *Sustainable Biotechnology*. Dordrecht: Springer Netherlands. 63–81.

Chaturvedi, V. and Verma, P. (2013). An overview of key pretreatment processes employed for bioconversion of lignocellulosic biomass into biofuels and value added products. *3 Biotech*. **3**(5), 415–431.

Chen, H. (2014). Chemical composition and structure of natural lignocellulose. In *Biotechnology of Lignocellulose: Theory and Practice*. Springer Netherlands. 25–71.

Cheng, K. K., Liu, Q., Zhang, J. A., Li, J. P., Xu, J. M. and Wang, G. H. (2010). Improved 2,3-butanediol production from corncob acid hydrolysate by fed-batch fermentation using *Klebsiella oxytoca*. *Process Biochemistry*. **45**(4), 613–616.

Cheng, K. K., Zhang, J. A., Chavez, E. and Li, J. P. (2010). Integrated production of xylitol and ethanol using corncob. *Applied Microbiology and Biotechnology*. **87**(2), 411–417.

Chum, H., Johnson, D., Black, S., Baker, J., Grohmann, K., Sarkanen, K., Wallace, K. and Schroeder, H. (1988). Organosolv pretreatment for enzymatic hydrolysis of poplars: I. Enzyme hydrolysis of cellulosic residues. *Biotechnology and Bioengineering*. **31**(7), 643–649.

Chun, B. W., Dair, B., Macuch, P. J., Wiebe, D., Porteneuve, C. and Jeknavorian, A. (2006). The development of cement and concrete additive: Based on xylonic acid derived via bioconversion of xylose. *Applied Biochemistry and Biotechnology*. **131**(1–3), 645–658.

Clark, T. and Mackie, K. (1987). Steam explosion of the softwood *Pinus radiata* with sulphur dioxide addition. I. Process optimisation. *Journal of Wood Chemistry and Technology*. **7**(3), 373–403.

Dale, B. (1982). A freeze-explosion technique for increasing cellulose hydrolysis [Biomass conversion]. *Macromolecules*. **12**(12), 31–43.

Dale, B. E., Leong, C. K., Pham, T. K., Esquivel, V. M., Rios, I. and Latimer, V. M. (1996). Hydrolysis of lignocellulosics at low enzyme levels: Application of the AFEX process. *Bioresource Technology*. **56**(1), 111–116.

Datta, R., Tsai, S. P., Bonsignore, P., Moon, S. H. and Frank, J. R. (1995). Technological and economic-potential of poly(lactic acid) and lactic-acid derivatives. *FEMS Microbiology Reviews*. **16**(2–3), 221–231.

de Vries, R. P. and Visser, J. (2001). Aspergillus enzymes involved in degradation of plant cell wall polysaccharides. *Microbiology and Molecular Biology Reviews*. **65**(4), 497–522.

Decker, S. R., Siika-Aho, M. and Viikari, L. (2008). *Enzymatic Depolymerization of Plant Cell Wall Hemicelluloses*. Wiley Blackwell. 352–373.

Deshpande, V., Lachke, A., Mishra, C., Keskar, S. and Rao, M. (1986). Mode of action and properties of xylanase and β-xylosidase from *Neurospora crassa*. *Biotechnology and Bioengineering*. Advances in Polymer Science. Springer Berlin Heidelberg. 186:1–67. **28**(12), 1832–1837.

Ebringerova, A., Hromadkova, Z. and Heinze, T. (2005). Hemicellulose. In T. Heinze (ed.), *Polysaccharides I*. Berlin: Springer Verlag. 1–67.

Edgar, W. M. (1998). Sugar substitutes, chewing gum and dental caries—A review. *British Dental Journal*. **184**(1), 29–32.

Emodi, A. (1978). Xylitol: Its properties and food applications. *Food Technology*. **32**(1), 20–32.

Fernandez-Bolanos, J., Felizon, B., Heredia, A., Rodriguez, R., Guillen, R. and Jimenez, A. (2001). Steam-explosion of olive stones: Hemicellulose solubilization and enhancement of enzymatic hydrolysis of cellulose. *Bioresource Technology*. **79**(1), 53–61.

Gilbert, H. J. and Hazlewood, G. P. (1993). Bacterial cellulases and xylanases. *Journal of General Microbiology*. **139**, 187–194.

Gírio, F. M., Fonseca, C., Carvalheiro, F., Duarte, L. C., Marques, S. and Bogel-Łukasik, R. (2010). Hemicelluloses for fuel ethanol: A review. *Bioresource Technology*. **101**(13), 4775–4800.

Goksungur, Y. (2004). Optimization of the production of chitosan from beet molasses by response surface methodology. *Journal of Chemical Technology and Biotechnology*. **79**(9), 974–981.

Goldstein, I. S. and Easter, J. M. (1992). An improved process for converting cellulose to ethanol. *Tappi Journal*. **75**(8), 135–140.

Gould, J. M. (1984). Alkaline peroxide delignification of agricultural residues to enhance enzymatic saccharification. *Biotechnology and Bioengineering*. **26**(1), 46–52.

Goyal, P., Kumar, V. and Sharma, P. (2007). Carboxymethylation of tamarind kernel powder. *Carbohydrate Polymers*. **69**(2), 251–255.

Granstrom, T. B., Izumori, K. and Leisola, M. (2007). A rare sugar xylitol. Part II: Biotechnological production and future applications of xylitol. *Applied Microbiology and Biotechnology*. **74**(2), 273–276.

Groleau, D., Laube, V. and Martin, S. (1985). The effect of various atmospheric conditions on the 2, 3-butanediol fermentation from glucose by *Bacillus polymyxa*. *Biotechnology Letters*. **7**(1), 53–58.

Grover, B., Garg, S. and Verma, J. (1990). Production of 2, 3-butanediol from wood hydrolysate by *Klebsiella pneumoniae*. *World Journal of Microbiology and Biotechnology*. **6**(3), 328–332.

Hahn-Hägerdal, B., Jeppsson, H., Olsson, L. and Mohagheghi, A. (1994). An interlaboratory comparison of the performance of ethanol-producing microorganisms in a xylose-rich acid hydrolysate. *Applied Microbiology and Biotechnology*. **41**(1), 62–72.

Hahn-Hägerdal, B. and Pamment, N. (2004). Microbial pentose metabolism. *Applied Biochemistry and Biotechnology*. **113**, 1207–1209.

Harholt, J., Ulvskov, P., Sorensen, I., Willats, W. and Petersen, B. (2010). Cell wall composition and glycosyltransferases involved in cell wall formation. *Annual Review of Plant Biology*. **61**, 263–289.

Held, M. A., Jiang, N., Basu, D., Showalter, A. M. and Faik, A. (2015). *Plant Cell Wall Polysaccharides: Structure and Biosynthesis*: Springer International Publishing Switzerland. 3–54.

Hofvendahl, K. and Hahn-Hägerdal, B. (2000). Factors affecting the fermentative lactic acid production from renewable resources. *Enzyme and Microbial Technology*. **26**(2–4), 87–107.

Holscher, T., Schleyer, U., Merfort, M., Bringer-Meyer, S., Gorisch, H. and Sahm, H. (2009). Glucose oxidation and PQQ-dependent dehydrogenases in *Gluconobacter oxydans*. *Journal of Molecular Microbiology and Biotechnology*. **16**(1–2), 6–13.

Honda, Y. and Kitaoka, M. (2004). A family 8 glycoside hydrolase from *Bacillus halodurans* C-125 (BH2105) is a reducing end xylose-releasing exo-oligoxylanase. *Journal of Biological Chemistry*. **279**(53), 55097–55103.

Hyvönen, L., Koivistoinen, P. and Voirol, F. (1982). Food technological evaluation of xylitol. *Advances in Food Research*. **28**, 373–403.

Ivanova, G., Rakhely, G. and Kovacs, K. L. (2009). Thermophilic biohydrogen production from energy plants by *Caldicellulosiruptor saccharolyticus* and comparison with related studies. *International Journal of Hydrogen Energy*. **34**(9), 3659–3670.

Iyer, P. V., Thomas, S. and Lee, Y. Y. (2000). High-yield fermentation of pentoses into lactic acid. *Applied Biochemistry and Biotechnology*. **84**(1), 665–677.

Jansen, N. B., Flickinger, M. C. and Tsao, G. T. (1984). Production of 2,3-butanediol from D-xylose by *Klebsiella pneumoniae* ATCC 8724. *Biotechnology and Bioengineering*. **26**(4), 362–369.

Jansen, N. B. and Tsao, G. T. (1983). Bioconversion of pentoses to 2,3-butanediol by *Klebsiella pneumoniae*. *Advances in Biochemical Engineering/Biotechnology*. **27**, 85–99.

Jeffries, T. W. (1994). Biodegradation of lignin and hemicelluloses. In C. Ratledge (ed.), *Biochemistry of Microbial Degradation*, Springer Dordrecht. 233–277.

Kaar, W. E. and Holtzapple, M. T. (2000). Using lime pretreatment to facilitate the enzymic hydrolysis of corn stover. *Biomass and Bioenergy*. **18**(3), 189–199.

Kadam, K. L., Chin, C. Y. and Brown, L. W. (2008). Flexible biorefinery for producing fermentation sugars, lignin and pulp from corn stover. *Journal of Industrial Microbiology & Biotechnology*. **35**(5), 331–341.

Kaparaju, P., Serrano, M., Thomsen, A. B., Kongjan, P. and Angelidaki, I. (2009). Bioethanol, biohydrogen and biogas production from wheat straw in a biorefinery concept. *Bioresource Technology*. **100**(9), 2562–2568.

Kapdan, I. K. and Kargi, F. (2006). Bio-hydrogen production from waste materials. *Enzyme and Microbial Technology*. **38**(5), 569–582.

Karimi, K. (ed.) (2015). *Lignocellulose-Based Bioproducts*. Enzymatic Hydrolysis Technologies for the Production of Biofuels. *Biofuel and Biorefinery Technologies*. Springer Switzerland. 155–173.

Kongjan, P. and Angelidaki, I. (2010). Extreme thermophilic biohydrogen production from wheat straw hydrolysate using mixed culture fermentation: Effect of reactor configuration. *Bioresource Technology*. **101**(20), 7789–7796.

Kormelink, F. and Voragen, A. (1993). Degradation of different [(glucurono) arabino] xylans by a combination of purified xylan-degrading enzymes. *Applied Microbiology and Biotechnology*. **38**(5), 688–695.

Koroneos, C. and Rovas, D. (2012). A review on exergy comparison of hydrogen production methods from renewable energy sources. *Energy & Environmental Science*. **5**(5), 6640–6651.

Kosaric, N. and Velikonja, J. (1995). Liquid and gaseous fuels from biotechnology—Challenge and opportunities. *FEMS Microbiology Reviews*. **16**(2–3), 111–142.

Koullas, D. P., Christakopoulos, P. F., Kekos, D., Koukios, E. G. and Macris, B. J. (1993). Effect of alkali delignification on wheat straw saccharification by fusarium–oxysporum cellulases. *Biomass & Bioenergy*. **4**(1), 9–13.

Laser, M., Schulman, D., Allen, S. G., Lichwa, J., Antal, M. J. and Lynd, L. R. (2002). A comparison of liquid hot water and steam pretreatments of sugar cane bagasse for bioconversion to ethanol. *Bioresource Technology*. **81**(1), 33–44.

Lee, G. N. and Na, J. (2013). Future of microbial polyesters. *Microbial Cell Factories*. **12**(54), 1–4.

Li, Z., Ji, X., Kan, S., Qiao, H., Jiang, M., Lu, D., Wang, J., Huang, H., Jia, H., Ouyuang, P. and Ying, H. (2010). Past, present, and future industrial biotechnology in China. *Biotechnology in China II: Chemicals, Energy and Environment*. **122**, 1–42.

Liepman, A. H., Nairn, C. J., Willats, W. G. T., Sorensen, I., Roberts, A. W. and Keegstra, K. (2007). Functional genomic analysis supports conservation of function among cellulose synthase-like A gene family members and suggests diverse roles of mannans in plants. *Plant Physiology*. **143**(4), 1881–1893.

Magee, R. J. and Kosaric, N. (1987). The microbial production of 2, 3-butanediol. *Advances in Applied Microbiology*. **32**, 89–161.

Mamman, A., Lee, J., Kim, Y., Hwang, I., Park, N., Hwang, Y., Chang, J. and Hwang, J. (2008). Furfural: Hemicellulose/xylose-derived biochemical. *Biofuels, Bioproducts & Biorefining*. **2**(5), 438–454.

Marinova, M., Mateos-Espejel, E. and Paris, J. (2010). From Kraft mill to forest biorefinery: An energy and water perspective. II. Case study. *Cellulose Chemistry and Technology*. **44**(1–3), 21–26.

Martinez, E. A., Silva, J. B. D. E., Giulietti, M. and Solenzal, A. I. N. (2007). Downstream process for xylitol produced from fermented hydrolysate. *Enzyme and Microbial Technology*. **40**(5), 1193–1198.

Maruyama, K., Goto, C., Numata, M., Suzuki, T., Nakagawa, Y., Hoshino, T. and Uchiyama, T. (1996). O-acetylated xyloglucan in extracellular polysaccharides from cell-suspension cultures of Mentha. *Phytochemistry*. **41**(5), 1309–1314.

McCleary, B. V. and Matheson, N. K. (1983). Action patterns and substrate-binding requirements of β-D-mannanase with mannosaccharides and mannan-type polysaccharides. *Carbohydrate Research*. **119**, 191–219.

McMillan, J. (1993). Xylose fermentation to ethanol. A review. National Renewable Energy Lab., Golden, CO.

McMillan, J. D. (1994). Pretreatment of lignocellulosic biomass. *Enzymatic Conversion of Biomass for Fuels Production*. **566**, 292–324.

Meier, H. and Reid, J. (1982). Reserve polysaccharides other than starch in higher plants. In: F. A. Loewus and W. Tanner (eds.), *Plant Carbohydrates I*, Springer Berlin Heidelberg. 13, 418–471.

Menon, V., Prakash, G. and Rao, M. (2010). Value added products from hemicellulose: Biotechnological perspective. *Global Journal of Biochemistry*. **1**(1), 36–67.

Menon, V. and Rao, M. (2012). Trends in bioconversion of lignocellulose: Biofuels, platform chemicals & biorefinery concept. *Progress in Energy and Combustion Science*. **38**(4), 522–550.

Miyafuji, H., Nakata, T., Ehara, K. and Saka, S. (2005). Fermentability of water-soluble portion to ethanol obtained by supercritical water treatment of lignocellulosics. *Applied Biochemistry and Biotechnology*. **121**, 963–971.

Mohagheghi, A., Dowe, N., Schell, D., Chou, Y. C., Eddy, C. and Zhang, M. (2004). Performance of a newly developed integrant of *Zymomonas mobilis* for ethanol production on corn stover hydrolysate. *Biotechnology Letters*. **26**(4), 321–325.

Moller, I., Sørensen, I., Bernal, A. J., Blaukopf, C., Lee, K., Øbro, J., Pettolino, F., Roberts, A., Mikkelsen, J. D. and Knox, J. P. (2007). High-throughput mapping of cell-wall polymers within and between plants using novel microarrays. *The Plant Journal*. **50**(6), 1118–1128.

Montane, D., Salvado, J., Torras, C. and Farriol, X. (2002). High-temperature dilute-acid hydrolysis of olive stones for furfural production. *Biomass & Bioenergy*. **22**(4), 295–304.

Moreira, L. R. and Filho, E. X. (2008). An overview of mannan structure and mannan-degrading enzyme systems. *Applied Microbiology and Biotechnology*. **79**(2), 165–178.

Motwani, M., Seth, R., Daginawala, H. F. and Khanna, P. (1993). Microbial-production of 2,3-butanediol from water hyacinth. *Bioresource Technology*. **44**(3), 187–195.

Moure, A., Gullón, P., Domínguez, H. and Parajó, J. C. (2006). Advances in the manufacture, purification and applications of xylo-oligosaccharides as food additives and nutraceuticals. *Process Biochemistry*. **41**(9), 1913–1923.

Nakashimada, Y., Marwoto, B., Kashiwamura, T., Kakizono, T. and Nishio, N. (2000). Enhanced 2,3-butanediol production by addition of acetic acid in *Paenibacillus polymyxa*. *Journal of Bioscience and Bioengineering*. **90**(6), 661–664.

Nigam, J. N. (2001). Ethanol production from wheat straw hemicellulose hydrolysate by *Pichia stipitis*. *Journal of Biotechnology*. **87**(1), 17–27.

Nigam, P. and Singh, D. (1995). Processes for fermentative production of xylitol—A sugar substitute. *Process Biochemistry.* **30**(2), 117–124.

O'Brien, D. J., Senske, G. E., Kurantz, M. J. and Craig, J. C. (2004). Ethanol recovery from corn fiber hydrolysate fermentations by pervaporation. *Bioresource Technology.* **92**(1), 15–19.

Oh, S. E., Van Ginkel, S. and Logan, B. E. (2003). The relative effectiveness of pH control and heat treatment for enhancing biohydrogen gas production. *Environmental Science & Technology.* **37**(22), 5186–5190.

Ohno, H. and Fukaya, Y. (2009). Task specific ionic liquids for cellulose technology. *Chemistry Letters.* **38**(1), 2–7.

Otieno, D. O. and Ahring, B. K. (2012). A thermochemical pretreatment process to produce xylooligosaccharides (XOS), arabinooligosaccharides (AOS) and mannooligosaccharides (MOS) from lignocellulosic biomasses. *Bioresource Technology.* **112**, 285–292.

Ou, M. S., Ingram, L. O. and Shanmugam, K. T. (2011). L: (+)-Lactic acid production from non-food carbohydrates by thermotolerant *Bacillus coagulans. Journal of Industrial Microbiology & Biotechnology.* **38**(5), 599–605.

Pan, W. Y., Perrotta, J. A., Stipanovic, A. J., Nomura, C. T. and Nakas, J. P. (2012). Production of polyhydroxyalkanoates by *Burkholderia cepacia* ATCC 17759 using a detoxified sugar maple hemicellulosic hydrolysate. *Journal of Industrial Microbiology & Biotechnology.* **39**(3), 459–469.

Pattra, S., Sangyoka, S., Boonmee, M. and Reungsang, A. (2008). Biohydrogen production from the fermentation of sugarcane bagasse hydrolysate by *Clostridium butyricum. International Journal of Hydrogen Energy.* **33**(19), 5256–5265.

Peng, F., Peng, P., Xu, F. and Sun, R.-C. (2012). Fractional purification and bioconversion of hemicelluloses. *Biotechnology Advances.* **30**(4), 879–903.

Perego, C. and Bianchi, D. (2010). Biomass upgrading through acid–base catalysis. *Chemical Engineering Journal.* **161**(3), 314–322.

Perego, P., Converti, A. and Del Borghi, M. (2003). Effects of temperature, inoculum size and starch hydrolyzate concentration on butanediol production by *Bacillus licheniformis. Bioresource Technology.* **89**(2), 125–131.

Perera, K. R. J., Ketheesan, B., Gadhamshetty, V. and Nirmalakhandan, N. (2010). Fermentative biohydrogen production: Evaluation of net energy gain. *International Journal of Hydrogen Energy.* **35**(22), 12224–12233.

Perttunen, J., Myllykoski, L. and Keiski, R. L. (2001). Lactic acid fermentation of hemicellulose liquors and their activated carbon pretreatments. In M. Hofman and P. Thonart (eds.), *Engineering and Manufacturing for Biotechnology.* Springer Finland. **4**(1), 29–38.

Petkowicz, C. de O., Reicher, F., Chanzy, H., Taravel, F. and Vuong, R. (2001). Linear mannan in the endosperm of *Schizolobium amazonicum. Carbohydrate Polymers.* **44**(2), 107–112.

Polizeli, M. L., Rizzatti, A. C., Monti, R., Terenzi, H. F., Jorge, J. A. and Amorim, D. S. (2005). Xylanases from fungi: Properties and industrial applications. *Applied Microbiology and Biotechnology.* **67**(5), 577–591.

Popper, Z. A. (2008). Evolution and diversity of green plant cell walls. *Current Opinion in Plant Biology*. **11**(3), 286–292.

Poutanen, K., Tenkanen, M., Korte, H. and Puls, J. (1991). Accessory enzymes involved in the hydrolysis of xylans. In G. F. Leatham and M. E. Himmel (eds.), *Enzymes in Biomass Conversion. ACS Symposium Series*, Vol. 460. New York: ACS. 426–436.

Powlowski, J., Mahajan, S., Schapira, M. and Master, E. R. (2009). Substrate recognition and hydrolysis by a fungal xyloglucan-specific family 12 hydrolase. *Carbohydrate Research*. **344**(10), 1175–1179.

Prade, R. A. (1996). Xylanases: From biology to biotechnology. *Biotechnology and Genetic Engineering Reviews*. **13**(1), 101–132.

Preston, R. (1968). Plants without cellulose. *Scientific American*. **218**, 102–108.

Rabea, E. I., Badawy, M. E. T., Stevens, C. V., Smagghe, G. and Steurbaut, W. (2003). Chitosan as antimicrobial agent: Applications and mode of action. *Biomacromolecules*. **4**(6), 1457–1465.

Reilly, P. J. (1981). Xylanases, structure and function. In A. Hollaender (ed.), *Trends in the Biology of Fermentation for Fuels and Chemicals*. Plenum, New York: Plenum. 111–129.

Rodriguez, L. C. and O'Connell, D. (2011). Biofuels: Balance the blend of food and fuel. *Nature*. **476**(7360), 283.

Sabiha-Hanim, S., Noor, M. A. M. and Rosma, A. (2011). Effect of autohydrolysis and enzymatic treatment on oil palm (*Elaeis guineensis* Jacq.) frond fibres for xylose and xylooligosaccharides production. *Bioresource Technology*. **102**(2), 1234–1239.

Saha, B. C. (2003). Hemicellulose bioconversion. *Journal of Industrial Microbiology & Biotechnology*. **30**(5), 279–291.

Saha, B. C. and Bothast, R. J. (1997). Microbial production of xylitol. *Fuels and Chemicals from Biomass*. **666**, 307–319.

Saha, B. C. and Bothast, R. J. (1999). Pretreatment and enzymatic saccharification of corn fiber. *Applied Biochemistry and Biotechnology*. **76**(2), 65–77.

Saha, B. C., Iten, L. B., Cotta, M. A. and Wu, Y. V. (2005). Dilute acid pretreatment, enzymatic saccharification, and fermentation of rice hulls to ethanol. *Biotechnology Progress*. **21**(3), 816–822.

Sahu, R. and Dhepe, P. L. (2012). A one-pot method for the selective conversion of hemicellulose from crop waste into C5 sugars and furfural by using solid acid catalysts. *ChemSusChem*. **5**(4), 751–761.

Salamanca-Cardona, L., Ashe, C. S., Stipanovic, A. J. and Nomura, C. T. (2014). Enhanced production of polyhydroxyalkanoates (PHAs) from beechwood xylan by recombinant *Escherichia coli*. *Applied Microbiology and Biotechnology*. **98**(2), 831–842.

Samala, A., Srinivasan, R. and Yadav, M. (2015). Comparison of xylo-oligosaccharides production by autohydrolysis of fibers separated from ground corn flour and DDGS. *Food and Bioproducts Processing*. **94**, 354–364.

Samanta, A. K., Senani, S., Kolte, A. P., Sridhar, M., Sampath, K. T., Jayapal, N. and Devi, A. (2012). Production and in vitro evaluation of xylooligosaccharides generated from corn cobs. *Food and Bioproducts Processing*. **90**(C3), 466–474.

Santos, J. C., Mussatto, S. I., Dragone, G., Converti, A. and Silva, S. S. (2005). Evaluation of porous glass and zeolite as cells carriers for xylitol production from sugarcane bagasse hydrolysate. *Biochemical Engineering Journal*. **23**(1), 1–9.

Scheller, H. V. and Ulvskov, P. (2010). Hemicelluloses. *Annual Review of Plant Biology*. **61**(1), 263–289.

Schmidt, A. S. and Thomsen, A. B. (1998). Optimization of wet oxidation pretreatment of wheat straw. *Bioresource Technology*. **64**(2), 139–151.

Schulze, E. (1891). Information regarding chemical composition of plant cell membrane. *Berichte der Deutschen Chemischen Gesellschaft*. **24**, 2277–2287.

Sedlak, M. and Ho, N. W. Y. (2004). Production of ethanol from cellulosic biomass hydrolysates using genetically engineered *Saccharomyces yeast* capable of cofermenting glucose and xylose. *Applied Biochemistry and Biotechnology*. **113**, 403–416.

Serrano, L., Egues, I., Alriols, M. G., Llano-Ponte, R. and Labidi, J. (2010). *Miscanthus sinensis* fractionation by different reagents. *Chemical Engineering Journal*. **156**(1), 49–55.

Shallom, D. and Shoham, Y. (2003). Microbial hemicellulases. *Current Opinion in Microbiology*. **6**(3), 219–228.

Silva, H. S. R. C., dos Santos, K. S. C. R. and Ferreira, E. I. (2006). Chitosan: Hydrosoluble derivatives, pharmaceutical applications and recent advances. *Quimica Nova*. **29**(4), 776–785.

Sims, I. M., Munro, S. L., Currie, G., Craik, D. and Bacic, A. (1996). Structural characterisation of xyloglucan secreted by suspension-cultured cells of *Nicotiana plumbaginifolia*. *Carbohydrate Research*. **293**(2), 147–172.

Singhania, R., Parameswaran, B. and Pandey, A. (2009). *Handbook of Plant-Based Biofuels*. Boca Raton, FL: CRC Press.

Sjoman, E., Manttari, M., Nystrom, M., Koivikko, H. and Heikkila, H. (2008). Xylose recovery by nanofiltration from different hemicellulose hydrolyzate feeds. *Journal of Membrane Science*. **310**(1–2), 268–277.

Sjöström, E. (1981). *Wood Chemistry: Fundamentals and Applications. Carbohydrate Research*. Academic Press, New York. 111(2), 24–25.

Smeets, E. M., Faaij, A. P., Lewandowski, I. M. and Turkenburg, W. C. (2007). A bottom-up assessment and review of global bio-energy potentials to 2050. *Progress in Energy and Combustion Science*. **33**(1), 56–106.

Spiridon, I. and Popa, V. I. (2008). Hemicelluloses: Major sources, properties and applications. In M. N. Belgacem and A. Gandini (eds.), *Monomers, Polymers and Composites from Renewable Resources*. Amsterdam: Elsevier Science Netherlands. 289–304.

Stephen, A. (1983). Other plant polysaccharides. In G. O, Aspinall (ed.), *The Polysaccharides*. New York: Academic Press. 97–193.

Su, B., Wu, M., Lin, J. and Yang, L. (2013). Metabolic engineering strategies for improving xylitol production from hemicellulosic sugars. *Biotechnology Letters*. **35**(11), 1781–1789.

Sun, L., Wang, X., Dai, J. and Xiu, Z. (2009). Microbial production of 2,3-butanediol from Jerusalem artichoke tubers by *Klebsiella pneumoniae*. *Applied Microbiology and Biotechnology*. **82**(5), 847–852.

Synowiecki, J. and Al-Khateeb, N. A. (2003). Production, properties, and some new applications of chitin and its derivatives. *Critical Reviews in Food Science and Nutrition.* **43**(2), 145–171.

Syu, M. J. (2001). Biological production of 2,3-butanediol. *Applied Microbiology and Biotechnology.* **55**(1), 10–18.

Tai, C., Li, S., Xu, Q., Ying, H., Huang, H. and Ouyang, P. (2010). Chitosan production from hemicellulose hydrolysate of corn straw: Impact of degradation products on *Rhizopus oryzae* growth and chitosan fermentation. *Letters in Applied Microbiology.* **51**(3), 278–284.

Tenkanen, M., Makkonen, M., Perttula, M., Viikari, L. and Teleman, A. (1997). Action of *Trichoderma reesei* mannanase on galactoglucomannan in pine kraft pulp. *Journal of Biotechnology.* **57**(1–3), 191–204.

Tenkanen, M., Thornton, J. and Viikari, L. (1995). An acetylglucomannan esterase of *Aspergillus oryzae*; purification, characterization and role in the hydrolysis of O-acetyl-galactoglucomannan. *Journal of Biotechnology.* **42**(3), 197–206.

Toivari, M. H., Nygard, Y., Penttila, M., Ruohonen, L. and Wiebe, M. G. (2012). Microbial D-xylonate production. *Applied Microbiology and Biotechnology.* **96**(1), 1–8.

Toivari, M. H., Ruohonen, L., Richard, P., Penttila, M. and Wiebe, M. G. (2010). *Saccharomyces cerevisiae* engineered to produce D-xylonate. *Applied Microbiology and Biotechnology.* **88**(3), 751–760.

Tyree, R., Clausen, E. and Gaddy, J. (1990). The fermentative characteristics of *Lactobacillus xylosus* on glucose and xylose. *Biotechnology Letters.* **12**(1), 51–56.

Vazquez, M., Alonso, J., Dominguez, H. and Parajo, J. (2000). Xylo-oligosaccharides: Manufacture and applications. *Trends in Food Science & Technology.* **11**(11), 387–393.

Vazquez, M., Alonso, J., Dominguez, H. and Parajo, J. (2002). Enzymatic processing of crude xylooligomer solutions obtained by autohydrolysis of eucalyptus wood. *Food Biotechnology.* **16**(2), 91–105.

Vazquez, M., Oliva, M., Tellez-Luis, S. J. and Ramirez, J. A. (2007). Hydrolysis of sorghum straw using phosphoric acid: Evaluation of furfural production. *Bioresource Technology.* **98**(16), 3053–3060.

Wang, L., Fan, X., Tang, P. and Yuan, Q. (2013). Xylitol fermentation using hemicellulose hydrolysate prepared by acid pre-impregnated steam explosion of corncob. *Journal of Chemical Technology and Biotechnology.* **88**(11), 2067–2074.

Willetts, A. (1984). Butane 2, 3-diol production by *Aeromonas hydrophila* grown on starch. *Biotechnology Letters.* **6**(4), 263–268.

Win, D. T. (2005). Furfural—Gold from garbage. *AU Journal of Technology.* **8**(4), 185-190.

Winkelhausen, E. and Kuzmanova, S. (1998). Microbial conversion of D-xylose to xylitol. *Journal of Fermentation and Bioengineering.* **86**(1), 1–14.

Wyman, C. E., Decker, S. R., Himmel, M. E., Brady, J. W., Skopec, C. E. and Viikari, L. (2005). Hydrolysis of cellulose and hemicellulose. In S. Dumitriu (ed.), *Polysaccharides: Structural Diversity and Functional Versatility.* Marcel Dekker, New York. **1**, 1023–1062.

Ximenes, E. (1998). In S. G. Pandalai (ed.), *Recent Research Developments in Microbiology*. Trivandrum: Research Signpost. 165–176.

Xu, F., Sun, R. and Zhan, H. (2003). Progress in non-wood hemicellulose research. *Transactions of China Pulp and Paper.* **18**(1), 145–151.

Xu, J., Ren, N., Su, D. and Qiu, J. (2010). Bio-hydrogen production from acetic acid steam-exploded corn straws by simultaneous saccharification and fermentation with *Ethanoligenens harbinense* B49. *International Journal of Energy Research*. **34**(5), 381–386.

Xu, Y., Chen, M., Yu, S., Yong, Q. and Yao, C. (2002). Study on enzymatic hydrolysis of xylan into xylo-oligosaccharide. *Chemistry & Industry of Forest Products*. **2**, 013.

Xu, Y., Wang, X., Zhu, J., Yong, Q. and Yu, S. and ShiYuan, Y. (2012). A new way for bioconversion of xylose in high efficiency. *China Biotechnology*. **32**(5), 113–119.

Xu, Y., Yong, Q., Yu, S. and Wang, X. (2012). *The Method of Efficient Conversion of Xylose to Xylonic Acid Using Whole Cell Biocatalysis*. Patent. CN102676608A, China.

Yang, S. (2008). *Plant Fiber Chemistry*. Beijing: China Light Industry Press.

Yin, Z. and Fan, R. (1999). The research progress of plant cell wall. *Bulletin of Botanical Research*. **19**(4), 407–414.

Yong, X., Yang, G., Mei C., and Shi-yuan Y. (2005). Effects and mechanism of xylo-oligosaccharides for layers. *Feed Industry*. **22**, 56–59.

Yoshida, S., Okano, K., Tanaka, T., Ogino, C. and Kondo, A. (2011). Homo-D-lactic acid production from mixed sugars using xylose-assimilating operon-integrated *Lactobacillus plantarum*. *Applied Microbiology and Biotechnology*. **92**(1), 67–76.

Yu, E. K. and Saddler, J. N. (1982). Enhanced production of 2,3-butanediol by *Klebsiella pneumoniae* grown on high sugar concentrations in the presence of acetic acid. *Applied and Environmental Microbiology*. **44**(4), 777–784.

Zavrel, M., Bross, D., Funke, M., Buchs, J. and Spiess, A. C. (2009). High-throughput screening for ionic liquids dissolving (ligno-)cellulose. *Bioresource Technology*. **100**(9), 2580–2587.

Zhang, H., Zhou, X., Xu, Y. and Yu, S. (2017). Production of xylooligosaccharides from waste xylan, obtained from viscose fiber processing, by selective hydrolysis using concentrated acetic acid. *Journal of Wood Chemistry and Technology*. **37**(1), 1–9.

Zhang, J. H., Yong, Q. and Yu, S. (2007). Metabolism of xylooligosaccharides by *Bifidobacterium adolescentis*. *Chemistry & Industry of Forest Products*. **27**(5), 1–5.

Zhang, Y., Sun, W., Wang, H. and Geng, A. (2013). Polyhydroxybutyrate production from oil palm empty fruit bunch using *Bacillus megaterium* R11. *Bioresource Technology*. **147**, 307–314.

Zhang, Y. D., Yu, G., Li, B., Mu, X. D., Peng, H. and Wang, H. S. (2016). Hemicellulose isolation, characterization, and the production of xylo-oligosaccharides from the wastewater of a viscose fiber mill. *Carbohydrate Polymers*. **141**, 238–243.

Zhou, X., Lü, S., Xu, Y., Mo, Y. and Yu, S. (2015). Improving the performance of cell biocatalysis and the productivity of xylonic acid using a compressed oxygen supply. *Biochemical Engineering Journal.* **93**, 196–199.

Zhu, J. and Zhuang, X. (2012). Conceptual net energy output for biofuel production from lignocellulosic biomass through biorefining. *Progress in Energy and Combustion Science.* **38**(4), 583–598.

Zhu, S. D., Wu, Y. X., Yu, Z. N., Zhang, X., Li, H. and Gao, M. (2006). The effect of microwave irradiation on enzymatic hydrolysis of rice straw. *Bioresource Technology.* **97**(15), 1964–1968.

7

PRODUCTION OF *GANODERMA LUCIDUM* POLYSACCHARIDES BY FERMENTATION

ZHEN-HUA WEI AND JIAN-JIANG ZHONG

Contents

7.1 Introduction

Mushrooms (higher fungi) are well-known as a rich source of polysaccharides with great functionality and structure diversity. *Ganoderma lucidum*, known as Lingzhi in Chinese and Reishi in Japanese, is a species of basidomycetes belonging to Ganodermaceae of Aphyllophorales, and is a famous traditional Chinese medicinal mushroom [1]. It has been used as a traditional medication for the prevention and treatment of various human diseases for more than 2000 years [2]. The name Lingzhi started to appear in ancient Chinese literature after the Han dynasty. According to the literature of modern traditional Chinese medicine, *G. lucidum* presents three characteristics for the prevention or treatment of human diseases. First, there is no toxicity and no apparent side effects to human beings when using *G. lucidum*. Second, it does not specifically effect one particular organ, but focuses on the macro effect on an entire human body. And third, is has been shown to improve the normal functioning of human organs [1].

Nowadays *G. lucidum* has gained wide recognition for its immunomodulation and anti-tumor functions, and its cultivation has been very popular in many regions of East Asia [1–4]. The phytochemistry research revealed that the most important active constituents of *G. lucidum* were polysaccharides and triterpenoids, including ganoderic acids (GAs). Polysaccharides show good immunoregulatory activities [5,6]. Some lanostane type triterpenes-enriched extracts of *G. lucidum* and a few purified ganoderic acids have been proven to suppress the growth of various types of tumor cell lines [7–10]. GAs have been demonstrated to possess many other important pharmacological activities such as anti-metastasis [9], anti-HIV [11,12], anti-hepatitis [13], and so on.

Commercial *G. lucidum* products are mostly obtained from field cultivated *G. lucidum*, while wild *G. lucidum* is very expensive and very limited. Even field cultivation of the mushroom is a time-consuming and labor-intensive process. It usually takes several months to obtain the fruiting body of *G. lucidum,* and it is difficult to control the product quality during its cultivation process. Because biotechnological production, which is relatively easy to control and optimize, has many advantages, the submerged cultivation of *G. lucidum* is viewed as a promising alternative for efficient production of its valuable metabolites and has received increasing attention around the world

[1,4,6,14–20]. Production of bioactive components from *G. lucidum* mycelia was stimulated by the significant development in bioprocessing technologies [18–20]. Based on such efficient production technologies, the anticancer activities and mechanism of polysaccharides and purified ganoderic acids were exploited further. The advances provide the technological and theoretical support for the medicinal application of *Ganoderma.*

7.2 Characteristics and Bioactivities of *Ganoderma lucidum* Polysaccharides

Mushroom polysaccharides have been recognized as among the list of the most potent antitumor and immune-stimulating compounds. Various mushroom polysaccharides, such as Lentinan derived from *Lentinula edodes* and Krestin (PSK) derived from *Coriolus versidor*, have been recommended for use as anticancer drugs [21,22].

As mentioned above, polysaccharides are one of the most important bioactive components in *G. lucidum. Ganoderma* polysaccharides contain active substances with remarkable bioactivities such as immunoregulation, anticancer, antioxidant, anti aging and blood circulation promotion by removing blood stasis. The recent research progress is significant and has garnered more attention.

7.2.1 *Structure of* G. lucidum *Polysaccharide*

The structure of *Ganoderma* polysaccharide determines its bioactivities. The chemical structures and properties of *Ganoderma* polysaccharides, such as the monosaccharide composition, type of glycosyl linkage, molecule weight and chain conformation, have been investigated by spectrum analysis, chemical analysis and chromatography technology. After the cleavage of all glycoidic linkage by hydrolysis, monosaccharide composition was detected by gas chromatography-mass spectrometry (GC-MS) method [23].

The monosaccharide composition of *Ganoderma* polysaccharide contains glucose, fructose, maltose, arabinose, xylose, rhamnose, galactose, and mannose. Yang et al. [24] studied the monosaccharide composition of polysaccharide from *G. lucidum* with the results of xylose, fructose, glucose, sucrose and maltose being detected in relative molar percentages of 0.4%, 14.4%, 12.8%, 0.7%, and 50.9%, respectively. Zhang et al. [25] purified and obtained a novel polysaccharide from *G. atrum*, which was mainly composed of glucose, galactose, mannose, and trigalacturonic acid (4.91: 1: 1.28: 0.71).

Miyazaki and Nishijima [26] reported a water-soluble branched arabinoxyloglucan from *G. ludicum*, which contained α, β-1-4)-, β-(1-6)- and β-(1-3)-D-glucopyranosyl in backbone and side-chains. Bao et al. [27] also isolated two heteroglucans (PL-1 and PL-4) and one glucan (PL-3) from fruit bodies of *G. lucidum* and demonstrated that PL-1 had a backbone consisting of α-(1-4)-D-glucopyranosyl residues and β-(1-6)-D-galactopyranosyl residues, while PL-4 was comprised of (1-3)-, (1-4)-, (1-6)-β-glucopyranosyl residues and (1-6)-β-mannopyranosyl residues. The glucan was highly branched with (1-3)-β-D-glucopyranosyl residues substituted at O-6 with (1-6)-glucosyl residues.

The relative molecular weight (Mw) of most *Ganoderma* polysaccharides ranged from 8×10^3 to 2×10^5 (Table 7.1). As reported [27], the Mw of two heteroglycans (PL-1, PL-4) and one glucan (PL-3) from *G. lucidum* was estimated to be 8.3×10^3, 2.0×10^5, and 6.3×10^4, respectively. The molecular weights of five heteropolysaccharides from *G. lucidum* were detected to be 6.1×10^4 to 18.4×10^4 by size-exclusion chromatography coupled with laser light scattering technology [28].

7.2.2 *Biosynthesis of* G. lucidum *Polysaccharide*

The research on *G. lucidum* polysaccharide has been focused on its structure, bioactivity, and fermentation technology. But the information on its biosynthesis fundamental research is very limited. It was reported that there was a similar polysaccharide biosynthetic pathway in different microorganisms [29,30]. Although the structures of various polysaccharides are different, their precursor biosynthesis pathway is basically the same. So the biosynthesis of *Ganoderma* polysaccharide was studied by using isotope tracer techniques, and its proposed pathway was shown in Figure 7.1.

Figure 7.1 indicates that glucose-6-phosphate which is transformed from glucose, has two metabolic branches. One is the transformation of glucose-6-phosphate to glucose-1-phosphate, which is a necessary step for the biosynthesis of *Ganoderma* polysaccharide. Another

Table 7.1 Source, Chemical Structure, and Molecule Weight (Mw) of Polysaccharides from *Ganoderma* genus

SPECIES	SOURCE	BACKBONE	MONOSACCHARIDE COMPOSITION	MW	REF.
G. lucidum	Fruiting bodies	β-(1-3)-, (1-6), (1-4)-, α-(1-4)- Arabinoxyloglucan	Glucose, xylose, arabinose	4×10^4	[26]
G. lucidum	Fruiting bodies	Water-soluble heteropolysaccharides	Glucose, galactose, mannose, arabinose, xylose, fucose,	8.2×10^4	[31]
G. tsugae	Fruiting bodies	Protein-containing glucogalactans and β-(1-3)-glucans	Galactose, glucose, manse, fucose	–	[32]
G. lucidum	Mycelia	Branched β-(1-3)-glucan	Glucose	–	[33]
G. lucidum	Fruiting bodies	Linear α-(1-3)-glucan	Glucose	–	[34]
G. lucidum	Spores	β-(1-3)-glucan	Glucose	1×10^4	[35]
G. lucidum	Extracellular	α-(1-4)-galactose	Galactose, mannose, glucose, arabinose, rhamnose	2.2×10^4	[36]
G. lucidum	Fruiting bodies	β-(1-4)-glucose	Glucose, galactose	2.5×10^6	[37]
G. lucidum	Fruiting bodies	α–(1–4)–, β–(1–6)–heteroglucan	Glucose, galactose, rhamonose	8.3×10^3	[38]
G. lucidum	Spores	β-(1-6)-glucan	Glucose	10.3×10^4	[39]

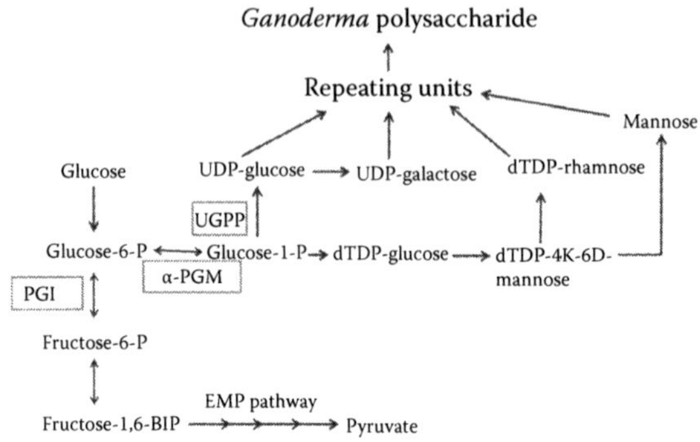

Figure 7.1 Proposed biosynthetic pathway of *Ganoderma* polysaccharide.

is transformation of glucose-6-phosphate to fructose-6-phosphate, and then it enters the glycolysis pathway [40]. The transformation of glucose-1-phosphate to uridine diphosphate glucose (UDP-glucose) is important for synthesis of *Ganoderma* polysaccharide which uses glucose and galactose as basic units. The transformation of glucose-1-phosphate to thymidine diphosphate glucose (dTDP-glucose or TDP-glucose) is important for biosynthesis of *Ganoderma* polysaccharide which utilizes rhamnose as a basic unit [31].

Phosphoglucomutase (PGM) and UDP glucose pyrophosphorylase (UGPP) are two important enzymes in the biosynthesis of polysaccharides [41–43]. PGM catalyzes the conversion of glucose-6-phosphate into glucose-1-phosphate, while UGPP catalyzes the synthesis of UDP-D-glucose, which is the donor of the monosaccharide unit in the biosynthesis of polysaccharides [41,42]. Both the intracellular polysaccharide (IPS) production and extracellular polysaccharide (EPS) production were improved by overexpression of PGM gene in *G. lucidum* [43]. Phosphoglucose isomerase (PGI), which catalyzes the transformation of glucose-6-phosphate to fructose-6-phosphate, also affects the synthesis of polysaccharide. PGI competes with PGM for the common substrate glucose-6-phosphate [41].

7.2.3 *Bioactivities of* G. lucidum *Polysaccharide*

Ganoderma has been extensively used in Chinese traditional medicine to promote human health, because of the contribution from its polysaccharide and other metabolites. The polysaccharide has been demonstrated to possess diverse and significant bioactivities such as immunomodulatory, anti-tumor, antioxidant, hepatoprotective, anti-ageing and anti-hypertensive activities [44].

7.2.3.1 Immunomodulatory Activity There is a general consensus that the immunomodulating effects of *G. lucidum* polysaccharide are extensive and important, with respect to enhancing the function of mononuclear phagocyte system, antigen-presenting cells, humoral immunity and cellular immunity [44].

You and Lin [45] found that *G. lucidum* polysaccharide peptide (GLPP), administered intraperitoneally, significantly increased the survival rate of macrophages injured by tert-butyl hydroperoxide *in vivo* and *in vitro*. It was also found that the phagocytosis and cytotoxicity of macrophages were significantly enhanced in cyclophosphamide-treated mice after the treatment of low-dosage (2.5 mg/kg) *G. lucidum* polysaccharide for 12 days [46]. A crude *G. ludicum* polysaccharide was extracted from solid fermentation of *Ganoderma* [47]; and, it was shown to strongly stimulate the proliferation of macrophages (158.02 ± 13.12%), the production of nitric oxide and phagocytosis (21.16 ± 1.65 μM), and protection of macrophage from doxorubicin [47]. It was demonstrated that polysaccharide PSG-1 from *G. atrum* could activate macrophages via toll-like receptor (TLR) 4-dependent signaling pathways, improving immunity and inhibiting CT26 tumor growth in mice, although PSG-1 did not kill CT26 cells directly [48]. Treatment with Gl-PS accelerated the recovery of splenic natural killer cells [49] and natural killer T (NKT) cells [50], and enhanced lymphokine-activated killer cells in immunosuppressed mice [46].

7.2.3.2 Anti-Tumor Activity Many studies have demonstrated that *Ganoderma* polysaccharides have strong anti-tumor effects. Treatment of mice with *G. lucidum* polysaccharide for 10 days can significantly reduce the tumor weight in a dose-dependent manner in S-180-bearing mice. Furthermore, the polysaccharide caused a significant cytotoxicity in human cervical cancer cell lines - CaSki and SiHa, human liver cancer cell lines - Hep3B and HepG2, and human colon cancer cell lines - HCT116 and HT29 *in vitro*, with marked apoptotic effects observed in CaSki, HepG2, and HCT116 cells [51]. The *G. lucidum* polysaccharide also exerted anti-tumor effects on solid tumors induced by Ehrlich's ascites carcinoma cells. *G. lucidum* polysaccharide (GLP) (100mg/kg) showed 80.8% and 77.6% reduction in tumor volume and tumor mass, respectively, when administered 24 h after tumor cell implantation. Furthermore, GLP with the same dosage, but when administered prior to tumor inoculation, showed 79.5% and 81.2% inhibition of tumor volume and tumor mass, respectively [49]. GLPP added into the cultured medium did not inhibit human lung carcinoma cells (PG) proliferation *in vitro*, but GLPP-treated serum potently inhibited PG cells proliferation *in vitro* and reduced the xenograft (PG cell line) in BALB/c nude mice *in vivo* [52]. It was reported that the mitogen-activated protein kinase (MAPK) pathway was simultaneously activated and involved in PSG-1-induced TNF-α secretion in RAW264.7 cells. Moreover, it was demonstrated that the phosphoinositide 3-kinase (PI3K)/Akt (protein kinase BB) pathway was stimulated and played an important role in the PSG-1 induced TNF-α secretion [53].

In fact, the structure and molecular weight played a crucial role in determining the antitumor activity of polysaccharides from *Ganoderma* [54]. The essential structure for the antitumor activity of *Ganoderma* polysaccharide might be a branched glucan core involving (1-3)-β-, (1-4)-β- and (1-6)–β-linkages [26].

7.2.3.3 Antioxidant Activity Spontaneous generation of reactive oxygen species (ROS) has widely been accepted as an unavoidable phenomenon that occurs in the progress of the energy metabolism of aerobic organisms. Despite its necessity in aerobic energy generation, an excess of ROS over the physiologically relevant level has been known to cause oxidative defects in cellular macromolecules, leading to a loss of normal cellular functions [55]. Accumulating evidences suggest that oxidative stress has been largely implicated in disease and various pathological conditions involved in aging. It has been well established that polysaccharide from *Ganoderma* exhibited a potent antioxidant activity [55–58].

Four polysaccharides were isolated from fruit bodies of *G. lucidum* [59]. The antioxidant activities of its four polysaccharides were evaluated. The results showed that GLP80 had the best reducing power, 2,2-diphenyl-1-picryhydrazyl (DPPH) radical scavenging ability and oxygen radical scavenging ability followed by

GLP, GLP60 and GLP40 [59]. A polysaccharide from *G. lucidum* (Gl-PS) also showed antioxidant effects by reducing ROS production and malondialdehyde content and increasing the activity of manganese superoxide dismutase in rat cerebral cortical neuronal cultures exposed to hypoxia/reoxygenation *in vitro*. These findings showed that Gl-PS exhibited remarkable protection against hypoxia/reoxygenation (H/R) injury in rat cortical neurons [60]. The antioxidant activity of PSG-1 was investigated *in vitro* by using DPPH free radical scavenging and self-oxidation of 1,2,3-phentriol assay. The results indicated that PSG-1 had strong scavenging abilities on DPPH free radical and superoxide anion radical compared to conventional antioxidants [61]. It was further indicated that PSG-1 could attenuate oxidative stress in D-galactose-induced aging mice, and the results suggested that PSG-1 has good potential as a novel anti-oxidant agent to promote human health [62].

7.2.3.4 Hypolipidemic Activity Two hypoglycemic principles, ganoderans B and C, isolated from the fruit bodies of *G. lucidum*, were shown to be peptidoglycans with each molecular weights of 7400 and 5800, respectively [63]. A hyperbranched proteoglycan from *G. lucidum* was found to inhibit the protein tyrosine phosphatase 1B (PTP1B) activity on a competitive mechanism *in vitro* [64]. The effect of different doses of *Ganoderma* polysaccharides on blood sugar was studied [65]. It was found that there was a significant effect of reducing blood sugar level under administration of more than 200 mg/kg *Ganoderma* polysaccharide [66].

7.2.3.5 Other Activities Besides the above important bioactivities, numerous reports indicate that *Ganoderma* polysaccharides possess other beneficial physiological activities. These include anti-inflammatory, anti-virus, anti-fatigue, and anti-radiation effects [66–69].

7.3. Why Production of *G. lucidum* Polysaccharides by Liquid Fermentation

As mentioned above, until now the commercial *Ganoderma* products are mostly obtained from field cultivation, but field cultivation is a time-consuming and labor-intensive process, and it is also very difficult to control the product quality during the cultivation. There is a compelling need to supply the market with a large amount of high-quality *Ganoderma* products. Nowadays, the submerged cultivation of *G. lucidum* has received much attention in Asia and beyond as a promising alternative to field cultivation for efficient production of its valuable metabolites such as polysaccharides and triterpenoids (especially ganoderic acids). It is obviously necessary and important to develop a process for simultaneous production of these valuable metabolites in large-scale bioreactor systems. There have been reports on liquid cultures of *G. lucidum* for production of polysaccharides [1,4–16] and ganoderic acids [2,3,15] in flasks and bioreactors. Based on this bioprocessing technology, the anticancer activities and mechanisms of polysaccharides and purified ganoderic acids were exploited further. Some related technological advancement will be overviewed as follows.

7.4. Strategies for Enhancing *G. lucidum* Polysaccharide Production in Liquid Fermentation Process

7.4.1 Breeding of Ganoderma *Cell Line*

The genus *Ganoderma* (the family: *Ganodermataceae*) involves different species that are mostly well known for their medicinal properties. The content of bioactive components such as polysaccharides and ganoderic acids vary significantly in different strains [70–72]. Therefore, screening the strains with high content and good fermentation property is necessary and critical for further application. It might provide a good source for searching for new compounds with possibly better bioactivities. Da et al. reported that there was a significant difference in triterpenoid content (including eight ganoderic acids and four ganoderic acids) between fruit bodies of *G. lucidum* and *G. sinense* [70]. The content of polysaccharides and GAs from the fruit bodies of 12 *Ganoderma* strains was determined in order to choose suitable strains for extraction and processing [71]. Five strains with high polysaccharide content and six strains with high triterpenoid content were identified [71].

Artificial selection, also known as selective breeding, is a traditional breeding method, which uses artificial means to choose superior strains from natural sources in order to biologically obtain new species and reproduce them selectively. The basic method of artificial selection includes the tissue separation and spore separation method for obtaining pure strains, followed by optimization of the screened strains, and then obtaining the best one [73]. During the mushroom production procedure, the tissue separation method is often employed to obtain the strains because the *Ganoderma* spore is hard to germinate [74]. Artificial selection is more commonly used in the breeding of edible mushrooms other than *Ganoderma*. As partial data are shown in Table 7.2, we analyzed the differences in biomass and the contents of polysaccharide and GAs in a number of *Ganoderma* strains [72]. The results showed that *G. lucidum* 5.26 had a high content of both polysaccharide and ganoderic acids.

Mutation breeding, which changes genes of strains and achieves genetic recombination for better characteristics, is a more effective method compared to artificial selection. The protoplast is usually chosen for mutation breeding [73], and the final goal is to obtain increased bioactive components such as polysaccharide [74] and triterpenoids [75]. Mutation breeding can not only increase strain mutation rate through a simple operation, but also provides genetic markers for further crossbreeding and cell fusion breeding. However, it also has some disadvantages. For example, the mutation generation is random and the work on selecting mutants is tedious, etc.

Genetic engineering is a rather new and more effective method in strain breeding. But the gene manipulation in *G. lucidum* is difficult due to the immature genetic tools and very low transformation efficiency. Zhang et al. [27] over-expressed rice OsUgp2 gene

Table 7.2 Screening of *Ganoderma* Strains

STRAINS	DCW (G/L)	IPS CONTENT (%)	TOTAL GA CONTENT (%)	EPS PRODUCTION (G/L)
G. lucidum 5.65	12.9 ± 0.4	17.3	0.88	1.00
G. lucidum 5.26	16.0 ± 0.4	17.2	2.04	0.82
G. lucidum 5.616	16.6 ± 0.4	17.9	1.76	0.52
G. formosanum 5.536	10.9 ± 0.6	15.3	1.68	0.53
G. tsugae 5.542	14.3 ± 1.0	13.3	0.96	1.17
G. applanatum 5.249	8.1 ± 1.4	14.1	0.98	0.53

Source: Modified from Wei ZH et al. (2014) *Bioprocess and Biosystems Engineering* 37: 1789–1797.

in *G. sinensis* and increased the content of intracellular and extracellular polysaccharide (EPS) in *G. sinensis*. Xu et al. [43] overexpressed α-phosphoglucomutase gene in *G. lucidum*. The effects of PGM gene overexpression on intracellular polysaccharide content, extracellular polysaccharide production and transcription levels of three genes encoding the enzymes involved in polysaccharide biosynthesis were all investigated. The maximum IPS content and EPS production in the engineered strain was increased by 40.5% and 44.3%, respectively, compared to the wild-type strain. It is expected that the development of genetic and metabolic engineering techniques, such as the powerful gene-editing tool CRISPR-Cas9, would greatly contribute to the new vigor to breed favorable and powerful *Ganoderma* cell lines.

7.4.2 Optimization of Fermentation Media

Developing fermentation media is an important step in improving production of bioactive metabolites. Due to the potentially large market demand for *Ganoderma* biomass, polysaccharide and GAs, the development of a low-cost, industrial medium formula is necessary and important.

Carbon source is an important factor for biomass, polysaccharide and GA production in the *G. lucidum* fermentation process. Lactose and glucose were widely used in *G. lucidum* fermentation as reported. Effects of carbon source and initial sugar concentration on the production of GAs and polysaccharide were studied [73]. It was found that sucrose as a carbon source was suitable for the extracellular polysaccharide (EPS) production although the cells could not grow very well. As shown in Table 7.3, lactose was beneficial for the cell growth and production of GAs and intracellular polysaccharide (IPS), thus, it was chosen as the carbon source. Sucrose greatly inhibited *G. ludicum* cell growth, but enhanced EPS production. When the initial lactose concentration exceeded 35 g/l, GA accumulation was decreased [75]. Glucose was found to be the best carbon source in *G. lucidum* 5.26 fermentation in our recent work [72]. It was found that sucrose greatly enhanced intracellular polysaccharide (IPS) content in fermentation of *G. lucidum* 5.26, but inhibited the cell growth. The mechanism was studied by analyzing the activities of enzymes related to polysaccharide biosynthesis. The

Table 7.3 Effects of Carbon Sources on the Cell Growth and Production of EPS, IPS and GAs

	SUCROSE	MALTOSE	LACTOSE	GLUCOSE
Maximum DW (g/L)	3.55 ± 0.40 (d 2)[a]	12.2 ± 1.1 (d 10)	12.3 ± 0.7 (d 12)	10.1 ± 0.5 (d 10)
Maximum EPS production (g/L)	0.63 ± 0.02	0.57 ± 0.02	0.53 ± 0.03	0.47 ± 0.01
Maximum IPS content (mg/100 mg DW)	12.21 ± 0.56	11.97 ± 0.27	12.25 ± 0.38	10.30 ± 0.15
Maximum IPS production (g/L)	0.25 ± 0.01	1.16 ± 0.05	1.43 ± 0.10	0.97 ± 0.06
Maximum GA content (mg/100 mg DW)	1.16 ± 0.07	1.28 ± 0.08	1.46 ± 0.06	1.42 ± 0.06
Maximum GA production (mg/L)	34.9 ± 2.5	118.7 ± 11.5	180.4 ± 6.1	143.5 ± 5.7

[a] $p < 0.05$.

data suggested that sucrose changed the metabolic flux of glucose 6-phosphate from glycolysis pathway to the synthesis of *Ganoderma* polysaccharide by regulating the activities of PGM and PGI. So, glucose and sucrose were used as a mixture carbon source. As a result, both the *Ganoderma* biomass and the polysaccharide production were enhanced.

The effects of nitrogen source and initial glucose concentration were studied in submerged fermentation of *G. lucidum* for simultaneous production of ganoderic acids and polysaccharides. The cells could not grow well when either yeast extract or peptone was used as the sole nitrogen source. However, a combined addition of 5 g/l of yeast extract and 5 g/l of peptone was optimal for the cell growth and polysaccharide production [15]. Under nitrogen-source limitation conditions, the cell concentration was low, but both polysaccharide production (1.61 g/L) and specific polysaccharide production (0.492 g/g-cell) were high [76]. Our work also showed that the production of four individual ganoderic acids, GA-Mk, GA-T, GA-S, and GA-Me, was 2.8-, 5.8-, 8.3-, and 5.1-fold higher in nitrogen limited conditions [77]. The transcriptional levels of biosynthetic genes encoding 3-hydroxy-3-methylglutaryl-CoA reductase, farnesyl pyrophosphate synthase, squalene synthase, lanosterol synthase, and a putative nitrogen regulator, AreA, were up-regulated under nitrogen limitation.

The work suggested that increased GA biosynthesis may result from higher expression of those genes [77].

Vitamins and minerals are important for *G. lucidum* cell growth and polysaccharide production, too [78]. Vitamin B_1 was widely used in *G. lucidum* fermentation medium [15,23]. Supplying $CaCO_3$ promoted both mycelial growth and polysaccharide production [76]. Our results showed that $Fe_2(SO_4)_3$ was important for *G. lucidum* mycelial growth and its polysaccharide production [72].

7.4.3 *Effect of Oxygen Supply and Shear Stress on* G. lucidum *Polysaccharide Production*

Because *G. lucidum* mycelia and pellets are shear-sensitive, and culture viscosity usually increases during fermentation, oxygen supply and shear stress might become serious problems in large-scale submerged cultures of *G. lucidum*. Oxygen affects cell growth, morphology, nutrient uptake, and metabolite biosynthesis. We investigated the impacts of initial volumetric oxygen transfer coefficient (K_La) and dissolved oxygen tension (DO) on cultivation of *G. lucidum* to obtain useful information for large-scale production of *Ganoderma* polysaccharide and ganoderic acids by the bioprocess.

The effect of initial K_La on *G. lucidum* cultures was studied by setting various initial K_La values [14]. All cell cultures were agitated at 200 rpm and the aeration rate was adjusted over a range of 220 to 3500 mL/min to produce the desired initial K_La values from 16.4 to 96.0 h^{-1}. The initial K_La level affected the biomass accumulation, and its peak value of 15.6 g/L was obtained at an initial K_La value of 78.2 h^{-1} (Table 7.4). The results indicated that the initial K_La value had a significant effect on cell growth during cultivation. The production and productivity of both extracellular polysaccharide (EPS) and intracellular polysaccharide (IPS) was relatively lower at an initial K_La value of 60.0 h^{-1} (Table 7.4). The data also showed that relatively higher initial K_La levels (78.2 and 96.0 h^{-1}) were favorable for polysaccharide production.

Although the highest biomass was obtained at an initial K_La value of 78.2 h^{-1}, the highest GAs production of 450 mg/L was obtained at 96.0 h^{-1}, because of the high GA content obtained in the case (Table 7.4). An increase in initial K_La increased the production and productivity of GAs, the amount of which, at an initial K_La value of

Table 7.4 Effects of Initial K_La Values on the Cell Growth and Production of EPS, IPS, and GAs

	INITIAL K_La (h^{-1})			
	16.4	60.0	78.2	96
Dry cell weight (DCW) (g/L)	11.8	14.1	15.6	13.6
EPS production (g/L)	1.0	0.7	0.9	0.9
EPS productivity (mg/(L d))	73	52	69	69
IPS production (g/L)	1.9	1.6	2.2	2.1
IPS productivity (mg/(L d))	189	161	217	207
GA production (mg/L)	246	280	339	450
GA productivity (mg/(L d))	24	27	33	44

Source: Modified from Zhong JJ, Tang YJ (2004) *Advances in Biochemical Engineering/Biotechnology* 87: pp. 25–59.

96.0 h^{-1},was 1.8-fold that at 16.4 h^{-1}. It is clear that an initial K_La value of 96.0 h^{-1} was optimal for both GAs production and productivity.

The DO gradient in the mycelia pellet can be calculated according to the following equation [14]:

$$D\left(\frac{d^2C}{dr^2}+\frac{2}{r}\frac{dC}{dr}\right)=\rho(Q_{O2})=\frac{\rho(Q_{O2})m^C}{K_O+C} \quad (7.1)$$

Where C is the DO concentration over the pellet cross sections at a distance; r from the pellet center is ρ density of cells in the pellet; D is the coefficient of oxygen diffusion along the liquid channels inside the mycelia pellet; Q_{O2} is the oxygen consumption rate per pellet volume; $(Q_{O2})_m$ is the maximum oxygen consumption rate per pellet volume; and K_O is oxygen constant. When DO was controlled around 10% of air saturation, its concentration was 2.44×10^{-2} mol/m^3, the radius of the mycelia pellet was about 1.9×10^{-4} m on average, the maximum specific oxygen uptake rate (SOUR) was measured as 1.85 mmol O_2 g^{-1} h^{-1}, the pellet density was 75 kg/m^3, D was assumed to be 1.9×10^{-9} m^2/s, and K_O was 4.47×10^{-5} mol/m^3 [14]. From Equation (7.1), the DO in the center of mycelia pellet was estimated to be zero, implying that the cells in the mycelia pellet were under severe oxygen limitation.

The results showed that DO levels affected the *G. ludicum* biomass and production of EPS, IPS and GAs [18]. EPS production and productivity obtained around 10% DO were higher than those around 25% DO (Figure 7.2b). The IPS content around 10% DO was also higher than that around 25% DO (Figure 7.2c). Therefore, oxygen limitation

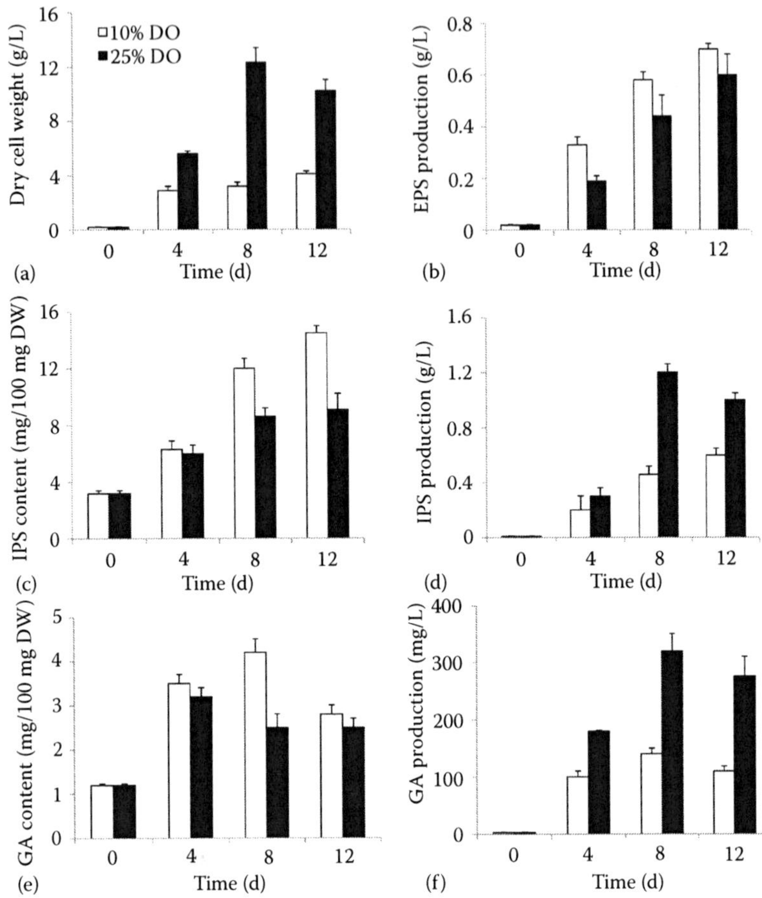

Figure 7.2 Effect of DO levels on (a) cell growth, (b) EPS production, (c) IPS content, (d) IPS production, (e) GA content, and (f) GA production in *G. lucidum* fermentation. Symbols for DO: blank bars, around 10% of air saturation; filled ones, around 25% of air saturation. (Data modified from Tang YJ, Zhong JJ (2003) *Enzyme and Microbial Technology* 32: 478–484.)

was suggested to be beneficial for the metabolic flux toward polysaccharide biosynthesis. The total IPS production and productivity were much higher at 25% DO than those at 10% DO, because a higher biomass density was obtained at 25% DO. The GAs content around 10% DO was higher than that around 25% DO (Figure 7.2e). The data indicated that a relatively lower DO, in which the cells in the center of the mycelia pellet could be under oxygen limitation, was beneficial for GAs biosynthesis. The results implied that oxygen limitation was also beneficial for metabolic flux towards the biosynthesis of GAs [18].

Hydrodynamic shear stress is well known as an important factor for shear sensitive bioprocesses such as cultures of cells of animals, insects, plants, and higher fungi. Mushroom (higher fungi) mycelial cells are generally sensitive to shear environment because of their morphological characteristics. Because investigation of the effect of shear on mushroom cultures was lacking, it was necessary to study the influence of hydrodynamic shear on cell growth and production of ganoderic acids and polysaccharide by *G. lucidum* cells in bioreactors [14].

Gong and Zhong [17] reported that hydrodynamic shear stress is an important factor for the batch fermentation of *G. lucidum*. As shown in Figure 7.3, the maximal cell density of 13.8 g/L and the intracellular polysaccharide production of 2.64 g/L were obtained under a low impeller tip speed (ITS, 0.51 m/s). The cell morphology of *G. lucidum* was also affected by ITS. The maximal mean projected area of dispersed hyphae increased much more at a lower ITS.

To achieve a highly efficient process for production of GAs and *Ganoderma* polysaccharide, the effect of shear stress on the fed-batch fermentation process was investigated by Tang et al. [79]. V_{tip} vaules of 1.234, 1.543, 1.853 and 2.161 m/s after cultivation of day 2 were selected to illustrate whether the shear stress was critical to the fed-batch fermentation of *G. lucidum*. As observed, the fermentation broth mixed well in the STR even at a low shear stress of 1.234 m/s, and there was no apparent cell sedimentation and no serious cell adhesion at the STR wall in all cases during fermentation. The results showed that cell growth was significantly influenced by shear stress. A low shear stress was better for

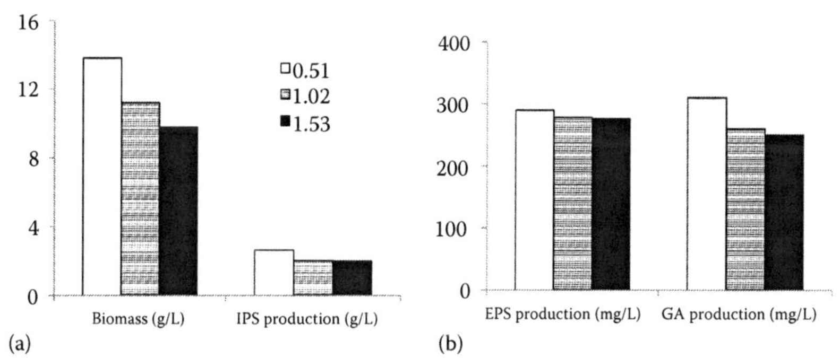

Figure 7.3 Results of biomass (a), IPS production (a), EPS production (b) and GA production (b) in fermentation of *G. lucidum* in a 5 L bioreactor at different impeller tip speeds. (Data modified from Gong HG, Zhong JJ (2005) *Chinese Journal of Chemical Engineering* 13: 426–428.)

G. lucidum cell growth. EPS production was obviously inhibited at a high V_{tip} of 2.161 m/s, and significantly higher at 1.543 m/s. The highest EPS production and productivity were attained at the V_{tip} of 1.543 m/s [79]. A different phenomenon was observed in the batch cultivation of *G. lucidum*, in which shear stress was reported to have no effect on EPS accumulation [17], while the maximal IPS production of 4.74 g/L was obtained on day 12 at the V_{tip} of 1.234 m/s because of the highest biomass and IPS content both obtained on day 12 [17].

7.4.4 Fed-Batch Fermentation

Fed-batch was a widely-adopted useful strategy in fermentation process intensification [75,80]. It was reported that a fed-batch fermentation process of *Tuber melanosporum* was developed for the efficient production of bioactive mycelia and *Tuber* polysaccharide by feeding of sucrose to maintain its concentration around 35.5 g/L when its residual level decreased to 10.5 g/L [80].

The biomass, production of extracellular polysaccharide (EPS) and intracellular polysaccharide (IPS) were enhanced compared to batch cultures [80]. A process for simultaneous efficient production of bioactive polysaccharide and ergosterol by fed-batch fermentation of medicinal mushroom *Agaricus brasiliensis* was developed by Zou [81]. A repeated fed-batch fermentation strategy with glucose as the carbon source successfully increased EPS and ergosterol production simultaneously [81].

A lactose fed-batch strategy was developed by Tang and Zhong with feeding of concentrated lactose solution [75,82]. On the basis of a preliminary study, feeding of lactose during cultivation was adopted to avoid the inhibitory effects of lactose on GAs biosynthesis while increasing final cell density, metabolite accumulation, and productivity [75]. As shown in Table 7.5, lactose feeding was tested on days 8, 10, and 12, when its residual level was 12.9, 6.0, and 3.3 g/L, respectively. The maximum cell density (by dry weight) of *G. lucidum* at different feeding times was almost the same, i.e. around 20 g/L, whereas it was only 15.8 ± 0.4 g L^{-1} (day 14) in batch cultures [75]. The EPS production under various feeding time was studied, and it had the best results when lactose was feeding at day 8. The EPS yield against lactose at feeding time of day 8, 10, and 12 was 19.85, 16.27, and 15.31 mg EPS/g lactose, respectively, while it was 15.98 mg EPS/g

Table 7.5 Results of the Cell Growth, EPS Production, EPS Yield, IPS Production and IPS Productivity at Different Feeding Time

FEEDING TIME	DRY CELL WEIGHT (G/L)	EPS PRODUCTION (G/L)	EPS YIELD (MG/G LACTOSE)	IPS PRODUCTION (G/L)	IPS PRODUCTIVITY (MG/L D)
Day 8	20.1	0.96	19.85	1.97	122.3
Day 10	19.7	0.76	16.27	2.01	142.9
Day 12	19.6	0.72	15.31	1.78	97.9
Control	15.8	0.57	15.98	1.57	86.7

Source: Data are modified from Tang YJ, Zhong JJ (2002) *Enzyme and Microbial Technology* 31: pp. 20–28.

lactose for the batch fermentation [75]. The IPS production reached its peak, i.e., 1.97, 2.01, and 1.78 g/L on day 16, 14, and 18 at feeding time of day 8, 10, and 12, respectively, and their IPS productivity was 122.3, 142.9, and 97.9 mg/L per day, respectively. In batch fermentation, the IPS production reached its peak (1.57 g/L) on day 18, and its productivity was 86.7 mg/L per day. The results indicated that fed-batch culture was favorable to the enhancement of both the production and productivity of IPS due to a relatively high cell density obtained under the condition, and the optimum feeding time for IPS production was on day 10 [75]. A pH-shift and DOT-shift integrated fed-batch fermentation process was developed for the production of ganoderic acid and polysaccharide by Tang et al., and it was successfully scaled up to 200 L bioreactor [16,79].

7.4.5 *Two-Stage Fermentation*

Investigations on the impact of pellet size on the cellular oxygen uptake and accumulation of GAs suggest the favorable effect of oxygen limitation on GA formation by the higher fungus *G. lucidum*. Interestingly, this well agreed to the two-stage fermentation process, which was by chance invented to enhance the GA production by combining conventional shaking fermentation with liquid static culture [4,83]. According to the results, a high cell density of 20.9 g of DW/L (DW = dry cell weight) was achieved through a 4-day shaking fermentation followed by a 12-day static culture [83]. A change in the cell morphology and a decrease in the sugar consumption rate were observed during the static liquid fermentation. GA production in the new two-stage process was considerably enhanced with its content increased from 1.36 (control) to 3.19 mg/100 mg of DW, which was much higher than previously observed. The higher GA biosynthesis was related to the cellular morphological changes (Figure 7.4) [84,85].

Tang et al. [82] studied the effect of liquid surface area per liquid volume (A_s) on EPS and IPS production in two-stage fermentation mode. The results showed that As had significant effects on EPS and IPS production in two-stage fermentation mode. The highest EPS production was 0.31, 0.32, and 0.26 g/L in 0.24, 0.90 and 1.53 cm^2/mL A_s condition, respectively. The IPS production at 0.90 and 1.53 A_s was enhanced 2 fold that of 0.24 A_s [82].

A simple unstructured kinetic model could describe the two-stage culture process as set up by Zhong's group [4,83]. The model described the dynamics of cell growth, lactose consumption and GA accumulation by *G. lucidum*. Parameter sensitivity analysis indicated that the model was sensitive to changes in maximum specific growth rate, the saturation constant, the biomass yield against lactose and the growth-associated product formation constant. The model was able to predict the two-stage culture process of *G. lucidum* in various laboratory-scale multi-layer static bioreactors with working volumes from 20 mL to 7.5 L. Because there is a great need for efficient large-scale production

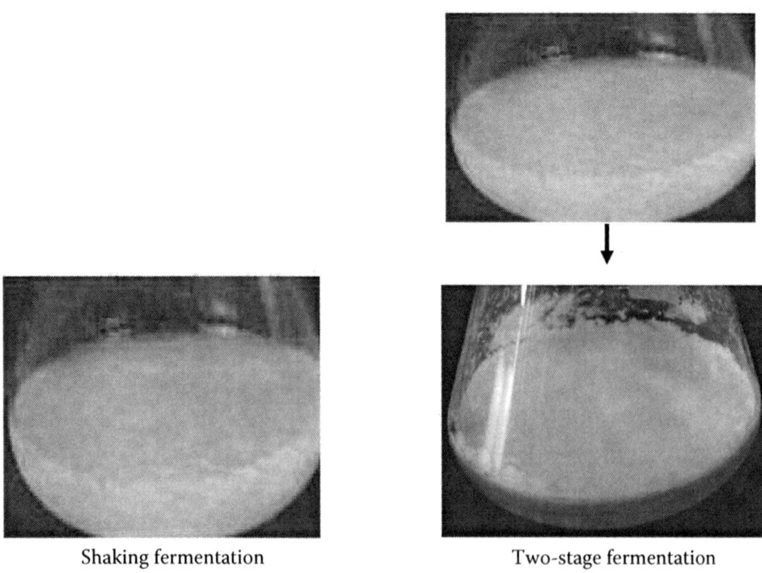

Figure 7.4 Morphology changes of *G. lucidum* by comparing its conventional shaking fermentation with its shaking-static two-stage cultivation

of GAs, a very useful antitumor metabolite, by the mushroom cultivation, the model identified might be useful in converting large-scale static culture processes to industrial application [83].

As different GAs have different bioactivities, it is critical to understand the kinetics of individual GA production during fermentation. The production of individual GAs between two-stage fermentation and shaking fermentation was compared by Xu et al. [84]. As shown in Figure 7.5, the content of individual GA and total GAs increased rapidly in the liquid static culture, and their maximum value was 6- to 25-fold that of shaking culture [84]. To understand the regulation of GA biosynthesis, investigation of the accumulation of intermediate (lanosterol) and by-product (ergosterol) and of the expression of three important biosynthetic genes was also conducted in liquid shaking and static cultures of *G. lucidum*. The transcript of squalene synthase (SQS), lanosterol synthase (LS) and 3-hydroxy-3-methylglutaryl coenzyme A reductase (HMGR) in liquid static culture was 4.3-, 2.1-, and 1.9-fold that of the shaking culture, respectively [84]. Higher GA content in liquid static culture was related to increased transcription levels of those genes especially SQS [84].

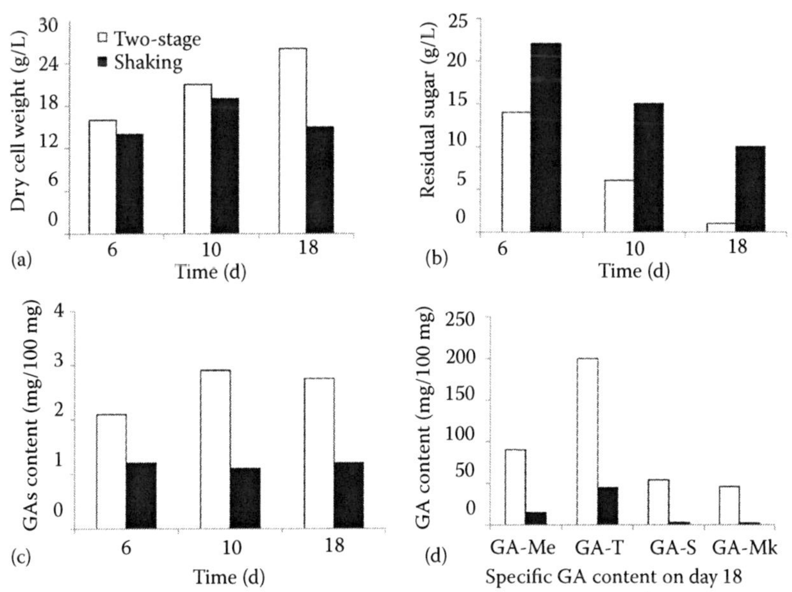

Figure 7.5 Results of the cell growth (a), sugar consumption (b), the content of GAs (c), and the content of GA-T, GA-Me, GA-Mk and GA-S (d) in liquid static and shaking cultures of *G. lucidum*. (Data are modified from Xu JW et al., (2010) *Applied Microbiology and Biotechnology* 85: 941–948.)

Subsequently, many strategies were developed in two-stage fermentation of *G. lucidum*, which further enhanced GA production. Effects of oxygen concentration within 21–100% in gaseous phase on the morphology and GA production by *G. lucidum* in liquid static culture were studied by Zhang and Zhong [85]. The results showed that the oxygen levels in the gaseous phase affected the spore formation and GA production, and a higher level of oxygen stimulated sporulation and GA synthesis. A high content of GAs was found in spores of cultured *G. lucidum*. The enrichment of oxygen in gas environment was favorable to the formation of GA-S and GA-Me [85]. These results indicated that oxygen enrichment into the two-stage culture provided favorable conditions for GA production by *G. lucidum*. The production of 4 individual GAs in two-stage culture of *G. lucidum* was enhanced under nitrogen-limiting conditions [77]. At 3 mM glutamine, a maximal content of GA-Mk, GA-T, GA-S, and GA-Me was 2.16, 11.76, 31.09, and 7.04 μg/mg cell dry weight, respectively, which was 2.8-, 5.8-, 8.3-, and 5.1-fold that obtained at 60 mM glutamine. The transcription levels of biosynthetic genes encoding HMGR, farnesyl pyrophosphate synthase, SQS, LS, and a putative nitrogen regulator, AreA, were up-regulated by 37-, 18-, 4.5-, 3.2-, and 13-fold, respectively, in nitrogen limitation conditions, suggesting that increased GA biosynthesis may result from higher expression of those genes [84]. The addition of 10 mM calcium ion to two-stage cultures of *G. lucidum* was proved as a very simple and useful strategy to enhance the production of total GAs and individual GA-Mk, -T, -S, and -Me as well [86]. Experiments using Ca^{2+} sensor inhibitors indicated the involvement of calcineurin signal in regulating GA biosynthesis. Quantitative gene transcription analysis revealed that the expression levels of genes of GA biosynthesis and Ca^{2+} sensor were up-regulated with calcium addition while down-regulated under the inhibitors addition, suggesting that higher GA production may result from higher expressions of those genes [86].

Although the production of polysaccharide was not greatly enhanced, two-stage fermentation is a useful technology for the production of *G. lucidum* biomass and GAs, which showed better antitumor activities [85–91].

7.4.6 Other Strategies

Fermentation pH was found to have significant effects on *G. ludicum* biomass, polysaccharide and GA production. Tang and Zhong studied the initial pH on production of GAs and polysaccharides by submerged fermentation of *G. lucidum* [92]. At an initial pH of 6.5, a maximum biomass was obtained, as well as a maximal GA production. Lowering the initial pH from 6.5 to 3.5 gradually led to a higher production of extracellular polysaccharide and a higher specific production of intracellular polysaccharide [92]. A pH-shift strategy, which changed pH from 3.0 to 4.5 on day 4, was developed by Tang et al. [79]. The strategy enhanced biomass, EPS, IPS and GA production significantly.

It was reported that the addition of some elicitors enhanced *Ganoderma* polysaccharide and GAs production. The addition of 254 mM methyl jasmonate solubilized in tween-20 that on day 6 enhanced GA yield by 45.3% [93]. The fungal elicitors extracted from *Tuber aestivum Vittad.* significantly stimulated the accumulation of extracellular polysaccharide [94]. *Sophora flavescens* water extract of 11 g/L promoted *G. lucidum* biomass and EPS production [95].

Other strategies were developed in the *G. lucidum* fermentation process, such as optimization of inoculation density [96], light irradiation [97], addition of oil [98], etc.

7.5 Scale-Up Study on *G. lucidum* Polysaccharide Fermentation

Bioprocess scale-up is a challenge to mushroom submerged fermentations, as it involves so many interreacted and, sometimes, competing factors. Although there have been a few reports on liquid cultivation of mushrooms using small bioreactors [16,75,79], there is a lack of reports on scale-up of mushroom cultivation processes [79]. The scale-up of the filamentous fungus *Glarea lozoyensis* to produce pharmaceutically important pneumocandin was successfully completed from pilot scale (0.07, 0.8, and 19 m^3) to production scale (57 m^3) by keeping K_La, oxygen uptake rate and oxygen transfer rate constant, which were effectively maintained across the scales, while the use of other criteria such as power per liquid volume (Pg/VL) or mixing time was less effective [99]. According to previous work [14], at an initial K_La value of 80.3 h^{-1} or

DO above 25%, cell growth and metabolite biosynthesis in a 5-L STR were comparable with or even better than those in shake flasks. Here, scale-up of *G. lucidum* cultures from a 5-L STR to a 30-L STR was first attempted on the basis of oxygen supply by Tang and Zhong [14].

Cultivations in a conventional 30-L turbine-impeller reactor (STR) (working volume 21 L) were conducted under different conditions by setting the initial K_La level at 80.3 h^{-1}, or controlling DO>25% of air saturation during cultivation, or changing agitation speeds between 200 and 300 rpm with DO>25%. As shown by the results summarized in Figure. 7.6, although EPS production was similar in both 5-L (control) and 30-L STR, maximum cell dry weight, IPS production, and GAs production were all much lower in 30-L STR than in 5-L STR. In addition, it was observed that fluid mixing was fine at an agitation speed of 200 rpm or 300 rpm. Therefore, initial K_La, DO, and mixing are not regarded as key factors in scale-up of *G. lucidum* cultures from 5-L to 30-L STR. It was considered that hydrodynamic

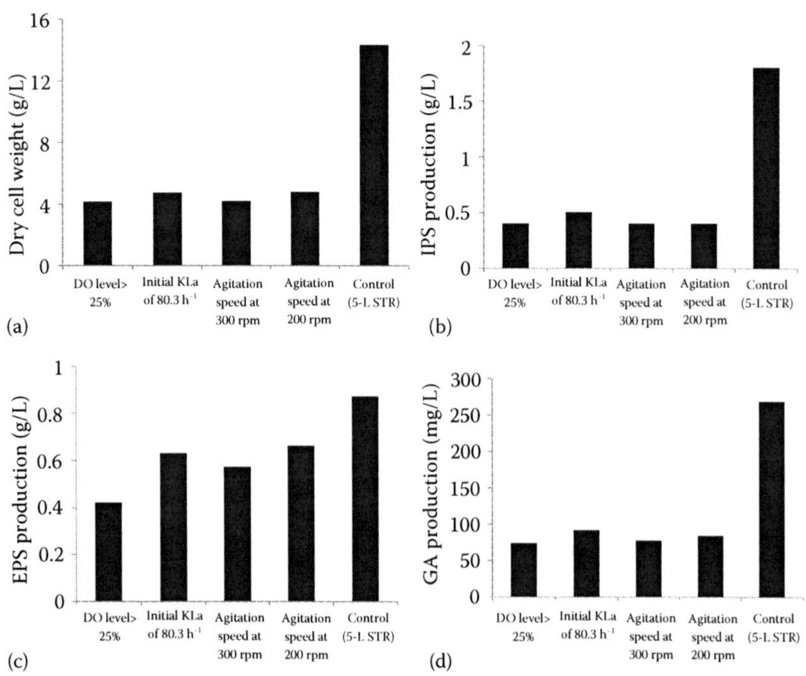

Figure 7.6 Summary of the accumulation of biomass (a), IPS (b), (EPS), and (GA) in various cultivations in a 30-L STR. (Modified from Zhong JJ, Tang YJ (2004) *Advances in Biochemical Engineering/ Biotechnology* 87: 25–59.)

shear might be critical for the bioprocess scale-up, because it was found to be significant to cell growth and metabolite production by the mushroom cultures as mentioned above.

An integrated scale-up strategy by combining pH-shift, DOT-shift, and lower shear stress was developed for the fed-batch fermentation process of *G. lucidum* by Tang et al. [79]. First, V_{tip} values of 1.234, 1.543, 1.853 and 2.161 m/s after cultivation of day 2 were selected to illustrate whether the shear stress was critical to the fed-batch fermentation of *G. lucidum*. The results showed that a low shear stress of 1.234 m/s had the best production of polysaccharides and GAs [79]. Based on the results, pH-shift, DOT-shift and lower shear stress were identified as key factors affecting the cell growth and the production of GAs and polysaccharide during the fed-batch fermentation of *G. lucidum*. Scaling up of *G. lucidum* fed-batch fermentation from 7.5- to 20-L STR, then to 200-L STR was successfully conducted based on the integrated scale-up criterion [79].

The *G. lucidum* fermentation process was scaled up in a 30-L centrifugal impeller bioreactor (CIB) [14], which had a low shear stress [100,101]. It is clear that the cultivation results in 5-L STR could be well reproduced in 30-L CIB. The *G. lucidum* fermentation process was scaled up to 300 L stirred tank reactor and airlift reactor by setting the initial $K_L a$ level at 92 h^{-1}. As shown in Figure 7.7, the biomass, EPS production and IPS production in a 300-L airlift reactor were much higher than that of a 300-L stirred tank reactor. The fermentation process in a small cultivation vessel was well reproduced in a 300-L airlift reactor.

Previous work in our laboratory demonstrated that the content and production of GAs could be significantly enhanced by means of a two-stage culture process combining conventional shaking cultivation (first stage) with static liquid culture (second-stage) [2,4,77,84]. It was essential to study the scale-up of this novel process from a flask to a bioreactor for potential commercial application of the mushroom culture. Tang and Zhong investigated the impact of liquid surface area per liquid volume (As) [83]. The formation of white aerial mycelia on the surface increased at larger As within the range investigated (0.24—1.53 cm^2/mL) and mycelial growth was significantly limited at an As of 0.24 cm^2/mL [83]. The maximum cell density in the liquid was 13.4, 16.0, and 13.4 g/L at As levels of 0.24, 0.90, and 1.53 cm^2/mL,

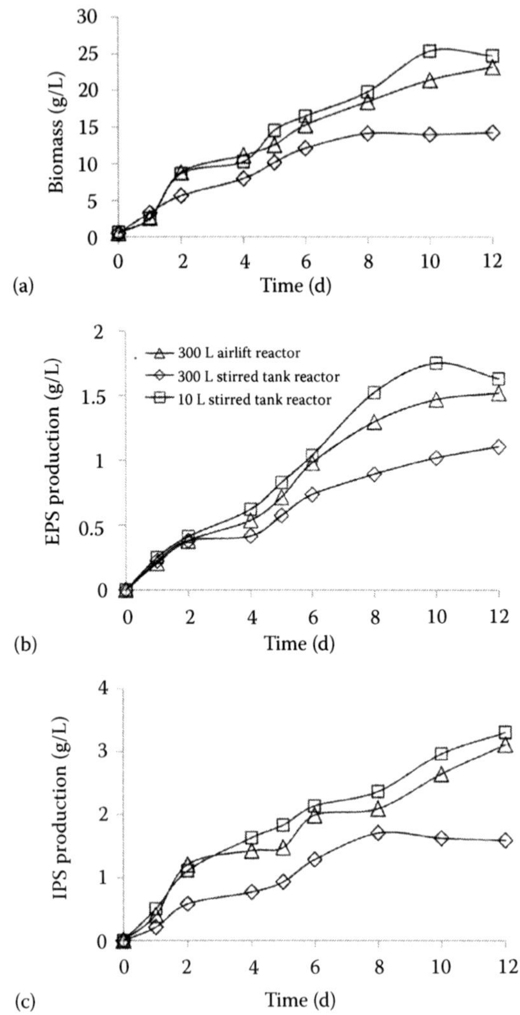

Figure 7.7 Time profiles of dry cell weight (a), EPS production (b), and GA production (c) during fermentation of *G. lucidum* in 10-L stirred tank reactor, 300-L stirred tank reactor and 300-L airlift reactor.

respectively, and the maximum cell density on the surface at As values of 1.53 and 0.90 cm^2/mL was about twice that of As 0.24 cm^2/mL [83]. The maximum GA production by cells in the liquid was 195, 318, and 236 mg/L at As values of 0.24, 0.90, and 1.53 cm/mL, respectively. The highest production and productivity of GAs obtained were 667 mg/L and 28.2 ± 0.3 mg/L/day at an As value of 0.90 cm^2/mL [83]. These results indicated that As significantly affected cell growth and GA accumulation. This information is also very important to the reactor design for the large-scale mushroom fermentation.

7.6 Perspectives and Concluding Remarks

Mushroom polysaccharides represent a structurally diverse class of biological macromolecules with wide-ranging physicochemical properties and bioactivities. *Ganoderma* is a common and important medicinal and nutritional mushroom used for the prevention and/or treatment of various types of diseases in East Asia, and its polysaccharide is found to be one of the most important components responsible for bioactivities of *Ganoderma*. More than 200 polysaccharides have been isolated from the fruiting bodies, spores, mycelia and fermentation broth of *Ganoderma* [102]. The major bioactive polysaccharides are (1-3)-, (1-6)-α/β glucan, glycoproteins and water-soluble heteropolysaccharides with different combinations of glucose, mannose, galactose, xylose, fructose and arabinose. The biosynthesis of *Ganoderma* polysaccharide was studied by using isotope tracer techniques. PGM and UGPP are two important enzymes in the biosynthesis of polysaccharides. *Ganoderma* polysaccharides exhibited diverse and potentially significant bioactivities such as immunomodulatory, antitumor, antioxidant, hypolipidemic activity, antivirus, anti-radiation activities, etc.

Nowadays the submerged cultivation of *G. lucidum* has received much attention as a promising alternative to the field cultivation for efficient production of its valuable metabolites, especially polysaccharide and GAs. Some strategies for enhancing *Ganoderma* polysaccharide by fermentation are summarized above, including strain breeding methods, modification and optimization of medium composition, as well as control and optimization of oxygen supply and shear stress especially for a large scale. Production of polysaccharide and GAs was greatly enhanced by fed-batch and two-stage fermentation processes. Scale-up of submerged cultures of *G. lucidum* was successfully realized by using a low-shear centrifugal impeller bioreactor and an airlift bioreactor. For liquid static cultures, both initial K_La level and liquid surface area per liquid volume were identified as key factors for the process scale-up. The information might be beneficial to other submerged cultivation processes of medicinal and edible mushrooms, which have wide application in the food and health-care sectors.

Metabolic engineering is a powerful biotechnological tool for optimizing production yields and directly manipulating the pathways to generate targeted products, which was successfully used in various strains for production of different bioactive compounds [103–105]. Metabolic engineering of *Ganoderma* species received a primary progress with enhanced ganoderic acid accumulation [31–34,106–110], and there surely remains a very promising area for further improvement in both polysaccharide and GA production. In the postgenomic era, the integration of transcriptomic, proteomic, and metabolomic tools should help elucidate the biochemical synthetic pathway and the regulatory mechanism of polysaccharide and GA biosyntheses. Based on the information from functional genomics studies, the rational design and manipulation of metabolic pathways with aid of computational biology and genome editing tools like CRISPR [106,108], together with the combination of engineering optimization of fermentation processes, could open up perspectives for higher yield of *Ganoderma* polysaccharide and GAs.

Acknowledgments

This work was supported by post-doctoral station of Hangzhou Wahaha Group Co., Ltd. Financial support to WZH from the Post-doc Foundation of Zhejiang Province is appreciated. JJZ thanks the National Natural Science Foundation of China (No. 21176153), National High Technology R&D Program (863 project No. 2012AA021701), National Basic Research Program (973 program No. 2012CB721006), and the Program of Shanghai Subject Chief Scientist from the Science and Technology Commission of Shanghai Municipality (No. 14XD1402600).

References

1. Zhou XW, Su KQ, Zhang YM (2012) Applied modern biotechnology for cultivation of *Ganoderma* and development of their products. *Applied Microbiology and Biotechnology* 93: 941–963.
2. Xu JW, Xu YN, Zhong JJ (2010) Production of individual ganoderic acids and expression of biosynthetic genes in liquid static and shaking cultures of *Ganoderma lucidum*. *Applied Microbiology and Biotechnology* 85: 941–948.

3. Xu P, Ding ZY, Zhu Q, Zhao CX, Zhang KC (2008) Improved production of mycelia biomass and ganoderic acid by submerged culture of *Ganoderma lucidum* SB using complex media. *Enzyme and Microbial Technology* 42: 325–331.
4. Fang QH, Zhong JJ (2002) Two-stage culture process for improved production of ganoderic acid by liquid fermentation of higher fungus *Ganoderma lucidum. Biotechnology Progress* 18: 51–54.
5. Lin KI, Kao YY, Kuo HK, Yang WB, Chou A, Lin HH, Yu AL, Wong CH (2006) Reishi polysaccharides induce immunoglobulin production through the TLR4/TLR2-mediated induction of transcription factor blimp-1. *Journal of Biological Chemistry* 34: 24111–24123.
6. Papinutti L (2010) Effect of nutrients, pH and water potential on exopolysaccharides production by a fungal strain belonging to *Ganoderma lucidum* complex. *Bioresource Technology* 101: 1941–1946.
7. Ruan WM, Popovich DG (2012) *Ganoderma lucidum* triterpenoid extract induces apoptosis in human colon carcinoma cells (Caco-2). *Biomedicine and Preventive Nutrition* 2: 203–209.
8. Chen CR, Yue QY, Wu ZY, Song XY, Tao SJ, Wu XH, Xu PP, Liu X, Guan SH, Guo DA (2010) Cytotoxic triterpenoids from *Ganoderma lucidum. Phytochemistry* 71: 1579–1585.
9. Chen NH, Liu JW, Zhong JJ (2011) Effect of ganoderic acid T on anti-invasion in vitro and anti-metastasis in vivo. *Pharmacological Reports* 62: 150–163.
10. Kim KC, Kim JS, Son JK, Kim IG (2007) Enhanced induction of mitochondrial damage and apoptosis in human leukemia HL-60 cells by the *Ganoderma lucidum* and Duchesnea chrysantha extracts. *Cancer Lett* 246: 210–217.
11. El-Mekkawy S, Meselhy MR, Nakamura N, Tezuka Y, Hattori M, Kakiuchi N (1998) Anti-HIV-1 and anti-HIV-1-protease substances from *Ganoderma lucidum. Phytochemistry* 49: 1651–1657.
12. Min BS, Nakamura N, Miyashiro H, Bae KW, Hattori M (1998) Triterpenes from the spores of *Ganoderma lucidum* and their inhibitory activity against HIV-1 protease. *Chemical and Pharmaceutical Bulletin* 46: 1607–1612.
13. Lin SB, Li CH, Lee SS, Kan LS (2003) Triterpene-enriched extracts from *Ganoderma lucidum* inhibit growth of hepatoma cells via suppressing proteinkinase C, activating mitogen-activated protein kinases and G2-phase cellcycle arrest. *Life Sciences* 72: 2381–2390.
14. Zhong JJ, Tang YJ (2004) Submerged cultivation of medicinal mushrooms for production of valuable bioactive metabolites. *Advances in Biochemical Engineering/Biotechnology* 87: 25–59.
15. Fang QH, Zhong JJ (2002) Submerged fermentation of higher fungus *Ganoderma lucidum* for production of valuable bioactive metabolites-ganoderic acid and polysaccharide. *Biochemical Engineering Journal* 10: 61–65.

16. Tang YJ, Zhang W, Zhong JJ (2009) Performance analyses of a pH-shift and DOT-shift integrated fed-batch fermentation process for the production of ganoderic acid and *Ganoderma* polysaccharides by medicinal mushroom *Ganoderma lucidum*. *Bioresource Technology* 100: 1852–1859.
17. Gong HG, Zhong JJ (2005) Hydrodynamic shear stress affects cell growth and metabolite production by medicinal mushroom *Ganoderma ludicum*. *Chinese Journal of Chemical Engineering* 13: 426–428.
18. Tang YJ, Zhong JJ (2003) Role of oxygen supply in submerged fermentation of *Ganoderma lucidum* for production of *Ganoderma* polysaccharide and ganoderic acid. *Enzyme and Microbial Technology* 32: 478–484.
19. Hsieh C, Tseng MH, Liu CJ (2006) Production of polysaccharides from *Ganoderma lucidum* (CCRC 36041) under limitations of nutrients. *Enzyme and Microbial Technology* 38: 109–117.
20. Zhu LW, Zhong JJ, Tang YJ (2008) Significance of fungal elicitors on the production of ganoderic acid and *Ganoderma* polysaccharides by the submerged culture of medicinal mushroom *Ganoderma lucidum*. *Process Biochemistry* 43: 1359–1370.
21. Zhang YY, Li S, Wang XH, Zhang L, Cheung CK (2011) Advances in lentinan: Isolation, structure, chain conformation and bioactivities. *Food Hydrocolloids* 25: 196–206.
22. Tsukagoshi S, Hashimoto Y, Fujii G, Kobayashi H, Nomoto K, Orita K (1984) Krestin. *Cancer Treatment Reviews* 11: 131–155.
23. Doco T, O'Neill MA, Pellerin P (2001) Determination of the neutral and acidic glycosyl-residue compositions of plant polysaccharides by GC–EI–MS analysis of the trimethylsilyl methyl glycoside derivatives. *Carbohydrate Polymers* 46(3): 249–259.
24. Yang Q, Wang SW, Xie YH, Sun JY, Wang JB (2010) HPLC analysis of *Ganoderma lucidum* polysaccharides and its effect on antioxidant enzymes activity and Bax, Bcl-2 expression. *International Journal of Biological Macromolecules* 46(2): 167–172.
25. Zhang H, Li WJ, Nie SP, Chen Y, Wang YX, Xie MY (2012) Structural characterization of a novel bioactive polysaccharide from *Ganoderma atrum*. *Carbohydrate Polymers* 88(3): 1047–1054.
26. Miyazaki T, Nishijima M (1981) Studies on fungal polysaccharides. XXVII. Structural examination of a water-soluble, anti-tumor polysaccharide of *Ganoderma lucidum*. *Chemical and Pharmaceutical Bulletin* 29(12): 3611–3616.
27. Bao XF, Wang XS, Dong Q, Fang JN, Li XY (2002) Structural features of immunologically active polysaccharides from *Ganoderma lucidum*. *Phytochemistry* 59(2): 175–181.
28. Wang JG, Ma ZC, Zhang LN, Fang YP, Jiang FT, Phillips GO (2011) Structure and chain conformation of water-soluble heteropolysaccharides from *Ganoderma lucidum*. *Carbohydrate Polymers* 86(2): 844–851.

29. Levander F, Radstrom P (2001) Requirement for the phosphoglucomutase in polysaccharide biosynsthesis in glucose- and lactose-utilizing *Streptococcus thermophilus*. *Applied and Environmental Microbiology* 67(6): 2734–2738.
30. Walling E, Dols-Lafargure M, Lonvaud-Funel A (2005) Glucose fermentation kinetics and expolysaccharide production by ropy *Pediococcus damnosus* IOEB8801. *Food Microbiology* 22(1): 71–78.
31. Ukai S, Yokoyama S, Hara C, Kiho T (1982) Structure of an alkali-soluble polysaccharide from the fruit body of *Ganoderma japonicum* Lloyd. *Carbohydrate Research* 105(2): 237–245.
32. Wang G, Zhang J, Mizuno T, Zhuang C, Ito H, Mayuzumi H et al. (1993) Antitumor activity polysaccharides from the Chinese mushroom Songshan lingzhi, the fruiting body of *Ganoderma tsugae*. *Bioscience Biotechnology and Biochemistry* 57(6): 894–900.
33. Sone Y, Okuda R, Wada N, Kishida E, Misaki A (1985) Structures and antitumor activities of the polysaccharides isolated from fruiting body and the growing culture of mycelium of *Ganoderma lucidum*. *Agricultural and Biological Chemistry* 49(9): 2641–2653.
34. Chen JH, Zhou JP, Zhang LN (1998) Chemical structure of the water-insoluble polysaccharide isolated from the fruiting body of *Ganoderma lucidum*. *Polymer Journal* 30(10): 838–842.
35. Bao XF, Liu CP, Fang JN, Li XY (2001) Structural and immunological studies of a major polysaccharide from spores of *Ganoderma lucidum* (Fr.) Karst. *Carbohydrate Research* 332(1): 67–74.
36. Li YQ, Fang L, Zhang KC (2007) Structure and bioactivities of a galactose rich extracellular polysaccharide from submergedly cultured *Ganoderma lucidum*. *Carbohydrate Polymers* 68(2): 323–328.
37. Huang SQ, Li JW, Li YQ, Wang Z (2011) Purification and structural characterization of a new water-soluble neutral polysaccharide GLP-F1-1 from *Ganoderma lucidum*. *International Journal of Biological Macromolecules* 48(1): 165–169.
38. Bao XF, Wang XS, Dong Q, Fang JN, Li XY (2002) Structural features of immunologically active polysaccharides from *Ganoderma lucidum*. *Phytochemistry* 59(2): 175–181.
39. Dong Q, Wang Y, Shi L, Yao J, Li J, Ma FL et al. (2012) A novel water-soluble β-D-glucan isolated from the spores of *Ganoderma lucidum*. *Carbohydrate Research* 353: 100–105.
40. Liu GQ, Zhao Y, Wang XL, Zhu CY (2011) Biosynthesis and fermentation control of polysaccharides from *Ganoderma lucidum*. *Mycosystema* 30(2): 198–205.
41. Tang YJ, Zhong JJ (2002) Exopolysaccharide biosynthesis and related enzyme activities of the medicinal fungus, *Ganoderma lucidum*, grown on lactose in a bioreactor. *Biotechnology Letters* 24(12): 1023–1026.
42. Gastebois A, Clavaud C, Aimanianda V, Latge JP (2009) *Aspergillus fumigatus*: Cell wall polysaccharides, their biosynthesis and organization. *Future Microbiology* 4(5): 583–505.

43. Xu JW, Ji SL, Li HJ, Zhou JS, Duan YQ, Dang LZ, Mo MH (2014) Increased polysaccharide production and biosynthetic gene expressions in a submerged culture of *Ganoderma lucidum* by the overexpression of α-phosphoglucomutase gene. *Bioprocess and Biosystems Engineering* 10.1007/s00449-014-1279-1.
44. Nie SP, Zhang H, Li WJ, Xie MY (2013) Current development of polysaccharides from *Ganoderma*: Isolation, structure and bioactivities. *Bioactive Carbohydrates and Dietary Fibre* 1: 10–20.
45. You YH, Lin ZB (2002) Protective effects of *Ganoderma lucidum* polysaccharides peptide on injury of macrophages induced by reactive oxygen species. *Acta Pharmacologica Sinica* 23(9): 789–791.
46. Zhu XL, Chen AF, Lin ZB (2007) *Ganoderma lucidum* polysaccharides enhance the function of immunological effector cells in immunosuppressed mice. *Journal of Ethnopharmacology* 111(2): 219–226.
47. Shi M, Yang YN, Hu XS, Zhang ZY (2014) Effect of ultrasonic extraction conditions on antioxidative and immunomodulatory activities of a *Ganoderma lucidum* polysaccharide originated from fermented soybean curd residue. *Food Chemistry* 155: 50–56.
48. Zhang SS, Nie SP, Huang DF, Li WJ, Xie MY (2013) Immunomodulatory effect of *Ganoderma atrum* polysaccharide on CT26 tumor-bearing mice. *Food Chemistry* 136: 1213–1219.
49. Soniamol J, Baby S, Varughese G, Kuttikkadan RA, Kainoor KJ (2011) Antitumor and anti-inflammatory fucogalactan from the fruiting bodies of *Ganoderma lucidum*. *Planta Medica* 74(14): 1730–1734.
50. Harris PJ, Henry RJ, Blakeney AB, Stone BA (1984) An improved procedure for the methylation analysis of oligosaccharides and polysaccharides. *Carbohydrate Research* 127(1): 59–73.
51. Gao YH, Tang W B, Dai XH, Gao H, Chen, GL, Ye JX et al. (2005) Effects of water-soluble *Ganoderma lucidum* polysaccharides on the immune functions of patients with advanced lung cancer. *Journal of Medicinal Food* 8(2): 159–168.
52. Cao QZ, Lin ZB (2004) Antitumor and anti-angiogenic activity of *Ganoderma lucidum* polysaccharide speptide. *Acta Pharmacologica Sinica* 25(6): 833–838.
53. Yu Q, Nie SP, Wang JQ, Yin PF, Li WJ, Xie MY (2012) Polysaccharide form *Ganoderma atrum* induces tumor necrosis factor-α secretion via phosphoinositide 3-kinase/Akt, mitogen-activated protein kinase and nuclear factor-Kb signaling pathway in RAW164.7 cells. *International Immunopharmacology* 14(4): 362–368.
54. Peng YF, Zhang LN, Zeng FB, Kennedy JF (2005) Structure and antitumor activities of the water-soluble polysaccharides from *Ganoderma tsugae* mycelium. *Carbohydrate Polymers* 59(3): 385–392.
55. Ames BN (1983) Dietary carcinogenesis and anticarcinogenesis, oxygen radicals and degenerative diseases. *Science* 221: 1256–1264.
56. Mau JL, Tsai SY, Tseng YH, Huang SJ (2005) Antioxidant properties of hot water extracts from *Ganoderma tsugae* Murrill. LWT—*Food Science and Technology* 38(6): 589–597.

57. Mau JL, Tsai SY, Tseng YH, Huang SJ (2005) Antioxidant properties of methanolic extracts from *Ganoderma tsugae*. *Food Chemistry* 93(4): 641–649.
58. Tseng YH, Yang JH, Mau JL (2008) Antioxidant properties of polysaccharides from *Ganoderma tsugae*. *Food Chemistry* 107(2): 732–738.
59. Kan YJ, Chen TQ, Wu YB, Wu JG, Wu JZ (2015) Antixidant activity of polysaccharide extract from *Ganoderma lucidum* using response surface methodology. *International Journal of Biological Macromolecules* 72: 151–157.
60. Zhao HB, Lin SQ, Liu JH, Lin ZB (2004) Polysaccharide extract isolated from *Ganoderma lucidum* protects rat cerebral cortical neurons from hypoxia/reoxygenation injury. *Journal of Pharmacological Sciences* 95(2): 294–298.
61. Chen Y, Xie MY, Nie SP, Li C, Wang YX (2008) Purification, composition analysis and antioxidant activity of a polysaccharide from the fruiting bodies of *Ganoderma atrum*. *Food Chemistry* 107(1): 231–241.
62. Li WJ, Nie SP, Xie MY, Yu Q, Chen Y, He M (2011) *Ganoderma atrum* polysaccharide attenuates oxidative stress induced by D-galactose in mouse brain. *Life Sciences* 88: 713–718.
63. Tomoda M, Gonda R, Kashimasa Y, Hikino H (1986) Glycan structures of ganoderans b and c, hypoglycemic glycans of *Ganoderma lucidum* fruit body. *Phytochemistry* 25(2): 2817–2820.
64. Pan D, Wang LQ, Chen CH, Hu BW, Zhou P (2015) Isolation and characterization of a hyperbranched proteoglycan from *Ganoderma lucidum* for anti-diabetes. *Carbohydrate Polymers* 117: 106–114.
65. Zhao J, Li P, Hu Y (2008) Effect of *Ganoderma lucidum* polysaccharides on reducing blood sugar in mice with diabetes. *Journal of North China Coal Medical College* 8(3): 287–288.
66. Pillai TG, Nair CK, Janardhan KK (2008) Polysaccharides isolated from *Ganoderm lucidum* occurring in southern parts of India, protects radiation induced damages both in vitro and in vivo. *Environmental Toxicology and Pharmacology* 26(1): 80.
67. Wagner R, Mitchell DA, Sassaki GL, Amazonas M (2004) Links between morphology and physiology of *Ganoderma lucidum* in submerged culture for the production of exopolysaccharide. *Journal of Biotechnology* 114(1–2): 153–164.
68. Wang SH, Liang CJ, Weng YW, Chen YH, Hsu HY, Chien HF, Tsai JS, Tseng YC, Li CY, Chen YL (2012) *Ganoderma lucidum* polysaccharides prevent platelet-derived growth factor-stimulated smooth muscle cell proliferation in vitro and neointimal hyperplasia in the endothelial-denuded artery in vivo. *Journal of Cell Physiology* 227(8): 3063–3071.
69. Eo SK, Kim YS, Lee CK, Han SS (2000) Possible mode of antiviral activity of acidic protein bound polysaccharide isolated from *Ganoderma lucidum* on herpes simplex viruses. *Journal of Ethnopharmacology* 72(3): 475–481.

70. Da J, Wu WY, Hou JJ, Long HL, Yao S, Yang Z, Cai LY, Yang M, Jiang BH, Liu X, Cheng CR, Li YF, Guo DA (2011) Comparison of two official Chinese pharmacopoeia species of *Ganoderma* based on chemical research with multiple technologies and chemometrics analysis. *Journal of Chromatography A* 1222: 59–70.
71. Fu LZ, Wu XQ, Li MY, Cheng JW, He L, Wu QQ, Li HB, Wei HL, Cao Q (2009) Analysis and evaluation of polysaccharides and triterpenes contents of fruit body of *Ganoderma lucidum* strains. *Edible Fungi China* 28: 38–40.
72. Wei ZH, Duan YY, Qian YQ, Guo XF, Li YJ, Jin SH, Zhou ZX, Shan SS, Wang CR, Chen XJ, Zheng YG, Zhong JJ (2014) Screening of *Ganoderma* strains with high polysaccharides and ganoderic acids contents and optimization of the fermentation medium by statistical methods. *Bioprocess and Biosystems Engineering* 37: 1789–1797.
73. Chen J, Su KM (2008) Proceeding of heredity, breeding and identification of edible fungi. *Edible Fungi* 27: 3–8.
74. Lin J, Zhou XW (1999) Artificial cultivation of organizational separation of *Ganoderma lucidum*. *Edible Fungi* 2: 10–11.
75. Tang YJ, Zhong JJ (2002) Fed-batch fermentation of *Ganoderma lucidum* for hyperproduction of polysaccharide and ganoderic acid. *Enzyme and Microbial Technology* 31: 20–28.
76. Hsieh C, Tseng MH, Liu CJ (2006) Production of polysaccharides from *Ganoderma lucidum* (CCRC 36041) under limitations of nutrients. *Enzyme and Microbial Technology* 38: 109–117.
77. Zhao W, Xu JW, Zhong JJ (2011) Enhanced production of ganoderic acids in static liquid culture of *Ganoderma lucidum* under nitrogen-limiting conditions. *Bioresource Technology* 102: 8185–8190.
78. Chang MY, Tsai GJ, Houng JY (2006) Optimization of the medium composition for the submerged culture of *Ganoderma lucidum* by Taguchi array design and steepest ascent method. *Enzyme and Microbial Technology* 38: 407–414.
79. Tang YJ, Zhang W, Liu RS, Zhu LW, Zhong JJ (2011) Scale-up study on the fed-batch fermentation of *Ganoderma lucidum* for the hyperproduction of ganoderic acid and *Ganoderma* polysaccharides. *Process Biochemistry* 46: 404–408.
80. Liu QN, Liu RS, Wang YH, Mi ZY, Li DS, Zhong JJ, Tang YJ (2009) Fed-batch fermentation of Tuber melanosporum for the hyperproduction of mycelia and bioactive *Tuber* polysaccharides. *Bioresource Technology* 100: 3644–3649.
81. Zou X (2006) Fed-batch fermentation for hyperproduction of polysaccharide and ergosterol by medicinal mushroom *Agaridus brasiliensis*. *Process Biochemistry* 41(4): 970–974.
82. Tang YJ (2003) Fermentation study on medicinal mushroom *Ganoderma lucidum* for production of ganoderic acid and *Ganoderma* polysaccharide, Ph.D. dissertation, East China University of Science and Technology, Shanghai, (in Chinese).

83. Tang YJ, Zhong JJ (2004) Modeling the kinetics of cell growth and ganoderic acid production in liquid static cultures of the medicinal mushroom *Ganoderma ludicum*. *Biochemical Engineering Journal* 21: 259–264.
84. Xu JW, Xu YN, Zhong JJ (2010) Production of individual ganoderic acids and expression of biosynthetic genes in liquid static and shaking cultures of *Ganoderma lucidum*. *Applied Microbiology and Biotechnology* 85: 941–948.
85. Zhang WX, Zhong JJ (2010) Effect of oxygen concentration in gas phase on sporulation and individual ganoderic acids accumulation in liquid static culture of *Ganoderma lucidum*. *Journal of Bioscience and Bioengineering* 109(1): 37–40.
86. Xu YN, Zhong JJ (2012) Impact of calcium signal transduction on the fermentation production of antitumor ganoderic acids by medicinal mushroom *Ganoderma ludicum*. *Biotechnology Advances* 30(6): 1301–1308.
87. Liu RM, Li YB, Zhong JJ (2012) Cytotoxic and pro-apoptotic effects of ganoderic acid derivatives on human cervical cancer cells in vitro. *European Journal of Pharmacology* 68: 23–33.
88. Liu RM, Zhong JJ (2011) Ganoderic acid Mf and S induced mitochondria mediated apoptosis in human cervical carcinoma HeLa cells. *Phytomedicine* 18: 349–355.
89. Chen NH, Zhong JJ (2011) p53 is important for the anti-invasion of ganoderic acid T in human carcinoma cells. *Phytomedicine* 18: 719–725.
90. Liu RM, Li YB, Zhong JJ (2012) Anti-proliferation and induced mitochondria-mediated apoptosis of ganoderic acid Mk from *Ganoderma lucidum* mycelia in cervical cancer HeLa cells. *Latin American Journal of Pharmacy* 31(1): 43–50.
91. Chen NH, Zhong JJ (2009) Ganoderic acid Me induces G1 arrest in wild-type p53 human tumor cells while G1/S transition arrest in p53-null cells. *Process Biochemistry* 44: 928–933.
92. Fang QH, Zhong JJ (2002) Effect of initial pH on production of ganoderic acid and polysaccharide by submerged fermentation of *Ganoderma lucidum*. *Process Biochemistry* 37: 769–774.
93. Ren A, Qin L, Shi L, Dong X, Mu DS, Li YX, Zhao MW (2010) Methyl jasmonate induces ganoderic acid biosynthesis in the basidiomycetous fungus *Ganoderma lucidum*. *Bioresource Technology* 101: 6785–6790.
94. Zhu LW, Zhong JJ, Tang YJ (2008) Significant of fungal elicitors on the production of ganoderic acid and *Ganoderma* polysaccharides by the submerged culture of medicinal mushroom *Ganoderma ludicum*. *Process Biochemistry* 43: 1359–1370.
95. Li YQ, Zhi ZC (2012) Promotion and inhibition to biomass and polysaccharides production of *Ganoderma lucidum* by *Sophora flavescens* extract. *African Journal of Microbiology* 6(2): 302–307.
96. Fang QH, Tang YJ, Zhong JJ (2002) Significance of inoculation density control in production of polysaccharide and ganoderic acid by submerged culture of *Ganoderma lucidum*. *Process Biochemistry* 37: 1375–1379.

97. Zhu LW, Zhong JJ, Tang YJ (2010) Multi-fed batch culture integrated with three-stage light irradiation and multiple additions of copper ions for the hyperproduction of ganoderic acid *Ganoderma* polysaccharides by the medicinal mushroom *Ganoderma lucidum*. *Process Biochemistry* 45: 1904–1911.
98. Huang HC, Chen CI, Hung CN, Liu YC (2008) Experimental analysis of the oil addition effect on mycelia and polysaccharide productions in *Ganoderma lucidum* submerged culture. *Bioprocess and Biosystems Engineering* 32: 217–224.
99. Pollard DJ, Kirschner TF, Hunt GR, Tong I-T, Stieber R, Salmon PM (2007) Scale up of a viscous fungal fermentation: Application of scale-up criteria with regime analysis and operating boundary conditions. *Biotechnology and Bioengineering* 96(2): 307–317.
100. Wang SJ, Zhong JJ (1996) A novel centrifugal impeller bioreactor. I. Fluid circulation, mixing, and liquid velocity profiles. *Biotechnology and Bioengineering* 51(5): 511–519.
101. Xia JY, Wang SJ, Zhang SL, Zhong JJ (2008) Computational investigation of fluid dynamics in a recently developed centrifugal impeller bioreactor. *Biochemical Engineering Journal* 38: 406–413.
102. Huie CW, Di X (2004) Chromatographic and electrophoretic methods for Lingzhi pharmacologically active components. *Journal of Chromatography B* 812(1–2): 241–257.
103. Upadhyaya BP, DeVeaux LC, Christopher LP (2014) Metabolic engineering as a tool for enhanced lactic acid production. *Trends in Biotechnology* 32(12): 637–644.
104. Weber T, Charusanti P, Musiol-Kroll EM, Jiang XL, Tong YJ, Kim HU, Lee SY (2015) Metabolic engineering of antibiotic factories: New tools for antibiotic production in actinomycetes. *Trends in Biotechnology* 33(1): 15–26.
105. Wang J, Xiao H, Qian ZG, Zhong JJ (2017) Bioproduction of antibody-drug conjugate payload precursors by engineered cell factories. *Trends in Biotechnology* 35(5): 466–478.
106. Xiao H, Zhong JJ (2016) Production of useful terpenoids by higher-fungus cell factory and synthetic biology approaches. *Trends in Biotechnology* 34: 242–255.
107. Xu JW, Xu YN, Zhong JJ (2012) Enhancement of ganoderic acid accumulation by overexpression of an N-terminally truncated 3-hydroxy-3-methylglutaryl coenzyme A reductase gene in the basidiomycete *Ganoderma lucidum*. *Applied and Environmental Microbiology* 78(22): 7968–7976.
108. Qin H, Xiao H, Zou G, Zhou ZH, Zhong JJ (2017) CRISPR-Cas9 assisted gene disruption in the higher fungus *Ganoderma* species. *Process Biochemistry* 56: 57–61.

109. Li HJ, He YL, Zhang DH, Yue TH, Jiang LX, Li N, Xu JW (2016) Enhancement of ganoderic acid production by constitutively expressing *Vitreoscilla* hemoglobin gene in *Ganoderma lucidum*. *Journal of Biotechnology* 227: 35–40.
110. Zhang DH, Li N, Yu X, Zhao P, Li T, Xu JW (2017) Overexpression of the homologous lanosterol synthase gene in ganoderic acid biosynthesis in *Ganoderma lingzhi*. *Phytochemistry* 134: 46–53.

8

Glucosamine and N-acetylglucosamine Production by Microbial Fermentation

Advances and Perspectives

YANFENG LIU AND LONG LIU

Contents

8.1 Introduction

Glucosamine (GlcN) and its derivative N-acetylglucosamine (GlcNAc) are ubiquitous amino sugars in almost all organisms, such as bacteria, fungi, and other higher cells (Hsieh et al. 2007; Sitanggang et al. 2010). In microorganisms, GlcN and GlcNAc are the building blocks of cell walls. In human cells, GlcN and GlcNAc constitute an important part, especially in joint and cartilage tissues. Therefore, providing more GlcN and GlcNAc was developed as a strategy for treatment of osteoarthritis and maintaining joint health (Anderson et al. 2005). Moreover, recent research indicates that GlcN and GlcNAc are effective for cell lifespan, which further expands potential application fields of GlcN and GlcNAc as pharmaceuticals and nutraceuticals (Weimer et al. 2014; Denzel et al. 2014). Therefore, a safe, sustainable, and green production method is needed to meet the increase demand of GlcN and GlcNAc.

Microbial fermentation is a key approach for many biofuels, industrially important chemical production, and pharmaceutically useful compound production (Peralta-Yahya et al. 2012; Stephanopoulos 2012; Chen and Nielsen 2013; Cho et al. 2015). It is a promising approach to substitute the unsustainable petroleum resources-based chemical production by using sustainable raw material. Development in metabolic engineering and synthetic biology expanded strategies and toolbox for cellular trait optimization. Increasing cell factories and fermentation processes were established for producing biochemical in green and sustainable approaches (Peralta-Yahya et al. 2012; Chen and Nielsen 2013; Cho et al. 2015). Systems biology facilitates systematical analysis of various networks for phenotype property regulation via genomics, transcriptomics, proteomics, and metabolomics (Otero and Nielsen 2010; Cho et al. 2015). Systematical characterization of cell factories via systems biology can identify engineering targets for further pathway engineering (Lee et al. 2007).

In this review, we will first summarize the applications and current production methods of GlcN and GlcNAc and their limitations. Next, the microbial fermentation approaches for GlcN and GlcNAc production are described and compared. Finally, prospective for further improving GlcN and GlcNAc production is discussed. The strategies used for optimizing GlcN and GlcNAc production may be useful for other biochemical production.

8.2 Applications and Current Production Methods of GlcN and GlcNAc

GlcN, 2-amino-2-deoxy-glucose, is a widely used nutraceutically and pharmaceutically important biochemical. As a dietary supplement, GlcN has been widely used to facilitate and maintain the health of the joint and cartilage for decades. In addition, GlcN is also used as a prescription medicine or over-the-counter preparation for the treatment of osteoarthritis (Subramanyam and Rao 1987).

Besides the known functions for maintaining joint and cartilage health, recent research demonstrated that GlcN can activate AMP-activated protein kinase for inhibiting glucose metabolism, which further facilitates mitochondria biogenesis for cell lifespan (Weimer et al. 2014). The GlcN derivative GlcNAc, which can be conveniently converted into GlcN by hydrolysis in mild conditions, also has positive effects on cell lifespan (Denzel et al. 2014). The mechanism is that GlcNAc can enhance the homeostasis of endoplasmic reticulum protein as the precursor of *N*-polysaccharide.

The commercial market for GlcN and GlcNAc has a broad prospect because of the wide ranging applications of GlcN and GlcNAc (Dostrovsky et al. 2011; Nakamura 2011). It is estimated that the current global sales of GlcN products have reached $2 billion. With an increasingly aging population and newly verified functions of GlcN, global demand for GlcN will further increase. It is estimated that the global production of GlcN will reach 46,600 t by 2017 (Muniyappa 2011). The main consumer market is the Asian-Pacific region, North America, and Europe, and China is the largest GlcN production and supply market (Byme 2010; Global Industry Analysts 2011).

The current GlcN production method is hydrolysis of chitin, which causes potential problems for various aspects of GlcN production, such as insufficient raw material supply, pollution of the environment, and potential risk of toxicity. First, the variability of quality and quantity of shellfish-derived chitin cannot meet the requirement of rapid market increase (Hsieh et al. 2007). Second, the process of acidic chitin hydrolysis with concentrated acids causes severe environmental pollution. Third, shellfish-derived GlcN may cause an allergic reaction in some consumers (Deng et al. 2005). Therefore, non-shellfish-derived and food-grade GlcN is the ideal product in the GlcN commercial market. Microbial fermentation has emerged as a promising approach for producing GlcN in a sustainable and environmentally friendly manner. Various production hosts were employed for GlcN production via process optimization and metabolic engineering, including filamentous fungi, *Escherichia coli*, and *Bacillus subtilis*.

8.3 Microbial Production of GlcN by Filamentous Fungi

GlcN and GlcNAc are building blocks of the cell wall of filamentous fungi, such as chitin and chitosan. Therefore, GlcN and GlcNAc can be attained by hydrolysis of the cell wall of filamentous fungi, which is a sustainable production approach and avoids the limitation of shellfish-based raw material supply. The following three wild-type filamentous fungi were used in the production of GlcN: *Rhizopus oligosorus*, *Monascus pilosus*, and *Aspergillus* sp. (Chen and Chiou 1999; Hsieh et al. 2007; Kim et al. 2001; Nwe and Stevens 2002, 2004; Sumbali 2005). Among the different fungi, *Aspergillus* sp. BCRC31742 achieved the highest GlcN production, which reached 5.48 g/L in optimized medium (Chang 2008). To further enhance GlcN production, methanol was identified as a stimulus to improve GlcN production, which further enhances GlcN production to 7.48 g/L (Sitanggang et al. 2010). Dissolved oxygen level was identified as a key factor for GlcN production via systematic investigation of effects of dissolved oxygen on GlcN production (Lin and Wu 2010). A two-stage dissolved oxygen control strategy was proposed, which led to the highest GlcN titer in *Aspergillus* sp. BCRC31742, reaching 14.37 g/L (Zhang et al. 2012). Despite GlcN production being significantly enhanced, the long fermentation period (5 to 7 days) of fungi reduced the economic competitiveness of this method. Furthermore, a hydrolysis step was also needed for GlcN production using fungi cell wall as raw material, which also caused environmental pollution. Ideally, GlcN should be secreted out of the cell as a monomer with high yield, which can significantly enhance the economic competitiveness and simplify the downstream isolation and purification process. In order to improve the cellular property of GlcN production host, a metabolic engineering approach was implemented to achieve extracellular GlcN production in *E. coli* and *B. subtilis*, respectively.

8.4 Microbial Production of GlcN and GlcNAc by *E. coli*

Because *E. coli* is the most characterized microorganism, it was chosen as the initial host for GlcN overproduction. As shown in Figure 8.1, the first step for amino sugar synthesis is that GlcN-6-P is synthesized from fructose-6-P catalyzed by glucosamine synthase (GlmS) using glutamine as amino donor. Feedback inhibition of GlcN-6-P for GlmS was the key mechanism for regulation of the balance of synthesis pathway. In addition, GlcN was a preferred carbon and nitrogen source for *E. coli*; extracellular GlcN can be taken up and further degraded via the amino sugar catabolic pathway. Therefore, the anabolism and catabolism of amino sugar are tightly

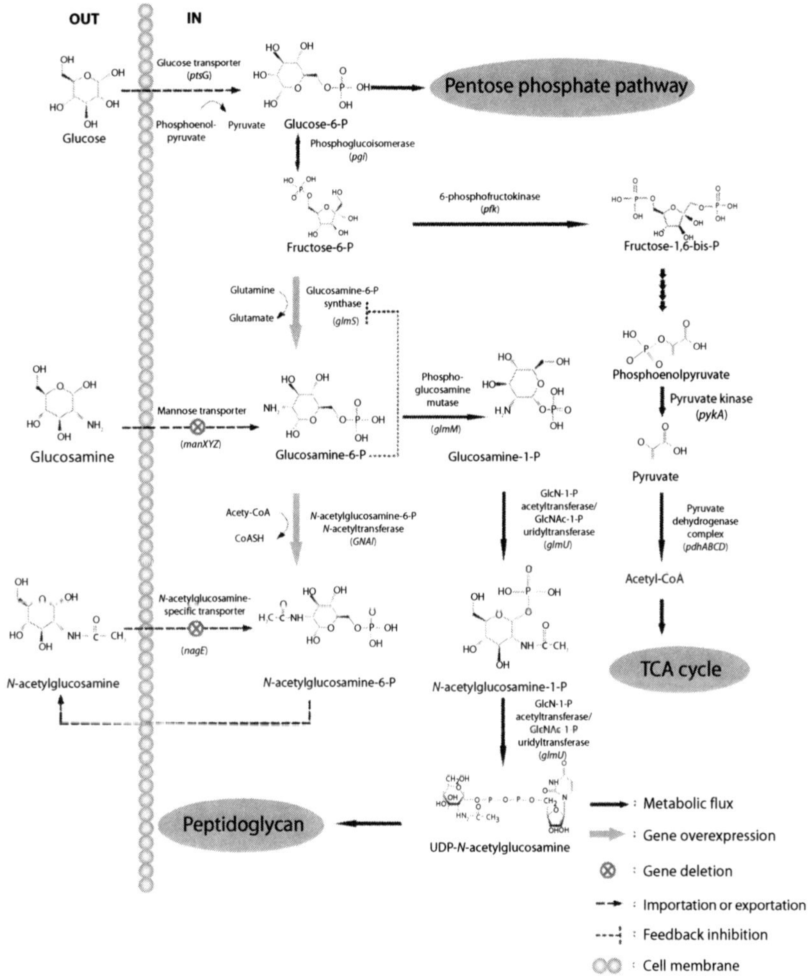

Figure 8.1 Biosynthetic pathways of GlcN and GlcNAc in *E. coli*. OUT, outside the cell; IN, inside the cell.

regulated. To overproduce GlcN, both anabolism and catabolism of amino sugar should be rationally engineered by metabolic engineering. First, GlcN accumulation was achieved by overexpression of feedback-resistant GlmS, obtained via directed evolution (Deng et al. 2006). Second, blockage of GlcN uptake pathway further improve GlcN titer, reaching 17 g/L. Finally, the GlcN pathway was further extended to GlcNAc by overexpression of GlcN-6-P acetyltransferase (GNA1), which led to improved GlcN titer, reaching 110 g/L (Deng et al. 2005). However, there are some drawbacks to using *E. coli* as a production host. First, the product of *E. coli* is not generally regarded as safe (GRAS), which limits the application fields of GlcN produced by *E. coli*. Second, *E. coli* is susceptible to phage contamination, which is an unsuitable characteristic for industrial production. Therefore, an ideal production host that can produce food-grade GlcN is needed for industrialization of microbial GlcN production.

8.5 Pathway Engineering of *B. subtilis* for GlcN and GlcNAc Production

B. subtilis is the model organism of Gram-positive bacterium, which has been intensively investigated for unraveling metabolism regulation mechanism and is widely engineered for biochemical production. Compared with *E. coli*, the products of *B. subtilis* are free of exotoxins and endotoxins, considered as GRAS products (Schallmey et al. 2004). Moreover, the fermentation process of *B. subtilis* is more favorable for industrial production, because *B. subtilis* can grow to high cell density with minimal medium and is not susceptible to the phage-infection problem. Therefore, *B. subtilis* is a more favorable production host for pharmaceutical and nutraceutical grade product fermentation in industrial conditions compared to *E. coli*. Finally, increasingly available genome engineering tools and gene expression systems enable various targets to be engineered in *B. subtilis*. Thus, *B. subtilis* holds several favorable advantages as a production host for GlcN and GlcNAc production in the pharmaceutical and nutraceutical fields.

The anabolic and catabolic pathways of the amino sugar in *B. subtilis* have been intensively studied (Mobley, Doyle et al. 1982; Vincent et al. 2004; Bertram et al. 2011). The structure of the amino sugar pathway in *B. subtilis* is the same as that in *E. coli*. The difference is that the GlcN-6-P triggered feedback regulation of GlmS is regulated via *glmS* ribozyme, which controls GlmS expression level based on GlcN-6-P binding (Mobley, Doyle et al. 1982).

To achieve the overproduction of GlcN and GlcNAc using *B. subtilis*, GlcNAc synthesis pathway and catabolic pathway were firstly engineered via rational pathway engineering (Liu YF et al. 2013). The encoding sequence of *B. subtilis* native GlmS was overexpressed via plasmid expression system, which eliminates feedback inhibition of GlmS expression by removing *glmS* ribozyme sequence from the upstream *glmS* encoding sequence. Next, the GlcN synthesis pathway can be extended to GlcNAc by overexpressing GlcN-6-P *N*-acetyltransferase (GNA1) to alleviate the inhibitory effects of GlcN on the growth of *B. subtilis.* Therefore, the GlcNAc synthesis pathway was strengthened by co-overexpression of GlmS and GNA1, resulting in 240 mg/L GlcNAc accumulation. To further facilitate GlcNAc accumulation, the entire GlcNAc catabolic pathway was blocked by knocking out all five genes involved in GlcNAc catabolism, including *nagP*, *gamP*, *gamA*, *nagA*, and *nagB*. GlcNAc titer reached 1.85 g/L, which was 7.7-fold of that without blocking GlcNAc catabolism (Liu YF et al. 2013).

In addition to the engineering strategies mentioned above, synthetic biology tools can be used to further improve GlcN and GlcNAc production by *B. subtilis.* Synthetic biology tools, like synthetic scaffolds and synthetic small regulatory RNAs (sRNA), have expanded the strategies for controlling various metabolic modules in different expression levels (Dueber et al. 2009; Na et al. 2013). Varying ratios of upstream and downstream pathway enzymes on scaffold result in various strengths of upstream and downstream reactions. Then, strain library can be generated by assembly of upstream and downstream reactions with different strength on scaffold. Finally, strains with balanced upstream and downstream modules, avoiding intermediate accumulation, can be attained by screening strain library. To this end, DNA-guided scaffolds were employed in engineered GlcNAc-producing *B. subtilis* for strengthening and balancing GlcNAc synthesis pathway (Liu et al. 2014b). Specifically, the key pathway enzymes GlmS and GNA1 were colocalized in DNA-guided scaffolds for strengthening the GlcNAc synthesis pathway. The relative activities of GlmS and GNA1 were balanced by controlling the ratio of GlmS and GNA1 at 1:2, which avoided intermediate accumulation and facilitated the final product GlcNAc synthesis. Finally, 2.5-fold increase in GlcNAc titer was attained by colocalizing pathway enzyme GlmS and GNA1 and controlling the ratio of GlmS and GNA1 at 1:2.

Modular pathway engineering has emerged as a promising strategy to optimize relative activities of product synthesis pathway and competition pathways, which artificially divides the metabolic pathway into various modules, and constructs artificially controlled modules with various expression levels and assemblies of multiple modules simultaneously for generating strain library (Ajikumar et al. 2010;

Juminaga et al. 2012). One round strain library generating and screening via modular pathway engineering can identify engineered strains with balanced metabolic fluxes. To optimize relative activities of GlcNAc synthesis pathway and other competition pathways, GlcNAc synthesis network was divided into three modules, including glycolysis module, GlcNAc synthesis module, and peptidoglycan synthesis module. Synthetic sRNAs were expressed to simultaneously inhibit expression of targeted genes, *pfk* and *glmM*, that were in the cross points of the above three modules, which controlled the flux distribution. Therefore, a strain library can be obtained by changing the inhibition efficiency of sRNAs for key genes of various branch modules. By controlling GlcNAc synthesis activity at a high level and glycolysis and peptidoglycan synthesis activity at low levels, a synthetic sRNA-based modular engineering strategy further improved GlcNAc production after rational metabolic pathway construction, and production reached 6.1-fold of that without modular pathway engineering (Liu et al. 2014a).

8.6 Conclusions and Perspectives

Microbial fermentation for GlcN and GlcNAc production has been developed which is a promising approach for sustainable production of safe GlcN and GlcNAc for use in food and medicinal fields. To further enhance the economic competitiveness and industrialize microbial GlcN and GlcNAc production, improved production and reduced medium cost are needed to be achieved. The following strategies can be implemented to further optimize GlcN and GlcNAc production.

First, dephosphorylation and exportation pathway of GlcN and GlcNAc should be identified and optimized for GlcN and GlcNAc production. Intracellular GlcN6P and GlcNAc6P should be first dephosphorylated and then exported for extracellular production. However, the enzymes that catalyze dephosphorylation and exportation are unknown. Identification of the enzymes that catalyze dephosphorylation and exportation steps can tell us what needs to be changed and how to implement it in dephosphorylation and exportation pathways.

Second, kinetics of synthetic pathway during GlcNAc synthesis should be systematically investigated. Dynamic metabolite change reflects the kinetics of product synthesis, which is the key feature of the synthetic pathway (Bujara et al. 2011; Link et al. 2014, 2015). Based on the dynamics of synthetic pathway, whether dephosphorylation and exportation are rate-limiting steps in GlcN and GlcNAc production or not, there are other limitations in GlcN and GlcNAc pathways that can be identified and further engineered. To this end, metabolomics can be used for monitoring metabolite dynamic changes during GlcNAc6P synthesis, dephosphorylation, and exportation. Systematically investigating the metabolite dynamic patterns can provide an informative description of the synthetic process and identify potential unknown regulation in the synthetic network.

Finally, multidimensional engineering strategies should be used for metabolic flux control. Metabolic fluxes were controlled by multilevel and multilayer strategies, such as gene expression and posttranslational modification, which ultimately affects pathway enzyme abundance and activity (Kochanowski et al. 2015). Metabolic fluxes are largely regulated by varying enzyme activities (Chubukov et al. 2013). However, previously implemented strategies for GlcN and GlcNAc production mainly focus on gene expression control for enzyme abundance regulation. Therefore, combining enzyme engineering with pathway engineering for posttranslational control can be employed for improving cellular properties of engineered strains for GlcN and GlcNAc production.

References

Ajikumar, P. K., Xiao, W. H., Tyo, K. E. J., Wang, Y., Simeon, F., Leonard, E., Mucha, O., Phon, T. H., Pfeifer, B. and Stephanopoulos, G. (2010). Isoprenoid pathway optimization for taxol precursor overproduction in *Escherichia coli*. *Science*. **330**(6000), 70–74.

Anderson, J. W., Nicolosi, R. J. and Borzelleca, J. F. (2005). Glucosamine effects in humans: A review of effects on glucose metabolism, side effects, safety considerations and efficacy. *Food and Chemical Toxicology*. **43**(2), 187–201.

Bertram, R., Rigali, S., Wood, N., Lulko, A. T., Kuipers, O. P. and Titgemeyer, F. (2011). Regulon of the N-acetylglucosamine utilization regulator NagR in *Bacillus subtilis*. *Journal of Bacteriology*. **193**(14), 3525–3536.

Bujara, M., Schumperli, M., Pellaux, R., Heinemann, M. and Panke, S. (2011). Optimization of a blueprint for in vitro glycolysis by metabolic real-time analysis. *Nature Chemical Biology*. **7**(5), 271–277.

Byme, J. (2010). Glucosamine market reaching maturity. Available at http://www.nutraingredients.com/Industry/Glucosamine-market-reaching-maturity. Last accessed August 5, 2017.

Chang, Y. (2008). Optimum production of GlcN using *Aspergillus* sp. Thesis, Department of Chemical Engineering and Materials Science, Yuan Ze University, Taiwan.

Chen, W. and Chiou, R. Y. (1999). A modified chemical procedure for rapid determination of glucosamine and its application for estimation of mold growth in peanut kernels and koji. *Journal of Agricultural and Food Chemistry*. **47**(5), 1999–2004.

Chen, Y. and Nielsen, J. (2013). Advances in metabolic pathway and strain engineering paving the way for sustainable production of chemical building blocks. *Current Opinion in Biotechnology*. **24**(6), 965–972.

Cho, C., Choi, S. Y., Luo, Z. W. and Lee, S. Y. (2015). Recent advances in microbial production of fuels and chemicals using tools and strategies of systems metabolic engineering. *Biotechnology Advances*. **33**(7), 1455–1466.

Chubukov, V., Uhr, M., Le Chat, L., Kleijn, R. J., Jules, M., Link, H., Aymerich, S., Stelling, J. and Sauer, U. (2013). Transcriptional regulation is insufficient to explain substrate-induced flux changes in *Bacillus subtilis*. *Molecular Systems Biology*. **9**, 709.

Deng, M. D., Grund, A. D., Wassink, S. L., Peng, S. S., Nielsen, K. L., Huckins, B. D., Walsh, B. L. and Burlingame, R. P. (2006). Directed evolution and characterization of Escherichia coli glucosamine synthase. Biochimie. **88**(5), 419–429.

Deng, M. D., Severson, D. K., Grund, A. D., Wassink, S. L., Burlingame, R. P., Berry, A., Running, J. A., Kunesh, C. A., Song, L. and Jerrell, T. A. (2005). Metabolic engineering of *Escherichia coli* for industrial production of glucosamine and N-acetylglucosamine. *Metabolic Engineering*. **7**(3), 201–214.

Denzel, M. S., Storm, N. J., Gutschmidt, A., Baddi, R., Hinze, Y., Jarosch, E., Sommer, T., Hoppe, T. and Antebi, A. (2014). Hexosamine pathway metabolites enhance protein quality control and prolong life. *Cell*. **156**(6), 1167–1178.

Dostrovsky, N. R., Towheed, T. E., Hudson, R. W. and Anastassiades, T. P. (2011). The effect of glucosamine on glucose metabolism in humans: A systematic review of the literature. *Osteoarthritis and Cartilage*. **19**(4), 375–380.

Dueber, J. E., Wu, G. C., Malmirchegini, G. R., Moon, T. S., Petzold, C. J., Ullal, A. V., Prather, K. L. and Keasling, J. D. (2009). Synthetic protein scaffolds provide modular control over metabolic flux. *Nature Biotechnology* **27**(8), 753–759.

Global Industry Analysts (2011). Global glucosamine market to reach 46.6 thousand metric tons by 2017, according to a new report by Global Industry Analysts, Inc. Available at http://www.prweb.com/releases/glucosamine/glucosamine_supplements/prweb8561248.htm. Last accessed August 5, 2017.

Hsieh, J. W., Wu, H. S., Wei, Y. H. and Wang, S. S. (2007). Determination and kinetics of producing glucosamine using fungi. *Biotechnology Progress*. **23**(5), 1009–1016.

Juminaga, D., Baidoo, E. E. K., Redding-Johanson, A. M., Batth, T. S., Burd, H., Mukhopadhyay, A., Petzold, C. J. and Keasling, J. D. (2012). Modular engineering of L-tyrosine production in *Escherichia coli*. *Applied and Environmental Microbiology*. **78**(1), 89–98.

Kim, W., Lee, W., Theodore, K. and Chang, H. (2001). Optimization of culture conditions and continuous production of chitosan by the fungi, *Absidia coerulea*. *Biotechnology and Bioprocess Engineering*. **6**(1), 6–10.

Kochanowski, K., Sauer, U. and Noor, E. (2015). Posttranslational regulation of microbial metabolism. *Current Opinion in Microbiology*. **27**, 10–17.

Lee, K. H., Park, J. H., Kim, T. Y., Kim, H. U. and Lee, S. Y. (2007). Systems metabolic engineering of *Escherichia coli* for L-threonine production. *Molecular Systems Biology*. **3**, 149.

Lin, B. C. and Wu, H. S. (2010). Production and separation of glucosamine using *Aspergillus* sp. BCRC 31742 cultivated in a fermentor. *Journal of Biotechnology*. **150**, S350.

Link, H., Christodoulou, D. and Sauer, U. (2014). Advancing metabolic models with kinetic information. *Current Opinion in Biotechnology*. **29**, 8–14.

Link, H., Fuhrer, T., Gerosa, L., Zamboni, N. and Sauer, U. (2015). Real-time metabolome profiling of the metabolic switch between starvation and growth. *Nature Methods*. **12**(11), 1091–1097.

Liu, L., Liu, Y., Shin, H.-d., Chen, R. R., Wang, N. S., Li, J., Du, G. and Chen, J. (2013). Developing Bacillus spp. as a cell factory for production of microbial enzymes and industrially important biochemicals in the context of systems and synthetic biology. *Applied Microbiology and Biotechnology*. **97**, 6113–6127.

Liu, Y. F., Liu, L., Shin, H. D., Chen, R. R., Li, J. H., Du, G. C. and Chen, J. (2013). Pathway engineering of *Bacillus subtilis* for microbial production of N-acetylglucosamine. *Metabolic Engineering*. **19**, 107–115.

Liu, Y. F., Zhu, Y. Q., Li, J. H., Shin, H. D., Chen, R. R., Du, G. C., Liu, L. and Chen, J. (2014a). Modular pathway engineering of *Bacillus subtilis* for improved N-acetylglucosamine production. *Metabolic Engineering*. **23**, 42–52.

Liu, Y. F., Zhu, Y. Q., Ma, W. L., Shin, H. D., Li, J. H., Liu, L., Du, G. C. and Chen, J. (2014b). Spatial modulation of key pathway enzymes by DNA-guided scaffold system and respiration chain engineering for improved N-acetylglucosamine production by *Bacillus subtilis*. *Metabolic Engineering*. **24**, 61–69.

Mobley, H. L., Doyle, R. J., Streips, U. N. and Langemeier, S. O. (1982). Transport and incorporation of N-acetyl-D-glucosamine in *Bacillus subtilis*. *Journal of Bacteriology*. **150**(1), 8–15.

Muniyappa, R. (2011). Glucosamine and osteoarthritis: Time to quit? *Diabetes/Metabolism Research and Reviews*. **27**(3), 233–234.

Na, D., Yoo, S. M., Chung, H., Park, H., Park, J. H. and Lee, S. Y. (2013). Metabolic engineering of *Escherichia coli* using synthetic small regulatory RNAs. *Nature Biotechnology*. **31**(2), 170–174.

Nakamura, H. (2011). Application of glucosamine on human disease—Osteoarthritis. *Carbohydrate Polymers*. **84**(2), 835–839.

Nwe, N. and Stevens, W. F. (2002). Production of fungal chitosan by solid substrate fermentation followed by enzymatic extraction. *Biotechnology Letters*. **24**(2), 131–134.

Nwe, N. and Stevens, W. F. (2004). Effect of urea on fungal chitosan production in solid substrate fermentation. *Process Biochemistry*. **39**(11), 1639–1642.

Otero, J. M. and Nielsen, J. (2010). Industrial systems biology. *Biotechnology and Bioengineering*. **105**(3), 439–460.

Peralta-Yahya, P. P., Zhang, F., del Cardayre, S. B. and Keasling, J. D. (2012). Microbial engineering for the production of advanced biofuels. *Nature*. **488**(7411), 320–328.

Schallmey, M., Singh, A. and Ward, O. P. (2004). Developments in the use of *Bacillus* species for industrial production. *Canadian Journal of Microbiology*. **50**(1), 1–17.

Sitanggang, A. B., Wu, H. S., Wang, S. S. and Ho, Y. C. (2010). Effect of pellet size and stimulating factor on the glucosamine production using *Aspergillus* sp BCRC 31742. *Bioresource Technology*. **101**(10), 3595–3601.

Stephanopoulos, G. (2012). Synthetic biology and metabolic engineering. *ACS Synthetic Biology*. **1**(11), 514–525.

Subramanyam, C. and Rao, S. (1987). An enzymatic method for the determination of chitin and chitosan in fungal cell walls. *Journal of Bioscience and Bioengineering*. **2**, 125–129.

Sumbali, G. (2005). *The Fungi*. Harrow: Alpha Science International.

Vincent, F., Yates, D., Garman, E., Davies, G. J. and Brannigan, J. A. (2004). The three-dimensional structure of the N-acetylglucosamine-6-phosphate deacetylase, NagA, from *Bacillus subtilis*. *Journal of Biological Chemistry*. **279**(4), 2809–2816.

Weimer, S., Priebs, J., Kuhlow, D., Groth, M., Priebe, S., Mansfeld, J., Merry, T. L., Dubuis, S., Laube, B., Pfeiffer, A. F., Schulz, T. J., Guthke, R., Platzer, M., Zamboni, N., Zarse, K. and Ristow, M. (2014). D-Glucosamine supplementation extends life span of nematodes and of ageing mice. *Nature Communications*. **5**, 3563.

Westers, L., Westers, H. and Quax, W. J. (2004). *Bacillus subtilis* as a cell factory for pharmaceutical proteins: A biotechnological approach to optimize the host organism. *Biochimica et Biophysica Acta*. **1694**, 299–310.

Zhang, J. X., Liu, L., Li, J. H., Du, G. C. and Chen, J. (2012). Enhanced glucosamine production by *Aspergillus* sp BCRC 31742 based on the time-variant kinetics analysis of dissolved oxygen level. *Bioresource Technology*. **111**, 507–511.

9

Functional Carbohydrates

Development, Characterization, and Biomanufacturing of Sugar Alcohols

GUOQIANG ZHANG
AND QINHONG WANG

Contents

9.1 Introduction

Sugar alcohols, known as polyols or polyhydric alcohols, are chemically defined as saccharide derivatives in which a ketone or aldehyde group is replaced by a hydroxyl group [1]. They are naturally present in small amounts in fruits and vegetables and are produced by microorganisms as carbohydrate reserves, reducing power storage, and osmoprotectants. Sugar alcohols are also recognized as a member of FODMAP (fermentable, oligo-, di-, monosaccharides, and polyols) [2].

Sugar alcohols have similar physicochemical properties and function due to their general chemical structure as formula of $HOCH_2(CHOH)_n CH_2OH$ (Figure 9.1) [3]. First, they have high thermal and chemical stability. For example, sorbitol, maltitol, and xylitol are stable up to 180°C without Maillard reaction. Second, sugar alcohols are hygroscopic. Sorbitol and maltitol solutions have excellent moisture-stabilizing effects. Additionally, they have high microbial stability. Xylitol, for example, is not normally recognized by transport mechanisms of cariogenic bacteria.

As a group, sugar alcohols are similar to sucrose to some extent, but they also have health-promoting properties such as lower caloric content, noncariogenicity, and low glycemic index and insulin response (Table 9.1) [4–6]. Sugar alcohols are relatively poorly absorbed and metabolized by the human body resulting in a low caloric content value. The slow ingestion of sugar alcohols results in the glycemic index and serum insulin responses being significantly lower than those following glucose or sucrose ingestion. These properties have clearly indicated that sugar alcohols can be regarded as suitable sweeteners in carbohydrate-controlled diets for obesity and diabetes [7–9]. Sugar alcohols can prevent the formation of dental caries. The bacteria in

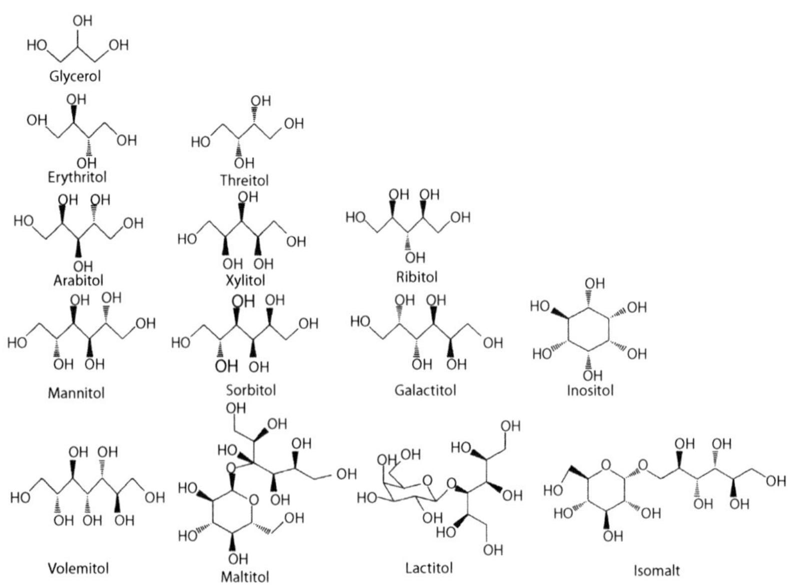

Figure 9.1 Chemical structures of primary sugar alcohols.

Table 9.1 Summary of Primary Sugar Alcohols and Sucrose

	SWEETNESS INDEX	CALORIES (g)	GLYCEMIC INDEX	CARIOGENICITY
Arabitol	0.7	0.2	–	Noncariogenic
Erythritol	0.8	0.2	0	Noncariogenic
Sorbitol	0.6	2.6	9	Noncariogenic
Mannitol	0.5	1.6	0	Noncariogenic
Xylitol	1.0	2.4	13	Anticariogenic
Maltitol	0.9	2.1	36	Noncariogenic
Lactitol	0.4	2.0	6	Noncariogenic
Isomalt	0.5	2.1	2	Noncariogenic
Sucrose	1.0	4.0	60	Cariogenic

Sources: Lee CK, *Food Chemistry* 1977, **2**:95–105; Mitchell H (ed.), *Sweeteners and Sugar Alternatives in Food Technology.* Oxford, UK: Blackwell Publishing, 2006; Salminen S, Halikainen A (eds.): *Sweeteners, Food Additives.* New York: Marcel Dekker, 2001.

dental plaque produce amounts of decay-promoting acid from sugars and starches. However, some sugar alcohols can avoid these chemical reactions because they are not normally recognized by transport mechanisms of cariogenic bacteria. In addition, some sugar alcohols have the potential to be outstanding building blocks for commodity chemicals and pharmaceuticals. For example, sorbitol can be converted into isosorbide and glycols as high value monomer intermediates for commercial applications [10].

The demand for sugar alcohols is increasing rapidly in the food, beverage, confectionery, and pharmaceutical industries because of their beneficial properties. It is well known that sugar alcohols are more widespread in plants and fungi and can be extracted by a series of operations as separation and purification [11]. However, the content of sugar alcohols is usually low, so it is not economically feasible to use the natural extraction method. At present, sugar alcohols are primarily produced by chemical hydrogenation from their corresponding sugars under high temperature and pressure. However, the chemical processes require highly purified substrates, extreme conditions, expensive chemical catalysts, and costly chromatographic purification steps, which lead to low final product yields and increasing the cost.

Along with the development of biotechnology, the biotechnological strategy is emerging as a promising alternative to offer a safer and more environmentally friendly process for sugar alcohols production (Figure 9.2). It has been reported that sugar alcohols are important

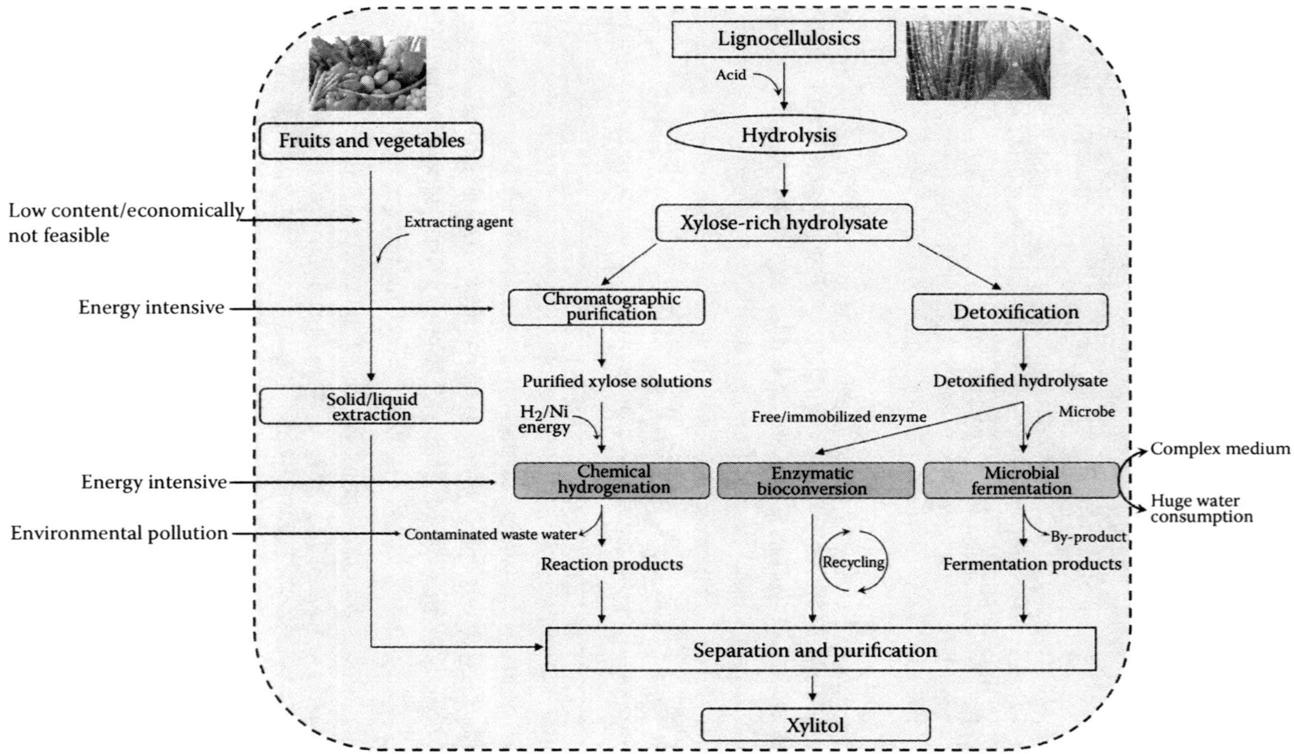

Figure 9.2 Sugar alcohol production via different methods (e.g., xylitol).

metabolites as osmoprotectants, carbohydrate reserves, translocatory compounds, and storages of reducing power in some microorganisms [12]. Several microbes with efficient capacities for sugar alcohol accumulation have been obtained through high-throughput screening and put into industrial use. For example, erythritol is mainly produced from glucose and sucrose by microbial fermentation in industry. Besides traditional strategies to improve sugar alcohol production through isolating strains and optimizing culture conditions, strain engineering has also been extensively developed to establish economic microbial production platforms [13].

Because of the aforementioned properties and developments, functional sugar alcohols are emerging as a sugar substitute and are widely used to provide functional food, nutraceuticals and pharmaceuticals. This chapter focuses on the role and biomanufacturing of primary sugar alcohols, such as erythritol, sorbitol, xylitol, arabitol, and mannitol.

9.2 Primary Sugar Alcohols

Currently, primary sugar alcohols such as erythritol, sorbitol, mannitol, arabitol, and xylitol are widely applied in food, medical, and chemical fields, and there is a rising demand for sugar alcohol products [14]. Though most sugar alcohol production still mainly depends on chemical synthesis in industry, sugar alcohol biomanufacturing is more and more competitive for its significant advantages.

9.2.1 Erythritol

Erythritol is a primary four-carbon sugar alcohol that has 60–80% of the sweetness of sucrose, and is intended for use as a low-calorie sweetener in food. It is widespread in nature as a metabolite or storage compound in seaweeds and fungi and is also a component of a number of fruits, such as grapes, melons, and pears [15]. It has been part of the human diet for thousands of years as it is present in fruits such as pears, melons, and grapes, as well as foods such as mushrooms and fermentation-derived foods such as wine, soy sauce, and cheese.

Erythritol occurs in the form of anhydrous crystals with a moderately sweet taste. It has the common chemical properties of sugar alcohols and thus has excellent heat and acid stability. Erythritol also has some different properties due to having a low molecular weight, such as higher osmotic pressure and lower water activity in solution. Erythritol is preferred over other sugar alcohols because it triggers a low or no glycemic response. It yields 0.2 cal g^{-1}, while the other sugar alcohols yield approximately 2 cal g^{-1}. Additionally, it is tolerated well as a natural food additive and results in minimal or no gastric discomfort or flatulence. The animal toxicological and clinical studies have consistently demonstrated the safety of erythritol when consumed on a daily basis in high amounts [16,17]. More than 90% of ingested erythritol is not metabolized by the human body and is excreted in the urine without changing blood glucose and insulin levels. A little amount of erythritol can be metabolized in some reversible metabolic reactions by dehydrogenation or phosphorylation in humans. Therefore, erythritol is a competitive candidate as a bulk sweetener in confectionery, chewing gum, beverages, and bakery products [18].

Since 1990, erythritol has been commercially produced and added into foods and beverages to provide sweetness, as well as to enhance their taste and texture. It is increasing in popularity, with a market of 20,000 metric tons per year by 2006. In addition, erythritol is also a cheap intermediate for pharmaceutical and chemical synthesis, such as acetonide and penthrite (Figure 9.3).

At present, chemical and fermentative processes have been introduced for large-scale production of erythritol. Unlike other sugar

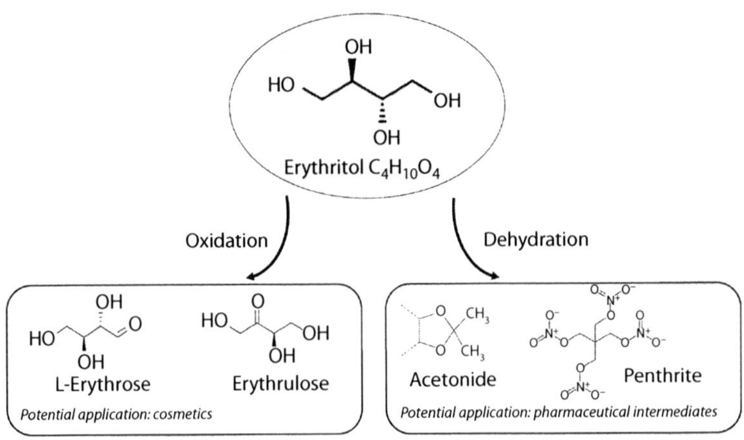

Figure 9.3 Chemistry to derivatives of erythritol.

alcohols, erythritol is produced industrially by biological fermentation processes from cornstarch, since the substrate erythrose is expensive for chemical hydrogenation. So far, erythritol is the only sugar alcohol produced commercially by fermentation rather than chemical method [19]. Large-scale production of erythritol uses glucose or sucrose as substrate for fermentative processes, using chemically and enzymatically hydrolyzed wheat and corn starches [20].

The improvement of erythritol production in biological processes is becoming more significant because of the growing demands in the food industry. In addition to condition optimization and fed-batch process, screening and reconstruction of robust microbial strains play a pivotal role in erythritol production [21,22]. Erythritol can be produced by microbial methods that utilize osmophilic yeasts and some bacteria (Table 9.2) [23–26]. In industrial manufacturing, the yield and productivity of erythritol produced by *Aureobasidium* sp. SN-G42 have been found to reach 47%, 2.0 g/L/h in 96 h in a 100,000 L fermentor with optimized medium components and oxygen transfer rate [27]. Additionally, it has been reported that scale-up studies for erythritol production by *Pseudozyma tsukubaensis* KN75 were carried out from a laboratory scale (7 L fermentor) to pilot (300 L) and plant (50,000 L) scales. Erythritol production at the pilot and plant scales was similar to that at the laboratory scale (245 g/L) corresponding to 2.86 g/L/h productivity and 61% yield,

Table 9.2 Erythritol Productions from Different Various Microorganisms

MICROORGANISM	SUBSTRATE	TITER (g/L)	PRODUCTIVITY (g/L/h)	YIELD (g/g)	REFERENCE
Ustilaginomycetes 618A-01	Glucose	100	0.19	39	[28]
M. tomentosa var. *pollinis*	Glucose	133	0.79	37	[29]
Trigonopsis variabilis	Glucose	46	ND	ND	[30]
Trichosporon sp. (fed-batch)	Glucose	149	1.86	45	[31]
Moniliella sp. 440	Glucose	116	0.8	39	[32]
Aureobasidium sp.	Glucose	175	1.82	44	[33]
Torula sp. (fed-batch)	Glucose	192	2.26	48	[34]
Candida magnoliae (fed-batch)	Glucose	200	1.2	43	[35]
Pseudozyma tsukubaensis (fed-batch)	Glucose	245	2.86	61	[36]

Note: ND represents not determined.

the highest erythritol yield ever reported by an erythritol-producing microorganism [36].

The erythritol in yeast and fungus species is synthesized from D-erythrose 4-phosphate after dephosphorylation and reduction reactions in pentose phosphate pathway. Several yeast species such as *Torula corallina*, *Candida magnoliae*, and *Trigonopsis variabilis* are known to produce erythritol, and the strains reach over 40% yields of erythritol from D-glucose (w/w) [30,33]. As shown in Figure 9.4, the production of erythritol in yeast results from the reduction and dephosphorylation of erythrose-4-phosphate. The erythrose-4-phosphate is from the splitting of fructose 6-phosphate through pentose phosphate pathway. It has been reported that the erythritol

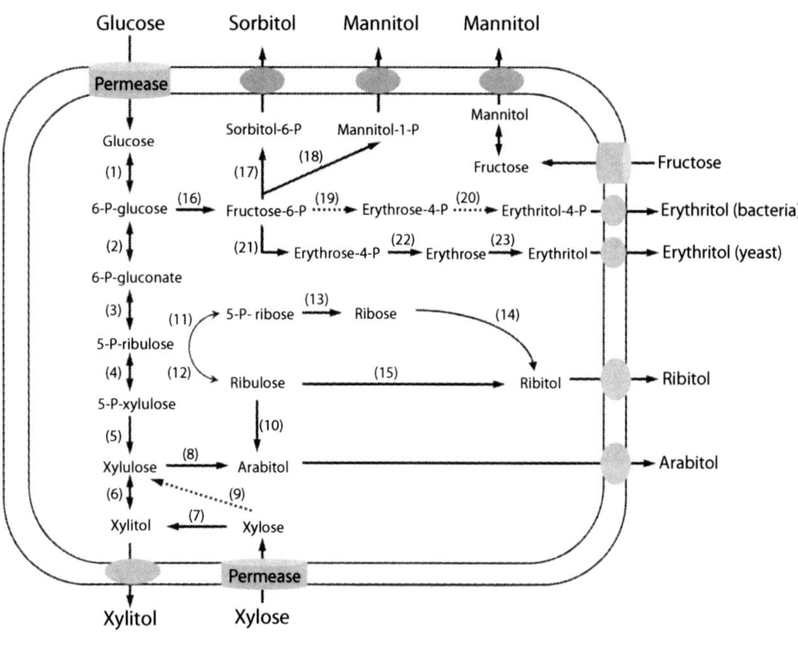

Figure 9.4 Proposed pathways for some sugar alcohol biosynthesis in eukaryotic microorganisms. (1) Glucose kinase, (2) 6-phospho-gluconolactonase, (3) 6-phosphogluconate dehydrogenase, (4) D-ribulose 5-phosphate 3-epimerase, (5) xylulose kinase, (6) xylitol dehydrogenase, (7) xylose reductase, (8) arabitol dehydrogenase, (9) xylose isomerase, (10) arabitol dehydrogenase, (11) D-ribulose 5-phosphate keol-isomerase, (12) ribulose phosphate phosphatase, (13) ribose phosphate phosphatase, (14) aldose reductase, (15) ribitol dehydrogenase, (16) phosphoglucose isomerase, (17) sorbitol-6-P dehydrogenase, (18) mannitol-6-P dehydrogenase, (19) phosphoketolase, (20) erythrose-4-P dehydrogenase, (21) pentose phosphate pathway, (22) erythrose phosphate phosphotransferase, and (23) erythrose dehydrogenase. The dotted line represents the different biosynthesis pathways of sugar alcohols that existed in bacteria.

formation pathway is also involved in bacteria *Leuconostoc oenos*, and the effect of oxygen on the regulation of erythritol production was also studied [26]. The erythritol biosynthesis pathway in *L. oenos* was shown to involve the isomerization of glucose 6-phosphate to fructose 6-phosphate by a phosphoglucose isomerase, the cleavage of fructose 6-phosphate by a phosphoketolase, the reduction of erythrose 4-phosphate by an erythritol 4-phosphate dehydrogenase, and, finally, the hydrolysis of erythritol 4-phosphate to erythritol by a phosphatase (Figure 9.4).

9.2.2 Xylitol

Xylitol is one of the most popular five-carbon sugar alcohols, which is naturally found in many fruits and vegetables. At present, the most significant application of xylitol is sweetener for diabetic patients for its low glycemic index. Other potential uses of xylitol are as an anti-cariogenic agent in toothpaste, as thin coatings of vitamin tablets, in chewing gum, soft drinks, beverages, and in bakery products [37–39]. Xylitol is currently approved for application in foods, pharmaceuticals, and oral health products in more than 50 countries [40]. Additionally, xylitol, as well as arabitol, has the potential to be an outstanding building block for commodity chemicals (Figure 9.5). The production of xylitol/arabitol and subsequent conversion to glycols has been

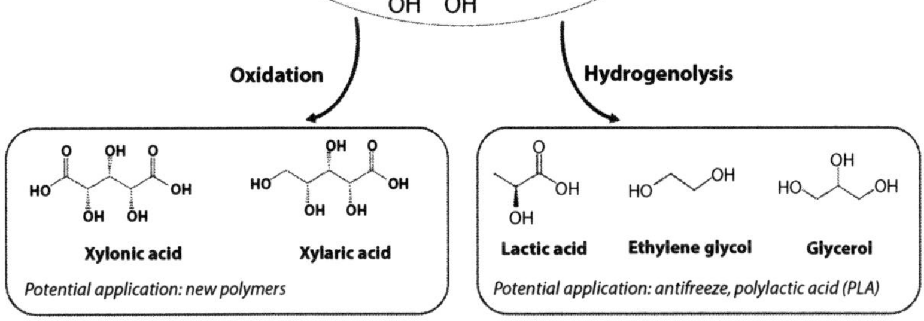

Figure 9.5 Chemistry to derivatives of arabitol and xylitol.

demonstrated and could be a first-generation process for the production of propylene glycol and ethylene glycol. Hence, they have been selected as one of the top 12 types of value-added building-block chemicals from biorefineries [10].

Xylitol is industrially produced by chemical hydrogenation of pure xylose. The chemical synthesis of xylitol starts with the extraction of xylose from hardwood or hemicellulose by acid-catalyzed hydrolysis. After impurity removal and purification, xylose-rich hydrolysate can be used for xylitol production through hydrogenation reaction at 80–140°C and hydrogen pressures up to 50 atmospheres in the presence of metal catalysts (Raney nickel). The obtained xylitol solution requires further purification by chromatography and then concentration and crystallization to form pure xylitol. The xylitol yield is about 50–60% of the xylan fraction; however, the xylitol production process is expensive due to the extensive separation and purification step [38,41,42]. Because of these drawbacks, making xylitol as a sweetener is still limited because of its high price and environmental impact. This has inspired researchers to develop different technologies to lower the production costs. Recently, many studies indicated that microorganisms are able to naturally synthesize xylitol from D-xylose or glucose, which makes its biotechnological production more and more attractive (Table 9.3). The biomanufacture of xylitol could be alternatively

Table 9.3 Xylitol Production from Different Various Microorganisms

MICROORGANISM	SUBSTRATE	TITER (g/L)	YIELD (g/g)	PRODUCTIVITY (g/L/h)	REFERENCE
Engineered *Bacillus subtilis*	D-glucose	–	0.23	–	[40]
Engineered *Corynebacterium glutamicum*	D-xylose	166	0.83	7.90	[43]
Engineered *Escherichia coli* PC07	D-xylose/ glucose	38	0.61	3.97	[44]
Engineered *Lactococcus lactis*	D-xylose/ glucose	54.4	0.68	2.70	[45]
Candida guilliermondii FTI-20037	D-xylose	77.2	0.74	0.80	[46]
Debaryomyces hansenii UFV-170	D-xylose	5.84	0.54	0.24	[47]
Saccharomyces cerevisiae	D-xylose/ cellobiose	93	0.58	1.0	[48]
Pichia anomala TIB-x229	D-xylose	172.3	0.86	1.80	[49]

performed by bacteria [50,51], filamentous fungi [52], yeast [49], and purified enzymes [53] from these microorganisms. Nevertheless, in order to make biomanufacturing exploitable at an industrial level, the process should be rapid, high yield, and adaptive for cheap culture media.

In view of potential alternatives to conventional chemical process, two biotechnological approaches—microbial fermentation and enzymatic bioconversion—seem promising (Figure 9.2). These biotechnological processes are highly attractive because they are able to produce a high-quality and cost-effective product. On a large scale, the biomanufacturing process is generally carried out in a bioreactor. The conditions in this process are generally very gentle with regards to temperature, pressure, and pH, when compared to those in a chemical process. In the fermentation process, different bacteria, fungi, and yeast can use xylose or hemicellulosic hydrolysate for xylitol production [42]. An advantage of the fermentation process over chemical procedures is the lower cost without the xylose substrate purification steps [37]. Other advantages of biotechnological production processes include high reaction specificity and selectivity with few by-products. Therefore, the implementation of biotechnological production methods can be considered an environmentally friendly production strategy [54].

Recently, yeasts have been proposed as the best xylitol producers among the microorganisms investigated [38,47]. The screening of different xylose-utilizing yeast has confirmed that the best xylitol producers belong to the genus *Candida*. In the fermentation process using yeast, the yield of xylitol obtainable from D-xylose is in a range of 65–85% of the theoretical value [38]. After growth in the presence of D-xylose, some yeast produces D-xylitol by the action of a D-xylose reductase (XR), and the other xylitol is converted into xylulose under xylitol dehydrogenase and then flows into pentose phosphate pathway [55]. The production of xylitol using bacteria and fungi has also been studied. A few bacteria, such as *Lactobacillus* and *Gluconobacter oxydans*, have been reported to produce xylitol [56]. D-xylose metabolic genes in bacteria showed a different catabolic pathway that D-xylose is catalyzed to xylulose by a D-xylose isomerase. Xylulose is then phosphorylated to xylulose-5-phosphate by a xylulose kinase, and this pentitol-phosphate is catalyzed via the phosphoketolase pathway (Figure 9.4) [57,58].

The fermentative xylitol production has been studied as an alternative, and its viability is dependent on the optimization of culture and process conditions, as well as the genetic character of the microorganisms. The production of xylitol through the fermentation process is affected by different factors, such as culture conditions, huge water consumption, and the type of process [47].

Xylitol production by biocatalysis is another attractive alternative to both fermentation and chemical processes. Compared to the fermentation process, the enzymatic approach employing XR for xylitol synthesis is expected to obtain a substantial increase in productivity. The enzymatic conversion of D-xylose into xylitol using XR of *Candida pelliculosa* coupled with the oxidoreductase system of *Methanobacterium* sp. has been reported, and the results showed that about 90% conversion could be achieved from xylose to xylitol at 35°C and pH 7.5 in 24 h [53,59].

Additionally, the choice of the biocatalysis system is another crucial point for the success of this bioprocess. Bioconversion of raw material to product is usually performed by microorganisms known as whole cell catalysis [49]. The continuous process is beneficial at the industrial level when immobilized cells are utilized to allow their catalytic activity to be continuously exploited, with consequent increase in volumetric productivity. In an effective continuous immobilized-cell system, xylitol formation is expected not only to minimize the carbon source loss but also to make the immobilized cell able to work as a resting cell biocatalyst. On the basis of these considerations, the D-xylose–xylitol conversion has been investigated in the yeast *Debaryomyces hansenii* UFV-170 as a first attempt for future development of an effective immobilized cell system for continuous xylitol production [47].

Since the demand for xylitol has grown rapidly during the past years, the availability of D-xylose has become a limiting step in xylitol manufacture. In order to make biomanufacturing feasible at an industrial level, the process should utilize alternative and cheap culture media such as glucose and biomass. For example, glucose is a common and relatively cheap material making it an attractive starting material for microbial bioconversions. Biotechnological production of D-arabitol and xylitol using glucose as a starting material has also been reported, but no industrially viable process has yet been developed for practical application due to low yields and interfering by-product [49,60].

9.2.3 Arabitol and Ribitol

Arabitol or arabinitol is a five-carbon sugar alcohol formed by the reduction of either arabinose or lyxose. Compared to its enantiomer, xylitol, little is known about the properties and application of D-arabitol. D-arabitol is, in fact, identified as one of the top 12 building block compounds based on biomass in 2004. Many important chemicals such as arabinoic acid, xylonic acid, propylene glycol, and ethylene glycol can be derived from arabitol [10]. In addition, D-arabitol can be used as a starting material for xylitol production, which has numerous applications in the food and pharmaceutical industries. Furthermore, D-arabitol itself is expected to show potential applications like xylitol as an alternative low-calorie sweetener. There are a few reports describing D-arabitol being nonmetabolizable in humans, which results in a very low caloric content of the molecule and therefore could offer another interesting sweetener alternative for weight management diets [61]. In the constantly growing markets of alternative sweeteners, there is certainly interest for new products such as arabitol. Thus, these properties stress the importance of industrial production of D-arabitol.

Ribitol, also known as adonitol, is an important isomer of arabitol, which is formed by the reduction of ribose. It occurs naturally in some plants such as *Adonis vernalis* and it is one of the components of teichoic acids in the bacteria cell wall [62]. Ribitol is used as the starting material to produce L-ribose, an expensive L-pentose with increasing utilization for the synthesis of L-nucleoside-based pharmaceutical compounds [63]. It also contributes to the chemical structure of riboflavin and flavin mononucleotide (FMN), which is a nucleotide coenzyme, present in the enzyme glycolate oxidase. There are fewer studies about ribitol metabolism and production. It has been reported that some yeasts can produce ribitol [64]; however, no references are found in the literature about ribitol producing bacteria.

D-arabitol can be produced by chemical hydrogenation of D-arabinose and by chemical reduction of lactones of arabinonic and lyxonic acids [65,66]. Since 1960, arabitol metabolism had been studied and many arabitol-producing microorganisms have been screened and engineered for polyol production. It is well known that many osmophilic yeast species such as *Zygosaccharomyces*, *Debaryomyces*,

Metschnikowia, *Candida*, *Pichia*, *Hansenula*, and others produce D-arabitol when grown on D-glucose or glycerol [49,67–71]. Most of these organisms have common tolerance to high concentrations of sugars and salts, and therefore, it has been revealed that the accumulation of sugar alcohols in yeast may be related to osmotic stress [72]. Two biosynthetic approaches for D-arabitol formation of the pentose phosphate pathway in fungi have also been reported [68,73]. For example, *Debaryomyces hansenii* converts D-glucose to D-ribulose-5-phosphate, dephosphorylate D-ribulose-5-phosphate, and then reduces D-ribulose to D-arabitol by NADP-dependent D-arabitol dehydrogenase (called D-ribulose forming pathway). On the other hand, other strains such as *Zygosaccharomyces rouxii* converts D-glucose to D-xylulose-5-phosphate, dephosphorylate D-xylulose-5-phosphate, and then reduces xylulose to D-arabitol by NAD-dependent D-arabitol dehydrogenase (D-xylulose forming pathway; Figure 9.4). Since D-arabitol is a by-product of yeast production, the typical yield is between 20% and 30% of consumed D-glucose. However, regardless of the growth conditions, the production of D-arabitol by yeast and fungi are always accompanied by the production of other sugar alcohols such as glycerol. There has been much research focusing on increasing D-arabitol production (Table 9.4). For

Table 9.4 Arabitol Production from Various Different Microorganisms

MICROORGANISM	SUBSTRATE	TITER (g/L)	YIELD (g/g)	PRODUCTIVITY (g/L/h)	REFERENCE
Engineered *Bacillus subtilis*	Glucose	100	0.38	0.19	[40]
Kodamaea ohmeri	Glucose	81.2	0.41	1.13	[60]
Zygosaccharomyces rouxii NRRL 27624	Glucose	83.4	0.48	0.35	[74]
Metschnikowia reukaufii AJ14787	Glucose	81.4	0.41	0.70	[70]
Hansenula polymorpha DSM 70277	Glucose	18.8	0.14	0.31	[75]
Debaryomyces hansenii ATCC 20212	Glucose	15.5	0.34	0.16	[67]
Endomycopsis chodati	Glucose	57	0.43	0.35	[76]
Pichia anomala TIB-*x*229	Glucose	37	0.37	0.61	[49]
Debaryomyces hansenii SBP-1	Glycerol	14	0.52	0.12	[77]
Candida quercitrusa NBRC1022	Glycerol	85.1	0.40	0.35	[78]

example, Nozaki et al. reported the production of 81.4 g/L D-arabitol by *Metschnikowia reukaufii* in 116 h with 0.7 g/L/h productivity, and with considerable amounts of by-products including ethanol, glycerol, and citric acid. After optimizing conditions, 206 g/L of D-arabitol was produced from glucose with a yield of 52% in 100 h [70]. Zhu et al. reported that the yeast *Kodamaea ohmeri* NH-9 produced 81.2 g/L D-arabitol from 200 g/L D-glucose in 72 h with a yield of 0.4 g/g and volumetric productivity of 1.1 g/L/h. Semicontinuous, repeated-batch fermentation was developed in shaker flasks to increase the D-arabitol production from D-glucose. Under repeated-batch culture, the highest volumetric productivity was 1.4 g/L/h [60].

As mentioned previously, xylitol can be produced by chemical reduction and biological process from xylose [49,79]. Although D-xylose is abundant and renewable, it is difficult to obtain qualified substrate from the hemicellulose fraction of biomass. To overcome this barrier, another strategy for xylitol production involves starting from D-glucose, and this has become an attractive alternative to the D-xylose-based processes. In 1969, Onishi and Suzuki described a three-step process for conversion of glucose to xylitol. In this process, D-glucose is converted into D-arabitol by an osmophilic yeast strain, *Debaryomyces hansenii*, followed by the oxidation of D-arabitol into D-xylulose with *Acetobacter suboxydans*. Finally, the D-xylulose was reduced into xylitol by *Candida guillermondii*. The yield of xylitol from glucose was about 11.6% [67]. Mayer et al. constructed a similar approach and replaced the reduction reaction step with an *in vitro* recombinant xylitol dehydrogenase from *Galactocandida mastotermitis* [80]. Besides, Suzuki et al. [56] reported a novel enzymatic catalysis method for conversion of D-arabitol to xylitol. In this process, D-arabitol was first oxidized to D-xylulose by a membrane-bound D-arabitol dehydrogenase and then reduced to xylitol by a NAD-dependent xylitol dehydrogenase [56]. Therefore, the development of efficient methods for production of D-arabitol is essential for a xylitol production process from glucose.

9.2.4 *Sorbitol*

Sorbitol, also referred to as D-glucitol, is a six-carbon sugar alcohol with a sweet taste, which occurs naturally in many stone fruits and berries from trees of the genus *Sorbus* [81]. It is obtained by the reduction of glucose, changing the aldehyde group to a hydroxyl group. This polyol has a relative sweetness of ~60% compared to sucrose, and also has applications in confectionery, chewing gums, diabetic foods, and a wide range of food products. Moreover, sorbitol works as a humectant in the food industry and modern cosmetics as it holds moisture, maintaining the proper balance and has a high refractive index sufficient for transparent formulations. Interestingly, it has been observed that sorbitol has a health benefit as well: it can increase the intestinal absorption of vitamin B12 and iron absorption in humans [82].

In addition, sorbitol has the potential to be an outstanding building block for commodity chemicals that can be produced from renewable biomass resources (Figure 9.6). For example, inexpensive sorbitol affords the potential for the production of isosorbide at low costs, which has been demonstrated to be a very effective monomer for raising the glass transition temperature of polymers. The conversion of sorbitol to glycols affords the opportunity to utilize a renewable resource for the production of a large-scale commodity chemical [10]. In addition, a mixture of sorbitol and potassium nitrate is successfully applied as an amateur solid rocket fuel, and it is identified as a potential key chemical intermediate for production of fuels from biomass resources [83].

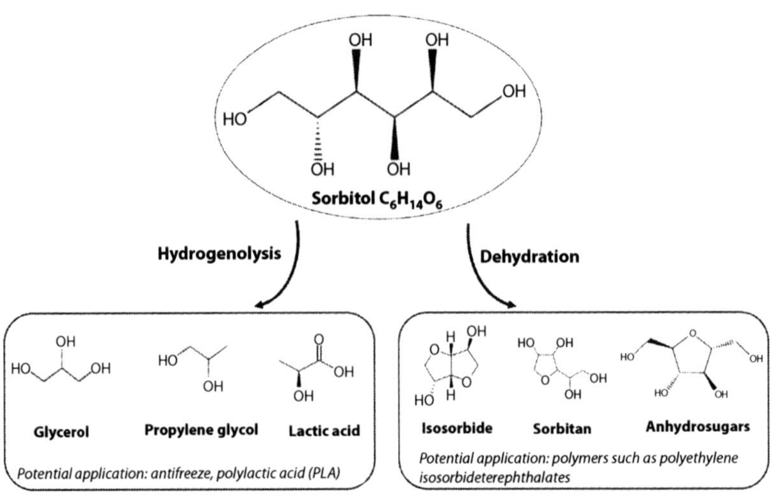

Figure 9.6 Sorbitol chemistry to derivatives.

The industrial production of sorbitol is traditionally performed by catalytic hydrogenation of D-glucose syrup, and about 80% of sorbitol is created by batch chemical method. Engelhard Company has demonstrated that the continuous production of sorbitol from glucose can be done using a ruthenium on carbon catalyst. The yields demonstrated were near 99% with very high productivity. This would make sorbitol an inexpensive feedstock for producing derivatives. Additionally, carbohydrate fractions in biomass such as cellulose could be used to produce sorbitol by sequential hydrolysis and hydrogenation in the presence of metal catalysts [84].

In 1984, it was first reported that the ethanol-producing bacterium *Zymomonas mobilis* could produce sorbitol from sucrose or mixtures of glucose and fructose [85]. In a typical batch fermentation of glucose and fructose by *Z. mobilis*, ethanol is the main product and sorbitol accumulation corresponds to 11% of the consumed substrate [86]. In this bioprocess, fructose and glucose were converted to sorbitol and gluconic acid by a periplasmic enzyme, glucose–fructose oxidoreductase. Besides *Z. mobilis*, only a few systems have been reported that might be able to produce sorbitol, such as *Candida boidinii* and *Candida famata* [87]. Though the biological process for sorbitol production has been proposed, the chemical process for sorbitol production is preferred, because of the relative low costs and high efficiency compared to a biotechnological process.

9.2.5 Mannitol

Mannitol or manna sugar is a naturally occurring six-carbon sugar alcohol. It is found in pumpkins, mushrooms, onions, and marine algae, especially brown seaweed. Brown algae contains about 10–20% mannitol in different growth periods. Manna, obtained by heating the bark of tree *Fraxinus ornus*, contains up to 50% mannitol and was the commercial source of mannitol for many years until the 1920s.

Like other sugar alcohols, mannitol has been widely used in the food, pharmaceutical, medical, and chemical industries [88]. Mannitol could either be added directly to foods, or the use of mannitol-producing bacteria, yeasts, or fungi might directly lead to "natural" mannitol-containing foods [89]. Mannitol is used in medicine as a powerful osmotic diuretic by increasing the formation of urine in order to prevent and treat acute renal failure. Hypertonic mannitol can enhance the transport of drugs across the blood–brain barrier for the treatment of life-threatening brain diseases [90]. Mannitol is the preferred drug for the treatment of acute glaucoma in veterinary medicine. It can dehydrate the vitreous humor and lower the intraocular pressure. Furthermore, mannitol is an important intermediate for chemical and pharmaceuticals synthesis (Figure 9.7). For example, mannitol nicotinate is a well-known vasodilator used in the treatment of hypertension and also a scavenger of hydroxyl radicals [91].

Since mannitol is found in a wide variety of natural products, including almost all plants, it can be directly extracted from natural products. Mannitol concentrations of plant exudates can range from 20% in seaweeds to 90% in the plane tree. The pure mannitol can be extracted and purified by supercritical or subcritical fluids methods.

At present, mannitol is commonly produced through the hydrogenation of fructose, either from starch or sucrose. In both starch and sucrose, the syrups are chromatographically purified to contain 90–95% fructose. The fructose is then hydrogenated using a nickel catalyst and hydrogen gas at a high temperature and pressure into a mixture of isomers sorbitol and mannitol. Mannitol is separated and purified by crystallization since it is less soluble than sorbitol. These production and purification processes make the manufacturing costs for mannitol relatively high.

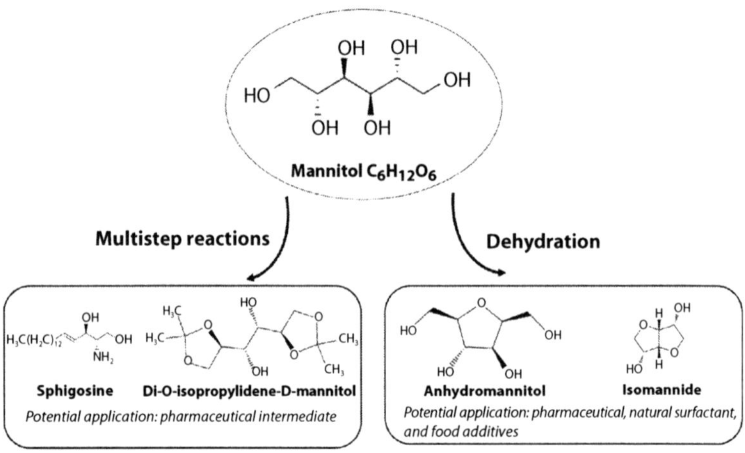

Figure 9.7 Chemistry to derivatives of mannitol.

Table 9.5 Mannitol Production from Various Different Microorganisms

MICROORGANISM	SUBSTRATE (g/L)	TITER (g/L)	PRODUCTIVITY (g/L/h)	REFERENCE
Lactobacillus fermentum	Fructose (100) + Glucose (50)	90	7.60	[92]
Lactobacillus intermedius B-3693	Fructose (250)	161	4.00	[93]
Lactobacillus sp. Y-107	Fructose (100)	73	0.61	[94]
Leuconostoc mesenteroides	Fructose (100) + Glucose (50)	90	3.75	[88]
Candida magnoliae	Fructose (250) + Glucose (50)	213	0.85	[95]
Torulopsis mannitofaciens	Glycerol (100)	31	0.18	[96]
Aspergillus candidus	Glucose (32)	22	0.08	[97]
Penicillium scabrosum	Sucrose (150)	60	0.21	[98]

Biotechnology has provided advantages to mannitol production. A fermentation process could have several advantages compared to the chemical synthesis, such as a high conversion rate of fructose to mannitol, fewer by-products that are difficult to remove, moderate reaction conditions, and no requirement of highly purified substrates. It has been reported that a number of homo- and heterofermentative lactic acid bacteria, osmophilic yeast species, and filamentous fungi can produce mannitol (Table 9.5). Mannitol production by lactic acid bacteria has been investigated over the past decade for its efficient production of mannitol from D-fructose. These bacteria convert fructose to mannitol with nearly 100% yields from a mixture of glucose and fructose. Glucose is converted to lactic acid and acetic acid, and fructose is converted to mannitol. Recently, Xue et al. described a mutant of *Lactobacillus brevis* produced 215 g/L mannitol from glycerol over 98 h under optimal conditions [99]. Further research is focusing on engineering more efficient mannitol pathways in lactic acid bacteria, as well as the use of other engineered microorganisms such as yeast and *E. coli* for mannitol production.

9.3 Nutritional and Health Benefits

9.3.1 Low Calorie and Glycemic Effect

In recent years, low sugar or low calorie is a top ranked market trend for the health food industry. Foods with low- or reduced-calorie sweeteners can have fewer calories than foods made with table sugar and other caloric sweeteners. Sugar alcohols are one type of reduced-calorie sweetener and are generally regarded as a safer and more natural sugar substitute than the artificial and frequently used sweeteners like aspartame or acesulfame K [100]. Sugar alcohols have been widely used in ice creams, cookies, puddings, candies, and chewing gum that is labeled as sugar-free. Sugar alcohols have less of an effect on blood glucose than other carbohydrates. Compared to artificial sweeteners, sugar alcohols are not completely calorie-free, and they are absorbed and digested to some extent. The absorption rate varies among sugar alcohols, from about 50% for xylitol to almost 100% for erythritol [101,102]. A handful of studies indicate that diabetic rats with xylitol supplementation can reduce body weight, blood glucose, and serum lipids, and increase glucose tolerance [103–105]. Therefore, sugar alcohols are a popular choice for weight loss due to their reduced calorie content, and for diabetics due to their low glycemic index.

9.3.2 Dental Caries

Dental caries is one of the most prevalent health issues affecting a large proportion of the world population, and it is also considered to be the most common form of chronic disease among school children. Regular sugars, such as sucrose and glucose, are suitable substrates for microorganisms that ferment carbohydrates and produce acids, resulting in dental caries lesions. Sugar alcohols are well-known substitutes for regular sugar because of their effect on dental health. They produce less acid from fermentation by oral microbiota. Large experiments have demonstrated xylitol to be more effective than fluoride at preventing dental caries [106]. Moreover, sugar alcohol can induce and increase the production of salivary enzymes leading to the growth inhibition of bacteria in plaque. Together, these properties are important in reducing dental caries incidence in patients who are using sugar alcohol food.

9.3.3 Other Health Benefits

In addition to the low calorie, dental, and hypoglycemic benefits already discussed, sugar alcohols also have other health benefits. For example, an interesting study found that collagen synthesis in the skin of aged rats increased after 10% xylitol supplementation over 20 months [107]. Preliminary rat studies have also shown that xylitol can increase mineral content and bone volume and prevent bone loss [108]. These studies suggest that sugar alcohols may have promise for preventing age-related decline in bone and skin health. Additionally, a few studies indicate that sugar alcohols may have an effect on bacteria distribution, and they found that xylitol promotes a shift from Gram-negative to Gram-positive bacteria, with reduced *Bacteroides* and increased *Bifidobacteria* [109].

9.4 Future Trends

On the basis of health trends among consumers, sugar alcohols are manifesting a vast potential application value and huge market demand in the future. Compared to traditional chemical production, the biotechnological strategy is becoming an attractive and efficient alternative method for industrial synthesis of sugar alcohols. In an attempt to enhance productivity, new bioprocesses such as enzymatic catalysis and bioconversion have been developed and investigated. Compared to the traditional fermentation method, these methods do not require strict sterility for the production process, and the products are usually much easier for downstream separation, which then reduces the cost for industrial applications. Enzymes are being used widely in industrial synthetic chemistry as excellent catalysts [110,111]. They are capable of catalyzing reactions with strict enantioselectivity and regioselectivity under moderate conditions. Moreover, bioconversion referring to the whole-cell catalysis has aroused more and more attention for its efficiency, convenience, and stability [13,49,112].

In biomanufacturing processes, microbial cells have been most commonly used for industrial purposes due to their diversity and ease of handling. The sugar alcohol-producing strains were previously screened and quantified by thin-layer chromatography, high-performance liquid chromatography, and *p*-iodonitrotetrazolium violet methods in previous studies [51,70,113,114]. However, these methods were time-consuming and not suitable for high throughput screening. Therefore, developing high throughput screening and selection of quality sugar alcohol-producing microbes id crucial for industrial application. For example, a colorimetric method for sugar alcohol detection [115,116] may be developed and optimized for high throughput assay of sugar alcohol in the future.

Though different sugar alcohol-producing microorganisms were screened and obtained, their yield is usually not high enough for commercial process by simple fermentation process optimization [93]. Therefore, it is necessary to develop efficient methods to further improve the strain performance, which is key to the industrial commercial production process. Traditional random mutagenesis, adaptive evolution, and genome shuffling have been considered as popular and efficient methods for strain engineering, and they could be applied for sugar alcohol-producing strain improvement [117,118]. Moreover, recent breakthroughs in engineering nucleases, such as zinc-finger nucleases, transcription activator-like effector nucleases, and clustered regulatory interspaced short palindromic repeat (CRISPR)/Cas-based RNA-guided DNA endonucleases, have offered some new opportunities for genome engineering strategies at the genome level aimed at allowing the accelerated evolution of biological systems and uncovering improved phenotypes with beneficial properties [119–121].

Overall, sugar alcohols appear to be safer and healthier than regular sugar and artificial sweeteners. With the increased availability and innovations of sugar alcohols in food technology and clinical medicine, sugar alcohol products are expected to become more acceptable and popular.

References

1. Zumbé A, Lee A, Storey D: Polyols in confectionery: The route to sugar-free, reduced sugar and reduced calorie confectionery. *Br J Nutr* 2001, 85:S31–S45.
2. Gibson PR, Shepherd SJ: Evidence-based dietary management of functional gastrointestinal symptoms: The FODMAP approach. *J Gastroenterol Hepatol* 2010, 25(2):252–258.
3. Axthammer QJ, Klapötke TM, Krumm B: Efficient synthesis of primary nitrocarbamates of sugar alcohols: From food to energetic materials. *Chem Asian J* 2016, 11(4):568–575.
4. Lee CK: Structural functions of taste in the sugar series: Taste properties of sugar alcohols and related compounds. *Food Chemistry* 1977, 2:95–105.
5. Mitchell H (ed.): *Sweeteners and Sugar Alternatives in Food Technology.* Oxford, UK: Blackwell Publishing; 2006.

6. Salminen S, Halikainen A (eds.): *Sweeteners, Food Additives*. New York: Marcel Dekker; 2001.
7. Payne AN, Chassard C, Lacroix C: Gut microbial adaptation to dietary consumption of fructose, artificial sweeteners and sugar alcohols: Implications for host microbe interactions contributing to obesity. *Obes Rev* 2012, 13(9):799–809.
8. Natah SS, Hussien KH, Tuominen JA, Koivisto VA: Metabolic response to lactitol and xylitol in healthy men. *Am J Clin Nutr* 1997, 65:947–950.
9. Nguyen NU, Dumoulin G, Henriet MT, Berthelay S, Regnard J: Carbohydrate metabolism and urinary excretion of calcium and oxalate after ingestion of polyol sweeteners. *J Clin Endocrinol Metab* 1993, 77:388–392.
10. Werpy T, Petersen G: Top Value Added Chemicals from Biomass Volume I: Results of Screening for Potential Candidates from Sugars and Synthesis Gas. Oak Ridge, TN: Department of Energy; 2004.
11. Lewis DH, Smith DC: Sugar alcohol in fungi and green plants. *New Phytol* 1967, 66(2):185–204.
12. Jennings DH: Polyol metabolism in fungi. *Adv Microb Physiol* 1984, 25:149–193.
13. Akinterinwa O, Khankal R, Cirino PC: Metabolic engineering for bio production of sugar alcohols. *Curr Opin Biotechnol* 2008, 19(5):461–467.
14. Bilanx M, Flourie B, Jaequemmim C, Messing B (eds.): *Handbook of Sweeteners*. Glasgow: Blackie Academic & Professional; 1991.
15. Bernt WO, Borzelleca JF, Flamm G, Munro IC: Erythritol: A review of biological and toxicological studies. *Regul Toxicol Pharmacol* 1996, 24:S191–S197.
16. Tokuoka K, Ishizuka H, Wako K, Taniguchi H: Comparison of three forms of erythrose reductase from an *Aureobasidium* sp. mutant. *J Gen Appl Microbiol* 1992, 38:145–155.
17. Yokozawa T, Kim HY, Cho EJ: Erythritol attenuates the diabetic oxidative stress through glucose metabolism and lipid peroxidation in streptozotocin-induced diabetic rats. *J Agric Food Chem* 2002, 50: 5485–5489.
18. Storey D, Lee A, Bornet F, Brouns F: Gastrointestinal tolerance of erythritol and xylitol ingested in a liquid. *Eur J Clin Nutraceuticals World* 2007, 61:349–354.
19. Lin SJ, Wen CY, Liau JC, Chu WS: Screening and production of erythritol by newly isolated osmophilic yeast-like fungi. *Proc Biochem* 2001, 36:1249–1258.
20. Aoki M, Pastore G, Park Y: Microbial transformation of sucrose and glucose to erythritol. *Biotechno Lett* 1993, 15:383–388.
21. Koh ES, Lee TH, Lee DY, Kim HJ, Ryu YW, Seo JH: Scale up of erythritol production by an osmophilic mutant of *Candida magnoliae*. *Biotechnol Lett* 2003, 25:2103–2105.
22. Rymowicz W, Rywi ńska A, Marcinkiewicz M: High-yield production of erythritol from raw glycerol in fed-batch cultures of *Yarrowia lipolytica*. *Biotechnol Lett* 2009, 31:377–380.

23. Lee JK, Koo BS, Kim SY: Fumarate-mediated inhibition of erythrose reductase, a key enzyme for erythritol production by *Torula corallina*. *Appl Environ Microbiol* 2002, 68:4534–4538.
24. Seo JH, Ryu YW, Jung SR, Kim SY: Fermentation processes for preparing erythritol by a high salt tolerant mutant of *Candida sp.* In: *US Patent* vol. 6287830B1; 2001.
25. Ryu YW, Park CY, Park JB, Kim SY, Seo JH: Optimization of erythritol production by *Candida magnoliae* in fed-batch culture. *J Ind Microbiol Biotechnol* 2000, 25:100–103.
26. Veiga-da-Cunha M, Santos H, Van Schaftingen E: Pathway and regulation of erythritol formation in *Leuconostoc oenos*. *J Bacteriol* 1993, 175:3941–3948.
27. Sawada K, Taki A, Yamakawa T, Seki M: Key role for transketolase activity in erythritol production by *Trichosporonoides megachiliensis* SN-G42. *J Biosci Bioeng* 2009, 4(5):385–390.
28. Hirata Y, Igarashi K, Ezaki S, Atomi H, Imanaka T: High-level production of erythritol by strain 618A-01 isolated from pollen. *J Biosci Bioeng* 1999, 87:630–635.
29. Hajny G, Smith J, Garver J: Erythritol production by a yeast-like fungus. *Appl Environ Microbiol* 1964, 12:240–246.
30. Kim SY, Lee KH, Kim JH, Oh DK: Erythritol production by controlling osmotic pressure in *Trigonopsis variabilis*. *Biotechnol Lett* 1997, 19:727–729.
31. Park JB, Seo BC, Kim JR, Park YK: Production of erythritol in fed-batch cultures of *Trichosporon* sp. *J Ferment Bioeng* 1998, 86: 577–580.
32. Lin S, Wen C, Huang C, Chu W: Erythritol-producing *Moniliella* strains. In: *US Patent*. vol. 20030008378A1; 2002.
33. Ishizuka H, Wako K, Kasumi T, Sasaki T: Breeding of a mutant of *Aureobasidium* sp. with high erythritol production. *J Ferment Bioeng* 1989, 68:310–314.
34. Oh DK, Cho CH, Lee JK, SY K: Increased erythritol production in fed-batch cultures of *Torula sp.* by controlling glucose concentration. *J Ind Microbiol Biotechnol* 2001, 26:248–252.
35. Koh ES, Lee TH, Lee DY, Kim HJ, Ryu YW, Seo JH: Scale-up of erythritol production by an osmophilic mutant of *Candida magnoliae*. *Biotechnol Lett* 2003, 25:2103–2105.
36. Jeya M, Lee KM, Tiwari MK, Kim JS, Gunasekaran P, Kim SY, Kim IW, Lee JK: Isolation of a novel high erythritol-producing *Pseudozyma tsukubaensis* and scale-up of erythritol fermentation to industrial level. *Appl Microbiol Biotechnol* 2009, 83:225–231.
37. Parajo JC, Dominguez H, Dominguez JM: Biotechnological production of xylitol. Part 1: interest of xylitol and fundamentals of its biosynthesis. *Bioresource Technol* 1998, 65:191–201.
38. Nigam P, Singh D: Processes for fermentative production of xylitol. *Process Biochemistry* 1995, 30:117–124.

39. Jaffe GM: Xylitol—A specialty sweetener. *Sugar Azucar* 1978, 93:36–42.
40. Povelainen M: *Pentitol Phosphate Dehydrogenases: Discovery, Characterization and Use in D-arabitol and Xylitol Production by Metabolically Engineered Bacillus subtilis*. Helsinki: University of Helsinki; 2008.
41. Parajo JC, Dominguez H, Dominguez JM: Biotechnological production of xylitol. Part 2: Operation in culture media made with commercial sugars. *Bioresource Technol* 1998, 65:203–212.
42. Nigam P, Singh D: Enzymes and microbial system involved in starch processing. *Enz Microb Technol* 1995, 17:770–778.
43. Sasaki M, Jojima T, Inui M, Yukawa H: Xylitol production by recombinant *Corynebacterium glutamicum* under oxygen deprivation. *Appl Microbiol Biotechnol* 2010, 86(4):1057–1066.
44. Khankal R, Chin JW, Cirino PC: Role of xylose transporters in xylitol production from engineered *Escherichia coli*. *J Biotechnol* 2008, 134(3–4):246–252.
45. Nyyssölä A, Pihlajaniemi A, Palva A, von Weymarn N, Leisola M: Production of xylitol from D-xylose by recombinant Lactococcus lactis. *J Biotechnol* 2005, 118(1):55–66.
46. Mussatto SI, Roberto IC: Xylitol production from high xylose concentration: Evaluation of the fermentation in bioreactor under different stirring rates. *J Appl Microbiol* 2003, 95(2):331–337.
47. Zhang G, Lin Y, He P, Li L, Wang Q, Ma Y: Characterization of the sugar alcohol-producing yeast Pichia anomala. *J Ind Microbiol Biotechnol* 2014, 41(1):41–48.
48. Oh EJ, Ha SJ, Rin Kim S, Lee WH, Galazka JM, Cate JH, Jin YS: Enhanced xylitol production through simultaneous co-utilization of cellobiose and xylose by engineered *Saccharomyces cerevisiae*. *Metab Eng* 2013, 15:226–234.
49. Sampaio FC, Chaves-Alves VM, Converti A, Lopes Passos FM, Cavalcante Coelho JL: Influence of cultivation conditions on xylose-to-xylitol bioconversion by a new isolate of *Debaryomyces hansenii*. *Bioresour Technol* 2008, 99(3):502–508.
50. Dhar KS, Wendisch VF, Nampoothiri KM: Engineering of *Corynebacterium glutamicum* for xylitol production from lignocellulosic pentose sugars. *J Biotechnol.* 2016, 230:63–71.
51. Rangaswamy S, Agblevor FA: Screening of facultative anaerobic bacteria utilizing D-xylose for xylitol production. *Appl Microbiol Biotechnol* 2002, 60(1–2):88–93.
52. Dahija JS: Xylitol production by *Petromyces albertensis* grown on medium containing D-xylose. *Can J Microbiol* 1991, 37(1):14–18.
53. Kitpreechavanich V, Hayashi M, Nishio N, Nagai S: Conversion of D-xylose into xylitol by xylose reductase from *Candida pelliculosa* coupled with the oxidoreductase system of methanogen strain HU. *Biotechnol Lett* 1984, 6(10):651–656.
54. Currell B, Van Dam Mieras RC (eds.): *Biotechnological Innovations in Chemical Synthesis*: Elsevier: Amsterdam, 1997.

55. Granstrom TB, Izumori K, Leisola M: A rare sugar xylitol. Part I. The biochemistry and biosynthesis of xylitol. *Appl Microbiol Biotechnol* 2007, 74:277–281.
56. Suzuki S, Sugiyama M, Mihara Y, Hashiguchi K, Yokozeki K: Novel enzymatic method for the production of xylitol from D-arabitol by *Gluconobacter oxydans*. *Biosci Biotechnol Biochem* 2002, 66(12):2614–2620.
57. Bor YC, Moraes C, Lee SP, Crosby WL, Sinskey AJ, Batt CA: Cloning and sequencing the *Lactobacillus brevis* gene encoding xylose isomerase. *Gene* 1992, 114:127–132.
58. Lokman BC, Heerikhuisen M, Leer RJ: Regulation of expression of the *Lactobacillus pentosus* xylAB operon. *J Bacteriol* 1997, 179:5391–5397.
59. Kitpreechavanich V, Nishio N, Hayashi M, Nagai S: Regeneration and retention of NADP (H) for xylitol production in an ionized membrane reactor. *Biotechnol Lett* 1985, 7(9):657–662.
60. Zhu HY, Xu H, Dai XY, Zhang Y, Ying HJ, Ouyang PK: Production of D-arabitol by a newly isolated *Kodamaea ohmeri*. *Bioprocess Biosyst Eng* 2010, 33(5):565–571.
61. Huck JHJ, Roos B, Jakobs C, van der Knaap MS, Verhoeven NM: Evaluation of pentitol metabolism in mammalian tissues provides new insight into disorders of human sugar metabolism. *Mol Gen Metab* 2004, 82:231–237.
62. Baur S, Marles-Wright J, Buckenmaier S, Lewis RJ, Vollmer W: Synthesis of CDP-activated ribitol for teichoic acid precursors in *Streptococcus pneumoniae*. *J Bacteriol* 2009, 191(4):1200–1210.
63. Woodyer RD, Wymer NJ, Racine FM, Khan SN, Saha BC: Efficient production of L-ribose with a recombinant *Escherichia coli* biocatalyst. *Appl Environ Microbiol* 2008, 74:2967–2975.
64. Toivari MH, Maaheimo H, Penttilä M, Ruohonen L: Enhancing the flux of D-glucose to the pentose phosphate pathway in *Saccharomyces cerevisiae* for the production of D-ribose and ribitol. *Appl Microbiol Biotechnol* 2009, 85:731–739.
65. Lemmens H, Coomans S, Röper H: Process for the production of arabinitol. European Patent EP0820979; 1998.
66. Fabre L, Gallezot P, Perrard A: Catalytic hydrogenation of arabinonic acid and lactones to arabitol. *J Catalys* 2002, 208:247–254.
67. Onishi H, Suzuki T: Microbial production of xylitol from glucose. *Appl Microbiol* 1969, 18:1031–1035.
68. Ingram JM, Wood WA: Enzymatic basis for D-arabitol production by *Saccharomyces rouxii*. *J Bacteriol* 1965, 89:1186–1194.
69. Hanssens L, Van Regenmortel A, Verachtert H: pH-dependent polyol production in *Moniliella tomentosa*. *Appl Microbiol* 1972, 24:831–833.
70. Nozaki H, Suzuki S, Tsuyoshi N, Yokozeki K: Production of D-arabitol by *Metschnikowia reukaufii* AJ14787. *Biosci Biotechnol Biochem* 2003, 67(9):1923–1929.
71. Groleau D, Chevalier P, Yuen TH: Production of polyols and ethanol by the osmophilic yeast *Zygosaccharomyces rouxii*. *Biotechnol Lett* 1995, 17:315–320.

72. Van Eck JH, Prior BA, Brandt EV: Accumulation of polyhydroxy alcohols by *Hansenula anomala* in response to water stress *J Gen Microbiol* 1989, 135:3505–3513.
73. Jovall P-A, Tunblad-Johansson I, Adler L: ^{13}C NMR analysis of production and accumulation of osmoregulatory metabolites in the salt-tolerant yeast *Debaryomyces hansenii*. *Arch Microbiol* 1990, 154:209–214.
74. Saha BC, Sakakibara Y, Cotta MA: Production of D-arabitol by a newly isolated *Zygosaccharomyces rouxii*. *J Ind Microbiol Biotechnol* 2007, 34(7):519–523.
75. Escalante J, Caminal G, de Mas C: Biomass production by a thermotolerant yeast: *Hansenula polymorpha*. *J Chem Technol Biotechnol* 1990, 48(1):61–70.
76. Hajny GJ: D-Arabitol production by *Endomycopsis chodati*. *Appl Microbiol* 1964, 12:87–92.
77. Koganti S, Kuo TM, Kurtzman CP, Smith N, Ju LK: Production of arabitol from glycerol: Strain screening and study of factors affecting production yield. *Appl Microbiol Biotechnol* 2011, 90(1):257–267.
78. Yoshikawa J, Habe H, Morita T, Fukuoka T, Imura T, Iwabuchi H, Uemura S, Tamura T, Kitamoto D: Production of D-arabitol from raw glycerol by *Candida quercitrusa*. *Appl Microbiol Biotechnol* 2014, 98(7):2947–2953.
79. Winkelhausen E, Kuzmanova S: Microbial conversion of D-xylose to xylitol. *J Ferment Bioeng* 1998, 86:1–14.
80. Mayer G, Kulbe KD, Nidetzky B: Utilization of xylitol dehydrogenase in a combined microbial/enzymatic process for production of xylitol from D-glucose. *Appl Biochem Biotechnol* 2002, 98–100:577–589.
81. Teo G, Suzuki Y, Uratsu SL, Lampinen B, Ormonde N, Hu WK, Dejong TM, Dandekar AM: Silencing leaf sorbitol synthesis alters long-distance partitioning and apple fruit quality. *Proc Natl Acad Sci* 2006, 103(49):18842–18847.
82. Kearsley MW, Deis RC: Sorbitol and mannitol. In H. Mitchell (ed.), *Sweeteners and Sugar Alternatives in Food Technology*. Oxford: Blackwell Publishing; 2006: 249–261.
83. Metzger JO: Production of liquid hydrocarbons from biomass. *Angew Chem Int Ed* 2006, 45(5):696–698.
84. Shrotri A, Tanksale A, Beltramini JN, Gurav H, Chilukuri SV: Conversion of cellulose to polyols over promoted nickel catalysts. *Catal Sci Technol* 2012, 2(9):1852–1858.
85. Barrow KD, Collins JG, Leigh DA, Rogers PL, Warr RG: Sorbitol production by *Zymomonas mobilis*. *Appl Microbiol Biotechnol* 1984, 20:225–232.
86. Viikari L: Formation of sorbitol by *Zymomonas mobilis*. *Appl Microbiol Biotechnol* 1984, 20:118–123.
87. Liu C, Dong H, Zhong J, Ryu DD, Bao J: Sorbitol production using recombinant *Zymomonas mobilis* strain. *J Biotechnol* 2010, 148(2–3): 105–112.

88. Soetaert W, Buchholz K, Vandamme EJ: Production of D-mannitol and D-lactic acid by fermentation with *Leuconostoc mesenteroides*. *Agro Food Ind Hi Tech* 1995, 6:41–44.
89. Efiuvwevwere BJO, Gorris LGM, Smid EJ, Kets EPW: Mannitol-enhanced survival of *Lactococcus lactis* subjected to drying. *Appl Microbiol Biotechnol* 1999, 51:100–104.
90. Miller G: Drug targeting. Breaking down barriers. *Science* 2002, 297(5584):1116–1118.
91. Shen B, Jensen RG, Bohnert HJ: Mannitol protects against oxidation by hydroxyl radicals. *Plant Physiol* 1997, 115:527–532.
92. von Weymarn FNW, Hujanen M, Leisola MSA: Production of D-mannitol by heterofermentative lactic acid bacteria. *Proc Biochem* 2002, 37:1207–1213.
93. Saha BC, Racine FM: Effects of pH and corn steep liquor variability on mannitol production by *Lactobacillus intermedius* NRRL B-3693. *Appl Microbiol Biotechnol* 2010, 87:553–560.
94. Yun JW, Kim DH: A comparative study of mannitol production by two lactic acid bacteria. *J Ferment Bioeng* 1998, 85:203–208.
95. Lee JK, Song JY, Kim SY: Controlling substrate concentration in fed-batch *Candida magnoliae* culture increases mannitol production. *Biotechnol Prog* 2003, 19:768–775.
96. Onishi H, Suzuki T: Microbial production of D-mannitol and D-fructose from glycerol. *Biotechnol Bioeng* 1970, 12:913–920.
97. Smiley KL, Cadmus MC, Liepins P: Biosynthesis of D-mannitol from D-glucose by *Aspergillus candidus*. *Biotechnol Bioeng* 1967, 9:365–374.
98. Hendriksen HV, Mathiasen TE, Alder-Nissen J, Frisvad JC, Emborg C: Production of mannitol by Penicillium strains. *J Chem Technol Biotechnol* 1988, 43:223–228.
99. Yue M, Cao H, Zhang J, Li S, Meng Y, Chen W, Huang L, Du Y: Improvement of mannitol production by *Lactobacillus brevis* mutant 3-A5 based on dual-stage pH control and fed-batch fermentations. *World J Microbiol Biotechnol* 2013, 10:1923–1930.
100. Bär A: Safety assessment of polyol sweeteners—some aspects of toxicity. *Food Chem* 1985, 16(3–4):231–241.
101. Wolever TMS, Piekarz A, Hollands M, Younker K: Sugar alcohols and diabetes: A review. *Can J Diabetes* 2002, 26(4):356–362.
102. Kroger M, Meister K, Kava R: Low-calorie sweeteners and other sugar substitutes: A review of the safety issues. *Compr Rev Food Sci Food Saf* 2006, 5(2):35–47.
103. Amo K, Arai H, Uebanso T, Fukaya M, Koganei M, Sasaki H, Yamamoto H, Taketani Y, Takeda E: Effects of xylitol on metabolic parameters and visceral fat accumulation. *J Clin Biochem Nutr* 2011, 49(1):1–7.
104. Islam MS: Effects of xylitol as a sugar substitute on diabetes-related parameters in nondiabetic rats. *J Med Food* 2011, 14(5):505–511.
105. Islam MS, Indrajit M: Effects of xylitol on blood glucose, glucose tolerance, serum insulin and lipid profile in a type 2 diabetes model of rats. *Ann Nutr Metab* 2012, 61(1):57–64.

106. Mäkinen KK: Sugar alcohol sweeteners as alternatives to sugar with special consideration of xylitol. *Med Princ Pract* 2011, 20(4):303–320.
107. Mattila PT, Pelkonen P, Knuuttila ML: Effects of a long-term dietary xylitol supplementation on collagen content and fluorescence of the skin in aged rats. *Gerontology* 2005, 51(3):166–169.
108. Mattila PT, Svanberg MJ, Jämsä T, Knuuttila ML: Improved bone biomechanical properties in xylitol-fed aged rats. *Metabolism* 2002, 51(1):92–96.
109. Tamura M, Hoshi C, Hori S: Xylitol affects the intestinal microbiota and metabolism of daidzein in adult male mice. *Int J Mol Sci* 2013, 14(12):23993–24007.
110. Ogawa J, Shimizu S: Industrial microbial enzymes: Their discovery by screening and use in large-scale production of useful chemicals in Japan. *Curr Opin Biotechnol* 2002, 13:367–375.
111. Schmid A, Dordick JS, Hauer B, Kiener A, Wubbolts M, Witholt B: Industrial biocatalysis today and tomorrow. *Nature* 2001, 409:258–268.
112. Kang HY, Kim YS, Seo JH, Ryu YW: Flocculation of an isolated flocculent yeast *Candida tropicalis* HY200 and its application for efficient xylitol production using repeated batch cultivation. *J Microbiol Biotechnol* 2006, 16:1874–1881.
113. Guo C, Zhao C, He P, Lu D, Shen A, Jiang N: Screening and characterization of yeasts for xylitol production. *J Appl Microbiol* 2006, 101(5):1096–1104.
114. Lin S, Wen C, Liau J, Chu W: Screening and production of erythritol by newly isolated osmophilic yeast-like fungi. *Process Biochem* 2001, 36(12):1249–1258.
115. Bok SH, Demain AL: An improved colorimetric assay for polyols. *Anal Biochem* 1977, 81(1):18–20.
116. Sánchez J: Colorimetric assay of alditols in complex biological samples. *J Agric Food Chem* 1998, 46(1):157–160.
117. Patnaik R, Louie S, Gavrilovic V, Perry K, Stemmer WP, Ryan CM, del Cardayré S: Genome shuffling of Lactobacillus for improved acid tolerance. *Nat Biotechnol* 2002, 20(7):707–712.
118. Zhang G, Lin Y, Qi X, Wang L, He P, Wang Q, Ma Y: Genome shuffling of the nonconventional yeast *Pichia anomala* for improved sugar alcohol production. *Microb Cell Fact* 2015,14:112.
119. Jiang W, Bikard D, Cox D, Zhang F, Marraffini LA: RNA-guided editing of bacterial genomes using CRISPR-Cas systems. *Nat Biotechnol* 2013, 31(3):233–239.
120. Wood AJ, Lo TW, Zeitler B, Pickle CS, Ralston EJ, Lee AH, Amora R, Miller JC, Leung E, Meng X et al.: Targeted genome editing across species using ZFNs and TALENs. *Science* 2011, 333(6040):307.
121. Zhang G, Lin Y, Qi X, Li L, Wang Q, Ma Y: TALENs-assisted multiplex editing for accelerated genome evolution to improve yeast phenotypes. *ACS Synth Biol* 2015, 4(10):1101–1111.

Index

D

I

J

K

L

M

N

O

P